CARTOGRAPHIC HUMANISM

CARTOGRAPHIC HUMANISM

The Making of Early Modern Europe

KATHARINA N. PIECHOCKI

THE UNIVERSITY OF CHICAGO PRESS
CHICAGO AND LONDON

The University of Chicago Press, Chicago 60637
The University of Chicago Press, Ltd., London
© 2019 by The University of Chicago
All rights reserved. No part of this book may be used or reproduced in any manner whatsoever without written permission, except in the case of brief quotations in critical articles and reviews. For more information, contact the University of Chicago Press, 1427 East 60th Street, Chicago, IL 60637.
Published 2019
Paperback edition 2021
Printed in the United States of America

30 29 28 27 26 25 24 23 22 21   1 2 3 4 5

ISBN-13: 978-0-226-64118-8 (cloth)
ISBN-13: 978-0-226-81681-4 (paper)
ISBN-13: 978-0-226-64121-8 (e-book)
DOI: https://doi.org/10.7208/chicago/9780226641218.001.0001

The University of Chicago Press gratefully acknowledges the generous support of Harvard University toward the publication of this book.

Library of Congress Cataloging-in-Publication Data

Names: Piechocki, Katharina N. (Katharina Natalia), author.
Title: Cartographic humanism : the making of early modern Europe / Katharina N. Piechocki.
Description: Chicago : The University of Chicago Press, 2019. | Includes bibliographical references and index.
Identifiers: LCCN 2019008632 | ISBN 9780226641188 (cloth : alk. paper) | ISBN 9780226641218 (e-book)
Subjects: LCSH: Europe—History—1492–1648. | Europe—Boundaries—History. | Europe—Maps—Early works to 1800. | Europe—In literature. | European literature—Renaissance, 1450–1600—History and criticism.
Classification: LCC D231 .P54 2019 | DDC 940.2—dc23
LC record available at https://lccn.loc.gov/2019008632

For Julianne and Karl-Heinz Meckmann, who in 1981 took
an unknown family of refugees under their roof—
and changed the course of my life.
And for my parents, courageous enough to migrate.

CONTENTS

List of Figures — ix
On Translations — xi

Introduction — 1

1. Gridding Europe's Navel: Conrad Celtis's *Quatuor Libri Amorum secundum Quatuor Latera Germanie* (1502) — 26
2. A Border Studies Manifesto: Maciej Miechowita's *Tractatus de duabus Sarmatiis* (1517) — 68
3. The *Alpha* and the *Alif*: Continental Ambivalence in Geoffroy Tory's *Champ fleury* (1529) — 107
4. Syphilitic Borders and Continents in Flux: Girolamo Fracastoro's *Syphilis sive Morbus Gallicus* (1530) — 148
5. Cartographic Curses: Europe and the Ptolemaic Poetics of *Os Lusíadas* (1572) — 185

Conclusion — 230

Acknowledgments — 235
Notes — 241
Index — 297

FIGURES

1 Michael Eitzinger, "Europa Virgo" (1588) 3
2 Albrecht Dürer, "Philosophia" (1502) 23
3 Nicolaus Cusanus, map of Central Europe (ca. 1480) 30–31
4 Erhard Etzlaub, "Romweg" map (1501) 33
5 Conrad Celtis, *Quatuor Libri Amorum*, frontispiece (1502) 38
6 Conrad Celtis, *Quatuor Libri Amorum*, "Phoebus Apollo" (1502) 46
7 *Rudimentum novitiorum*, world map (1475) 51
8 Johannes Schnitzer von Armsheim, Ptolemaic world map (1482) 52
9 Conrad Celtis, *Quatuor Libri Amorum*, "Barbara Codonea" (1502) 54
10 Benedetto Bordone, *Isolario*, Crimean Peninsula (1528) 77
11 Martin Waldseemüller, *Geographia*, "European Sarmatia" (1513) 81
12 Maciej Miechowita, *Descriptio Sarmatiarum*, frontispiece (1521) 91
13 Augustin Hirschvogel, map of Muscovy (1549) 105
14 Ptolemy, *Cosmographia*, "Europa" (1490) 108
15 Guillaume Postel, *Cosmographiae Disciplinae Compendium*, "Iapetia" (Europa), (1561) 110
16 Pomponius Mela, *De Orbis situ*, frontispiece (1522) 123
17 Geoffroy Tory, *Champ fleury*, Arabic alphabet (1529) 131
18 Charles de Bovelles, "Liber de Nichilo" (1510) 135
19 Geoffroy Tory, *Champ fleury*, Letters "I" and "O" (1529) 138
20 Johannes Augustinus Pantheus, *Ars et theoria transmutationis metallicae cum Voarchadumia*, title page (1550) 167
21 Sebastian Münster, *Cosmography*, Ptolemaic hemispheric map (1564) 194
22 Sebastian Münster, *Cosmography*, modern world map (1564) 195
23 Martin Waldseemüller, *Universalis Cosmographia* (1507) 212

ON TRANSLATIONS

All translations into English are my own unless otherwise noted. While I kept the original spelling of early modern primary sources, I modernized the Latin "u" and "v" and the tildes (~) which substitute for "n" or "m."

INTRODUCTION

But we must begin with the map.
—Marshall Hodgson, *Rethinking World History*[1]

In the late thirteenth century, Hereford Cathedral received a precious gift from a newly appointed cleric: an enormous and carefully crafted *mappa mundi*, a map of the world drawn on a single ox hide five feet in diameter, to be mounted on the cathedral wall. The largest medieval *mappa mundi* to have survived in its entirety,[2] it imagines the world as a vibrant *oikoumene*, a shared inhabitation of animals, monsters, and humans alike. The world is represented as a vast continuous body of land surrounded by a thin rim of water, the circular *okeanos*. The Hereford map is distinguished by the near indiscernibility of its continents and the absence of bodies of water, echoing the biblical credo that Noah's three sons peopled "the whole earth."[3] The land mass, brought together into a circular shape, restricts rivers to areas that now resemble veins marking the human body. The only clearly recognizable body of water, the Mediterranean Sea, studded with circles and triangles representing its numerous islands, is compressed into the center of the hide, while stylized towns and villages dot the world-continent. Oriented to the east, the map is organized according to Christian principles. On the map's top, Christ's outstretched arms embrace the entire inhabited world, while four rivers flow from the isle of Paradise, located at the world's easternmost boundary. The Red Sea, in a vivid red, allows the onlooker to locate the Arabian Peninsula, while the Pillars of Hercules to the west, inscribed on the bottom of the map, signal the mouth of the Mediterranean Sea and the tentative separation between Europe and Africa. The map's most striking feature, however, is the inversion of the toponyms "Europa" and "Africa": the word "Africa," written in large golden letters, prominently stretches across the entire region north of the Mediterranean Sea, while the land flanking the southern Mediterranean coast and extending to the south bears the name "Europa." Scholars have long noticed this blatant mismatch and suggested that this inversion could be a limner's mistake.[4] A slip of pen or not, the toponymic incongruity is quite telling: around 1300, the map is an artifact in which Europe is but one among other comparable geographical units and may even figure as provincial and interchangeable rather than central to this assembly.

An allegorical map made nearly three hundred years later tells a quite different story of Europe. In 1588, Michael Eitzinger, the Austrian jurist, proto-journalist, and designer of the *Leo Belgicus* map that famously imagined the seventeen provinces of the Low Countries in the shape of a lion, published a rare pocket atlas entitled *De Europae Virginis, Tauro insidentis, topographica atque historica descriptione, liber* (Book on the Topographical and Historical Description of Europe as a Virgin, sitting on a Bull),[5] the frontispiece of which presents a circular map with Europe in the form of Europa, the mythological princess firmly seated on a tamed bull (fig. 1).[6] The well-delineated anthropomorphic and zoomorphic contours of the virgin and the animal here intertwine in a joint representation of a clearly defined continent of Europe. The Iberian Peninsula, "Lusitania coronata" (crowned Lusitania),[7] forms Europa's head, while Italy is imagined as Europa's right arm, firmly holding a scepter. England and Scotland constitute the virgin's left arm, while France, Germany, Poland, Hungary, and Greece cover Europa's body like a long robe. Bohemia, protected by the circular Hercynian Forest, symbolizes the virgin's chaste womb, while the continent's orographic and hydrographic features align to evoke the folds of her dress. A closer examination reveals her seductively exposed left leg as the Baltic Sea, which separates the European mainland from the bull's croup, Scandinavia. Here, a distinctly delineated Europe is made up of bodies of land and water alike.

From a misnamed strip of land on the Hereford Map, almost indistinguishable from other parts of the *oikoumene*, by the second half of the sixteenth century Europe had become the sharply delineated continent of "proud Europe."[8] Eitzinger's contemporary, Luís de Camões, characterized Europe as such in his epic poem, *Os Lusíadas* (1572). Unlike the medieval *mappa mundi*, which imagined Europe as part of the *oikoumene*, and much smaller than Asia, Eitzinger's frontispiece extends Europe alone across the map's surface: only a sliver of Africa's northern coast protrudes from the map's margin, while Asia has entirely disappeared. The first poets to sing the myth of Europa, Moschus and Ovid, had imagined the daughter of the Phoenician king Agenor as a terrified maiden, whom the adulterous Jupiter, disguised as a

Figure 1. "Europa Virgo," in Michael Eitzinger, *De Europae Virginis, Tauro Insidentis* (Cologne: Gottfried von Kempen, 1588), n.p. [47.Jj.69]. Photograph: © Österreichische Nationalbibliothek, Wien.

white bull, abducted from the shores of Asia all the way to Crete at the limits of Europe. In contrast, on Eitzinger's map Europa has assumed the posture of a ruler who dominates the bull and the world alike, pushing the other continents to the margins of the map. In Eitzinger's world view, Europa (no longer an Eastern migrant) has turned into a powerful sovereign, controlling—indeed, embodying—land and sea alike. On Eitzinger's map, all connection to and memory of Europa's Asian origin have been strategically erased, as have all other continents.

The story of Europe's transformation from a derivative of Asia to a sovereign of the terraqueous globe is the story the following chapters will tell. It is not so much a tale of a linear progression of the rise of Europe in early modern times—all too often inflected by a retrospective vision which smooths out the wrinkles of history—as it is an examination of the different cartographic, poetic, and linguistic attempts to imagine Europe as an autonomous continent detached from the land masses surrounding it. *Cartographic Humanism* does not glorify Europe, but offers, rather, a necessary corrective to existing studies of the early modern period which tend to either take the existence of a monolithic and consolidated Europe for granted or reduce their focus of inquiry to Western European countries. Provincializing Europe, as Dipesh Chakrabarty has long called for, considering it as a "figur[e] of imagination whose geographical referents remain somewhat indeterminate" as it powerfully operates "as a silent referent in historical knowledge,"[9] demands a return to the question of what early modern Europe actually means and meant—for us and Renaissance humanists alike. It requires a critical intervention in the cartographic and poetic processes which propelled Europe's rise as a continent and an idea. The hypothesis underlying this book is that the new imagining of Europe as an increasingly sovereign continent was driven by the rise of a novel humanistic discipline: cartography.[10]

What is Europe, where are Europe's borders, and what gives Europe a special status? These questions may seem drawn from today's headlines, but they were essential for the Renaissance humanists who form the subject of this book. The question about Europe's borders is as urgent now as it was in the early modern period, when the idea of Europe as a continent was first emerging.[11] If "border zones, diaspora, and postcolonial relations are daily phenomena of contemporary life,"[12] as Walter Mignolo argues with an eye to the twenty-first century, this book delves into the period when continental boundaries were first drawn and when the question of natural and arbitrary boundaries was formulated in a radically new light. Early modern

Europe is a particularly pressing case that calls for a meticulous examination. Its rise to an ideologically charged metaphor, an idea that came to stand for a "universal and secular vision of the human"[13] and for a "modern world-system"[14]—especially from the sixteenth century on—has had both extraordinary and devastating consequences of global dimensions. While for Mignolo, the investigation of the "darker side of the Renaissance" presupposes "the need to make a cultural and political intervention by inscribing postcolonial theorizing into particular colonial legacies,"[15] *Cartographic Humanism* turns to spatial and geographic thinking[16] as an indispensable prerequisite for the examination of the origins of colonies and for the understanding of the formation of (what was perceived as) an increasingly autonomous Europe: from the unprecedented impact of Ptolemy's *Geography* (*Geōgraphikē Hyphēgēsis*, literally "Geographical Guidance"),[17] a second-century-CE Greek manuscript on geography brought to Italy in 1397 by Manuel Chrysoloras and the first to employ longitudinal and latitudinal lines, to the consequential invention of the cylindrical projection method by Gerhard Mercator for his 1569 world map. *Cartographic Humanism* identifies the increasingly complex spatial imagery and outstanding impact of the map, especially (but not exclusively) with the rise of the Ptolemaic coordinate system in the fifteenth and sixteenth centuries, as a vital, and frequently overlooked, driving force in the making of Europe.

All too often equated with "the West,"[18] Europe is an astonishingly unexamined continent commonly taken as a fixed and stable category, immune to time or history, a monolith within a broader understanding of "the world."[19] In its broad geographic, literary, and linguistic scope, this book functions as a critical wedge that explores Europe's contours from the dual perspective of cartography and literature[20] as it charts new and vibrant itineraries across parts of Europe never before considered together: Germany, Poland, France, Italy, and Portugal. It lingers on the interstitial geographic zones that the course of early modernity has morphed into continental boundaries. Advocating for the necessity to continuously historicize the idea of a continent, subjecting it to careful, interdisciplinary scrutiny, this book argues for a more nuanced investigation of the early modern formation of continental borders than scholars have heretofore undertaken. During the early modern period, far from consolidated, the blurred, messy, and confusing contours of Europe were being only negotiated. They demand to be disentangled, in particular given our insufficient—and often misleading—use of the term "Renaissance Europe." *Cartographic Humanism* traces the formation of Europe back from a metaphor and idea to what Roberto Esposito has recently called

Europe's "bare geographic given" (*nudo dato geografico*).[21] It thereby offers a long overdue corrective to our understanding of what we mean when we say "Europe," in the past but with an eye to the present.

CONTINENTS

Continental awareness in ancient Greece first emerged not as an opposition between Europe and Asia, but as a productive tension between the Aegean islands and the surrounding mainland (Greece and Asia Minor). The Aegean islands (often island-states) maintained a particularly close relationship with powerful coastal city-states on the mainland such as Athens (Greece) and Miletus (Asia Minor), benefiting from a robust and efficient economic and political network that united them, among others, through an elaborate system of daily ferrying (*porthmeutike*).[22] Before ancient Greek mariners called *Europe* and *Asia* the mainland divided by the waterway (the Aegean Sea and the Black Sea, up to the Sea of Azov),[23] the "continent" referenced the Eurasian mainland in opposition to the archipelago. The Greek term for mainland, *epeiron* (literally "without boundaries"), stood in stark contrast to the boundedness of an island, *nesos*.

The close, if antithetical, relationship between mainland and island was generative of an important network between the Greek island-states "able to overcome their insular boundaries and acquire possessions on the mainland"[24] and mighty city-states on the mainland such as Athens and Miletus whose power extended to the islands. The Greeks used the term *peraias* (from περαία, literally "the land across")[25] for the close nexus between island and mainland. For the Greek geographer Strabo the word *peraias* (περαίας, "continent opposite")[26] describes the Iberian mainland across from the city-island of Cádiz—itself stemming from the Latin "Gades," meaning "boundary"—in the vicinity of the mythical Pillars of Hercules, which, in turn, separate Europe from Africa. And when Homer refers, in the catalog of ships in the *Iliad*, to a coastal stretch on the mainland (ἤπειρον), he specifies that it lies "on the shores opposite (ἀντιπέραι) the isles"[27] (ruled by Odysseus). Continental thinking, then, first emerged out of the at once antithetical and productive bond between island and mainland.

Since the beginnings of pre-Socratic philosophy in Ionian Miletus, geography and philosophy unfolded as twin disciplines. If the opposition between mainland and island functioned as a practical tool of maritime transportation and geographic location, it equally operated as the basis of philosophical thought and conceptual definition. Anaximander, the first Greek philosopher, was equally the geographer who, "according to tradition, six to seven centuries before Christ, first dared to represent the inhabited Earth on a tablet."[28] In Ionian thought, the geographic practice of defining entities (islands) and imagining borderless space (mainland) was enmeshed with the foundational logical operations which drive philosophical thought: the act of defining concepts and imagining the infinite. The affinity between *epeiron* and Anaximander's *apeiron* has since its distant beginnings in Milesian Ionia intimately entangled philosophy and geography in their joint pursuit of defining in(de)finite concepts and spaces.[29] The intertwined preoccupation of philosophy and geography concerning definitions and the infinite has led Strabo to confidently open his *Geography* with the claim that "the science of geography, which I now propose to investigate, is, I think, quite as much as any other science, a concern of the philosopher."[30]

When early modern humanists embarked on the endeavor to explore the idea of the continent, they also mobilized the opposition between mainland and island. With the first circulation of Ptolemy's geographic work in Europe, the cartographic genre of the *isolario* (Island Book) suddenly emerged. The *Liber insularum arcipelagi* by Cristoforo Buondelmonti, the first Island Book, appeared around 1420 in Rhodes and Constantinople. Using the word "archipelago" in its etymological sense of "primordial sea" ("archi-" and "pelagus") for the Aegean Sea (not its islands), the genre is unique in that it not only describes but also depicts the islands of the Aegean. But alongside islands proper, Buondelmonti included important coastal cities and places such as Constantinople, Gallipoli, the shores of the Dardanelles, Mount Athos, and Athens among his islands.[31] An iconic image in Buondelmonti's *Liber insularum* is the conceptual and visual insulation of Constantinople (a city nowadays literally bridging two continents), timidly connected to the mainland through a narrow stretch of land. Buondelmonti's *isolario* recreates, as it were, the ancient Greek space of the *peraiai*, the very tension between island and coastline as the core of emerging continental thinking.

To imagine coastal cities as islands was a phenomenon well known to the ancient Greeks, for whom islands signified an indispensable tool for and symbol of imperial consolidation. Herodotus argues that island powers can make "claims to sea power, but mainland powers cannot."[32] Thucydides opens the first book of his *Peloponnesian War* by underscoring the significance of islands for colonial purposes, claiming that "there was no warfare on *land* that resulted in the acquisition of an empire."[33] At the height of the Athenian empire, the city of Athens was described as "the island of Athens."[34] In contrast to the more vulnerable mainland, islands, in Greek thought, have long symbolized political power, strength, and civilization alike. From the fifteenth century on, the popular cartographic genre of the *isolario* engaged with those imperial ambitions of the colonizing powers. When in 1528 the Italian cartographer Benedetto Bordone crafted the first *isolario* on a global scale, he juxtaposed the island-city of Venice with the recently conquered (and by then brutally destroyed) insular city of Tenochtitlan (Mexico City), which, albeit landlocked, was located in the midst of a lake.[35]

In the world of ancient Greek seafaring and empire building, then, the opposition between Greeks and so-called "barbarians" did, at first, not follow a continental divide between Europe and Asia, but rather hinged upon the structural difference between island and mainland within the regional framework of the Aegean world.[36] Because that binary productively spanned the maritime experience of sailors (and merchants) and theoretical philosophical concerns, the continental division which slowly emerged between Europe and Asia was at first felt as utterly arbitrary and impractical. However, by the time Herodotus was writing *The Histories*, the threefold division of the *oikoumene*, with Africa (Libya)—considered in those days to be significantly smaller than Asia—no longer tacitly subsumed under Asia, was widely established. But Herodotus marveled about the randomness with which bodies of water were inadequately used as artificial continental dividing lines—as in the case of the Nile which, far from separating Africa from Asia, visibly traverses Egypt's landlocked provinces.[37] Strabo, writing during the delicate transitional period from the Roman Republic to the Roman Empire when the memory of the beginnings of the threefold continental division was almost entirely erased, subjected to critical scrutiny the tacit, and yet so consequential, slippage between the Greek world of the Aegean Sea and the threefold division of the *oikoumene* which according to Eratosthenes (whose geographic thought Strabo's *Geography* attempts to reconstruct) was devoid of "any practical result":[38]

> The question, then, is whether the "first men" who divided these three continents by boundaries . . . were those "first men" who sought to divide by boundaries their own country from that of the Carians, which lay opposite; or, did the latter have a notion merely of Greece, and of Caria and a bit of territory that is contiguous thereto, without having, in like manner, a notion of Europe or Asia, or of Libya, whereas the men of subsequent times, travelling over what was enough of the earth to suggest the notion of the inhabited world—are these the men, I say, who made the division into three parts [τρια διαιρουντες]? And who, when speaking of three parts and calling each of the parts a continent, does not at the same time have a notion of the integer of which he makes his divi-

sion into parts? But suppose he does not have a notion of the inhabited world [οικουμενην], but should make his division of some part of it—of what part of the inhabited world, I ask, would anyone have said Asia was a part [μερος], or Europe, or a continent [ἤπειρον] in general?[39]

Strabo is puzzled by the weighty substitution of scales which occurred over time, the transition from the world of the Aegean Sea centered on small-scale regions, which he refers to as χωρῶν (*choron*), spanning Greece and Asia Minor (Caria) to the larger scale of the tripartite inhabited world (οικουμενην, *oikoumene*), whose full extension was still unknown to the Greeks (Ptolemy would later term this opposition "chorography" and "cosmography"). Mindful of the leap between the scales and the arbitrariness of continental divisions, Strabo concludes that "the Greeks named the three continents [τρεις ηπειρους] wrongly, because they did not look out upon the whole inhabited world, but merely upon their own country and that which lay directly opposite, namely, Caria, where Ionians and their immediate neighbors now live."[40] Overconfidently substituting one scale for another, Strabo surmises, the Greeks ended up arbitrarily dividing a geographic body that they had no sufficient knowledge of and that was by definition indivisible (*epeiron*). Equipped with sole local expertise, the Greeks set out to divide the entire world (*oikoumene*) instead.

When Jacopo Angelo da Scarperia first translated Ptolemy's *Geography* from Greek into Latin (ca. 1409), he, quite confusingly, rendered the Greek word *oikoumene* as "continens"—thus conflating the entire known world (*oikoumene*) made up of *three* continents with one of its constituent parts: the continent. For the continent understood as a part of the *oikoumene* Scarperia used the Latin word "pars" (part):

Diuturnius tempus eorum notitiam semper certiorem facit, circaque Cosmographiam hoc animadvertendum videtur cum concessum sit ex traditionibus vario in tempore editis, non unas nostri *continentis partes* ob excessum sue magnitudinis nondum ad nostram pervenisse noticiam aliquas vero non quemadmodum sese habent ob peragrantium negligentiam nobis minus diligenter traditas, alias autem esse que nunc aliter quam antea sese habent, sive ob corruptiones sive ob mutationes in quibus pro parte corruisse cognite sunt.

(The passage of time always makes far more accurate research possible; and such is the case with cosmography, too. For the consensus of the very reports that have been made at various times is that many parts of our *oikoumenē* have not reached our knowledge because its size has made them inaccessible, while other [parts] have been described falsely because of the carelessness of the people who undertook the researches; and some [parts] are themselves different now from what they were before because features have ceased to exist or have changed.)[41]

Following his Greek models, Ptolemy symbolically separates the parts of the known world by natural bodies of water (the Tanais River as a boundary between Europe and Asia and the Nile between Asia and Africa). But unlike his ancient Greek predecessors, for whom the *oikoumene* is girded by the endless ocean, Ptolemy inverts the relationship between water and land, expounding that "all the waters of the globe are enclosed by a sole continental mass."[42] The continent in the *Geography* is not only a single world-continent made of Europe, Africa, and Asia (as it was for the Greek geographers), but, in contrast to his Greek predecessors, a continuous land mass which hugs the world's oceans, turning them into landlocked bodies of water (such is, for instance, Ptolemy's description of the Indian Ocean).

If the threefold division of the *oikoumene* into separate continents contrasts with the idea of a continuous land mass, a limitless *epeiron* which, with Ptolemy, more readily isolates bodies of water than bodies of land, such paradox was conveyed by the polysemy of the Latin participle "continens," derived at once from two verbs, "continuare" and "continere." On the one hand, "continuare" means "mak[ing] continuous in space, join[ing] together, connect[ing]" and thus conveys the idea of territorial continuity.[43] This etymology circulated widely and proved influential in medieval encyclopedic texts such as Isidore of Seville's *Etymologies*. In his succinct definition of "continent," Isidore writes: "Continens p[er]petua terra nec ullo mari discreta, quem greci pyron [sic] vocant" (A mainland [*continens*] is a stretch of continuous land uninterrupted by any sea, what the Greeks call ηπειρος).[44] By the fifteenth century, as evidenced in this 1489 edition of the *Etymologies*, the Greek word *epeiron* had already faded so much from collective memory that—with the Greek alphabet in printed books only in its infancy—the typesetter clumsily attempted to recreate the word *epeiron* with the Latin letters "pyron," making the Greek etymology of the word "continent" wholly unrecognizable. The etymology of "continere," on the other hand, encapsulated an impressive (and contradictory) range of meanings. From "hold[ing] together, connect[ing], link[ing], join[ing]" (similar to the verb "continuare"), it flipped into its very opposite, denoting "to enclose, bound . . . keep within certain bounds, limit, confine."[45] Pulling in two directions at once, the verb "continere" alone evokes the image of the continent as self-contradiction: at once continuous and self-contained, the continent challenges any attempt at visual representation. This twofold understanding of "continens" resurfaced with unprecedented urgency during a cartographically informed period which, just like the Latin participle, pulled in two different directions: a Renaissance humanism eager to define and delineate new geographic entities such as continents and an early modernity geared toward expansion, exploration, and colonization of open, seemingly infinite, spaces.

EUROPOIESIS

The rise of the idea of Europe is frequently attributed to the Fall of Constantinople in 1453, which, according to traditional explanatory models, prompted the awakening of a new sense of European belonging under the sign of Christianity. Enea Silvio Piccolomini, later Pope Pius II (1458–64), is readily cited as having "famously equated the medieval term *Respublica christiana* with 'Europe.'"[46] John Marino has recently reiterated the claim that "Pius is credited with being among the first to use the adjective 'europeus' ('European'). From this vast humanist learning, the Sienese pope called for a united, Christian Europe to launch a crusade against the Ottoman Turks, recounted his broad international diplomatic experience, and described in chorographic texts on *Europa* (1458) and *Asia* (1461) the similarities and differences between the two continents, observing in his *Germania* (1457–58) that 'the inhabitants of Asia are always considered inferior to the inhabitants of Europe.'"[47]

But the identification of early modern Europe with Christianity is a starkly simplified and misleadingly reductive model for what was an exceedingly lengthy and highly complex process of geographical, cultural, and linguistic formation. Even in the case of Piccolomini, Denys Hay points to a certain built-in ambiguity in the equation of Europe and Christianity: "Pius II, who provided important illustrations of the trend towards a European spirit, used the phrase *Respublica Christiana* frequently in his writings—more frequently . . . than he used the word Europe. And if by Christendom he meant Europe, that was precisely the ambiguity which was to persist."[48] After all, medieval *mappae mundi* had posited an even more universal claim by equating the entire *oikoumene*—not just Europe—with Christianity, a claim later taken up by Columbus and other early modern travelers who sailed to the New World in the company of biblical prophecies and the firm belief to amass enough gold to launch yet another crusade to "liberate" Jerusalem.[49]

The tendency of scholars of the Renaissance, in past and present, to promote the coincidence between continental delineation and religious denomination has retrospectively ossified Europe as a religious monolith, eclipsing the complex wealth of cultural, linguistic, and religious articulations of a continent only in the making. Marino's contention that "Renaissance humanism identified Europe with Christendom at the very moment when the term 'Europe' first found wide acceptance"[50] is thwarted by the concomitant cartographic enterprise of the humanists themselves: not only is Piccolomini's *De Europa* (first printed in 1509), whose opening chapters are dedicated, perhaps surprisingly to modern readers, to eastern and central Europe,[51] a geographic rather than religious treatise,[52] but it was "the geographer" Giovanni Boccaccio who was the first to advance, in his *Comento alla Divina Commedia* (*Commentary on the Divine Comedy*), the adjectival neologism "Europico."[53] While neither Boccaccio's adjective nor the neologism "europianus" (used by the Polish humanist Maciej Miechowita in 1517)[54] gained currency (which only a retrospective eye can measure), the cartographic drive with which early humanists were forging the idea and the contours of a continent, an entity that has since become "the most elementary of our many geographical concepts,"[55] was unprecedented.

Cartographic Humanism plots an alternative model of defining Europe by privileging cartographic over religious thinking.[56] Without denying the importance of the manifold religious articulations that shaped Europe and without promoting an exclusively secular vision of the continent, this book argues that a shift to geography and cartography can yield a radically new and complex vision of a continent slowly taking shape. At a time when emerging nation-states were competing with empires, city-states popped up within local and regional structures, and an increasingly global world order was on the rise, Renaissance humanists revised and collated ancient and medieval geographic sources to shape a vision of Europe as a newly formed entity between the local (chorographic) and the global (cosmographic) scale. Along the way, their careful geographic investigation yielded a new and long-lasting cartographic terminology. Philologically creative and poetically versed, Renaissance humanists supplanted a vague geographic nomenclature from previous eras (*pars mundi, regio, figuratio*) and advanced a rich new cartographic lexicon, including the words "topography," "continent," and "European," that registered their rising interest in Europe as an autonomous continent and a new epistemological category. I call this process *europoiesis*.[57]

The geographic, poetic, and ideological formation of Europe as a continent took shape at a particularly momentous time: when Renaissance humanists started to think, for the first time, with a complex network of maps in mind. The new mobilization of ancient and medieval geographic sources, which—paired with the novel tools that the printing press offered—gave rise to cartography, substantially changed the humanists' engagement with Europe's heretofore blurred contours. While the European continent had been in the making since antiquity, it was in the fifteenth and sixteenth centuries that it was transformed from a poorly defined part of the world, its borders permeable,[58] to a sharply delineated, hegemonic, and metaphysically charged continent. From the early fifteenth century onward, humanists, politicians, and navigators all explored and exploited geography and map making at once to represent a continent continuously in flux and to spearhead the image of a self-contained Europe. Inextricably linked to philology, poetics, and translation, cartography became, on the brink of modernity, a humanistic discipline that would promote what I call continental thinking[59] (the unprecedented imagining of a new geographic unit, the continent) while propelling the rapid rise of humanism itself.

The powerful intervention of cartography as a new humanistic discipline transformed Europe from "a mapless world"[60]—its uncertain borders entangled in the ambiguous etymology of the Latin "continens"—into an autonomous continent. From the late fifteenth to the late sixteenth century in particular, significant cartographic events recurred with unparalleled frequency and import, often closely intertwined with—if not foreshadowing—key moments in global history: the troublesome contact with a heretofore unknown continent, a "New World" that Europeans would claim for centuries to have "discovered" in the wake of Columbus's crossing of the Atlantic Ocean (1492); the creation of the first (extant) terrestrial globe (1492); the Treaty of Tordesillas (1494), which for the first time cast a straight line as an arbitrary border upon the surface of the globe; the European rounding of the Cape of Good Hope by Vasco da Gama (1498); the circumnavigation of the globe by Magellan (1519–22); and the publication of Copernicus's *De revolutionibus orbium coelestium* (*On the Revolution of the Heavenly Spheres*, 1543), which for the first time challenged the Ptolemaic universe, to name but a few.

The accelerating speed with which (European) regions were explored while Europe's borders were multiplied was made possible by the rise of cartography, not only as a groundbreaking humanistic discipline but also as the only discipline capable of engaging with the fundamental human question, "Where am I?" Cartography crystallized as an increasingly urgent epistemological practice that allowed humans to locate and to be located, to define and to be defined, to include and exclude through the growing practice of drawing arbitrary territorial boundaries. At no point was this more important than in the period now recognized as the brink of early modernity, when Renaissance humanists set out to consciously detach Europe from other parts of the *oikoumene*—first Africa, later Asia and America—by strategically erasing the traces of a shared past and substantially complicating the conditions of possibility for a common future.

Traditionally understood as the revival of classical antiquity, (European) Renaissance humanism is often disassociated from spatial concerns and studied through its most classicizing lenses: philological research, particularly the legacy of antiquity and classical sources; the rise of vernacular languages to the status of ancient languages (mostly Latin and Greek); and the formation of nation-states grounded in a civic humanism that would look back to the ancient world for political models.[61] Frequently associated with western Europe (indeed, Italy), Renaissance humanism is typically "oriented toward the Mediterranean and often assum[es] the continuity of the classical tradition." In contrast, early modernity commonly "turn[s] toward the Atlantic and the Pacific and advocate[s] the perspective of the colonies."[62] Many scholars have, of course, long recognized the cross-fertilization between both strands and the impossibility of drawing clear-cut lines. But it is precisely cartography, Mignolo suggests, which plays a central role in its capacity to cut across these two distinct pathways:

> Alphabetic writing, Western historiography, and cartography became part and parcel of a large frame of mind in which the regional could be universalized and taken as a yardstick to evaluate the degree of development of the rest of the human race. . . . Cartography does not suddenly emerge as a rational enterprise from previous irrational organization of space. . . . The geometric rationalization of space that took place during the sixteenth century, in the context of a much larger epistemic transformation in Western Europe, did not replace the ethnic ones. It redistributed them in significant ways.[63]

Cartographic Humanism emphatically mobilizes cartography as a tool of self-reflection and bends it back to its intra-European workings capable of demasking "the West" (a misleading *pars pro toto* for Europe) and enabling a geographically more balanced view of Europe. It redefines cartography's role *within* Europe by at once challenging the all-too-facile equation between Europe and the West and underscoring cartography's pivotal role for the emergence of Renaissance humanism itself. More than a connector between Renaissance humanism and early modern studies, cartography emerges here in a twofold articulation: as an indispensable heuristic instrument to apprehend the making of the European continent and as a catalyst for poetic, philological, and translational production—the core of Renaissance humanism which, here decentralized,[64] unfolds as a vibrant and complex network connecting Europe's emerging contours in all its cardinal directions.

DRAWING LINES

Four elements of cartographic import were critical in propelling the making of Europe among early modern humanists. Besides the already mentioned impact of Ptolemy's *Geography*, the first geographic treatise to determine place and space by using arbitrary lines, independent of the territory they traversed, there was a concomitant resurgent interest in non-Ptolemaic ancient and medieval geographic texts and models of mapping, such as the Greek and Latin geographies of Strabo, Pliny the Elder, and Pomponius Mela, the Macrobian zonal map, the medieval *mappa mundi*, and the navigational chart—competing forms of imagining space that significantly affected the convoluted process of defining Europe's borders and establishing cartography as a new discipline. Europe's limits were also reimagined through extensive (and problematic) contact with new territories and its inhabitants and, lastly, through the invention and dissemination of the printing press, which fixed on the page not only words but toponyms and territorial boundaries as well. These four distinct yet intertwined cartographic pathways, explored in greater detail below, each share the centrality of language, translation, and philology in their specific spatial manifestations.

Continental thinking crystallized in the first years of the fifteenth century amid a renewed interest in Ptolemy's *Geography*, whose six central books are correlated in number with what Ptolemy considered to be the respective size of the three parts of the *oikoumene*.[65] What sets Ptolemy's work apart from other (previous or contemporary) geographic treatises is, as has been mentioned, his unprecedented use of longitudinal and latitudinal coordinates to provide the exact location of major places of the entire world (as far as it was known to him) in the form of a catalog of place-names.[66] This technique—alongside Ptolemy's new concept of the scale—allowed humanists and scientists to develop a new mode of conceiving and measuring space. In contrast to the medieval *mappa mundi*, modeled according to Christian principles which privileged symbolic places organized around a firm religious center such as Jerusalem (David Woodward has termed this cartographic method "center-enhancing"), and in contrast to the "route-enhancing" practice of plotting space characteristic of (medieval) navigational charts which devises itineraries away from set points of departure, Ptolemy's

Geography proposed an "equipollent-coordinate"[67] spatial model. His decentralized idea of gridding and reckoning space allowed him to allot equal importance to all points and locales on the surface of the earth. This spatial outlook was, Franco Farinelli surmises, the most democratic one, in fact, one that ushered in the condition of possibility for the development of democracy writ large: "Where is democracy born from," Farinelli asks, "if not from the same cartographic grid [*matrice*] which births, together with the first geometric city, the territory?"[68]

The rise of arbitrary boundaries, disconnected from the physical territory and grounded in the alleged impartiality and universality of mathematics, has long been hailed as "the quintessential modernity of Renaissance cartography."[69] One of the first concrete early modern geopolitical impacts of Ptolemy's coordinate system was a series of papal bulls and treaties issued in the wake of New World "discoveries," the Treaty of Tordesillas (1494) being the most famous, which demanded that a straight line (alternately called *linea* or *raya* in the Spanish original) be drawn on a map across the Atlantic, from the North to the South Pole. What appears to be a continental divide between Europe and the New World on maps, is, in fact, an intra-European line of demarcation: it signals the division of the globe into a Spanish and a Portuguese sphere of influence. As the treaty stresses, the straight line ought to be drawn "trezientas e setenta leguas delas dichas yslas del Cabo Verde por rrota derecha ala parte del poniente" (370 leagues west of the Cape Verde Islands) across the Atlantic Ocean. Independent from the physical landscape it would pass, be it on water or land, "se ponga la linea de la dicha partiçion, figurandose del dicho polo Artico al dicho polo Antartico" (one shall cast the line of said partition, imagined from the Arctic to the Antarctic pole).[70] Proved too difficult to implement and defend *in situ*,[71] the *linea* exerted its geopolitical power, impacting (early modern) territorial and colonial divisions exclusively on the map.

As the example of the *linea* shows, cartography's central role in early modern Europe was a mixed blessing. More akin to a *pharmakon*—present and poison alike—it ushered in democracy's demise while establishing its principles. It is a great paradox of Renaissance humanism that its cartographic underpinnings promoted a concept of space which, unlike its medieval counterparts (bound up with the individual experience of time and inextricably entangled with the embodied practice of travel),[72] disengaged the human body (deemed an unreliable yardstick to measure space) from spatial measurement: with the abstract coordinate system on the rise, the physical human presence on the ground was no longer felt to be needed as an essential element to reckon space (the medieval traveler's step was substituted by early modern volumetric sight and the method of triangulation).[73] Robert D. Sack argues in *Homo Geographicus* that human interactions and social formations "cannot be adequately comprehended unless the geographic is seen as a primary strand tying them together."[74] But during the Renaissance, the human body became a growing impediment for what evolved as a more rational, objective, and efficient manner of producing territorial lines. The physical human body was substituted by an allegorized body: popular mythological figures such as Atlas turned into powerful narratives of cartographic production itself in which "the mapmaker *embodies* the world,"[75] while newly conceived allegorical visualizations of Europe (as Princess Europa) started circulating, in particular from the first half of the sixteenth century on, increasingly eclipsing the physical body as a previously necessary tool for cartographic thinking.

The early modern period, intimately tied to its image of a boisterous human being taking center stage, was, paradoxically, the first period to take the human out of the (humanistic) equation. Cartography is an incisive case in point: a layer of abstract lines and dots, the equipollent map became the dominant interface through which the world was captured, measured, and contemplated. This is when, as Farinelli suggests, the world was transformed into its own model, ushering in what Martin Heidegger would call, centuries later, "the age of the world-picture."[76] The idea of a grid map upon which national and continental boundaries are drawn as arbitrary lines "independent[ly] of the geographical content beneath"[77] disconnected cartographic borders from the physical landscape, from humans and their habitat. This act of tracing arbitrary territorial lines, which was widely adopted in the early modern period and beyond, ultimately prepared the ground for cartographic decisions with dramatic global consequences, from the Treaty of Tordesillas in 1494 to the Berlin Congo Conference in 1884–85[78]—and well into the present. *Cartographic Humanism* speaks to the continuous oscillation between the presence of the human body in and its withdrawal from a cartographically inflected humanism.

THE FIRST PRINTED MAP

From the mid-fifteenth century on, the process of mapping Europe as a continent with clear-cut boundaries was both accelerated and complicated by the momentous invention of the printing press. Konrad Sweynheym, among Germany's very first printers, not only introduced the new technique to

Italy by setting up the first printing press in Subiaco, but also printed, together with Arnold Buckinck, one of the earliest editions of Ptolemy's *Geography* (Rome, 1478)—and the second edition containing maps.[79] The first printed map, however, was published in the imperial city of Augsburg in 1472. It was not a Ptolemaic map but an illustration of Isidore of Seville's seventh-century *Etymologies*: a diagrammatic woodcut *orbis terrarum* (T-O) map,[80] not unlike the Hereford Map, with symbolically conceived continents.[81] Here, the three known parts of the *oikoumene* (which the map isolates by bodies of water) are matched with the three sons of Noah: Asia is peopled by Shem, Africa by Ham, and Europe by Japheth.[82] But the correspondence between continent and Noachian population, all too often assumed to stem from the authority of the Bible and to coincide with the biblical episode of the flood, developed only over the course of the first thousand years of Christianity. The vision of historians of early Christianity such as Josephus and Church fathers such as Jerome and Augustine of how the world was populated stood in stark contrast to the way that their fifteenth- and sixteenth-century editors, commentators, and illustrators depicted it. The early Church fathers located Noah's sons with no regard to the regions early modern humanists started to identify as continents. Josephus, for instance, argued that Japheth inhabited a region straddling Europe and Asia, Ham's descendants populated the border region of Africa and Asia, while Shem's offspring spread across the region of Asia.[83] The *Etymologies* of Isidore of Seville also follow Josephus's complex "Afrasian, Eurasian, and Asian designations," as Benjamin Braude points out.[84]

When the first printed map of 1472 visualized the Isidorian description of the parts of the world, it did so by dramatically reinforcing continental dividing lines and the consequential distribution of peoples. The complex Noachic geography that had been in place for over a millennium is supplanted on the T-O map by a clear-cut continental division: each son is associated with one continent. What during Isidore's own lifetime had been a united *oikoumene* is imagined on this map as clearly disconnected continents, ruled by stylized geometry and laws of abstraction. Here, the parts of the world are subject to reduced geometric shapes and the dictates of simple lines. The printing press (in its infancy and operating with limited means) required a clear, easy, and straightforward distribution of lines and toponyms for its creation of maps. Paradoxically, perhaps, it reinforced the image of geographic immobility and cartographic symmetry in the very moment when the *oikoumene* was bound to lose its religious and geographic center, Jerusalem, becoming progressively decentralized.[85] The standardization of toponyms and geographic lines on maps proved progressive and conservative alike—so much so that scholars now compare the fifteenth-century cartographic correlation of Noah's three sons with the three continents to "a movement from medieval polyphony to modern monophony in the understanding of the Bible"[86] and its geographic spaces. The layout of the printed page reduced the complexity of what was, beyond the map, the messy reality of the physical world. Continents crystalized when necessarily simplified territorial lines, following the constraints of (early) printing, were fixed on the printed page.

SPACE

Preceding Ptolemy's *Geography* by two hundred years, Strabo's *Geography* was central for the introduction of a new concept of space, one that privileged a standardized metrical interval for the reckoning of distances. Farinelli summarizes this mathematical concept of space, in contrast to place, in the following way:

> Space . . . is a word which derives from the Greek *stadion*. For the ancient Greeks, the stadium was the unit for measuring distances, and thus signified literally a standard linear metrical interval. Deriving from it is the fact that, within space, all the parts are equivalent to one another, in the sense that they submit to the same abstract rule which takes no account of their qualitative differences. Such a rule is represented by the scale, which from the sixteenth century begins to appear systematically on maps . . . , and indicates the relationship between linear distances in drawing and those that exist in reality. Place, on the contrary, is a part of the Earth's surface which does not match any other and which cannot be exchanged with any other without everything changing. . . . In space, instead, every part can be substituted by another without anything being altered, exactly as when two things which have the same weight are moved from one side to the other of a set of scales without the equilibrium being compromised.[87]

This understanding of space goes back to Strabo (a crucial source for Renaissance cartographers), who was acutely aware of the necessity of introducing a standardized concept of space when measuring territories. He writes:

> The measurement of the length of the inhabited world [οικουμενης μετρησις] is made along a line parallel to the equator, because the inhabited world, in its length, stretches in the same way the equator does; and

in the same way, therefore, we must take as the length of each of the continents the space that lies between two meridians. Again, the measure [μετρα] employed for these lengths is that by stadia [σταδιασμοι]. . . . But no one employs rules and measures that are variable for things that are non-variable, nor reckonings that are made relative to one position to another for things that are absolute and unchanging.[88]

Aligning the contours of the *oikoumene* with the straight line of the equator, Strabo strikingly privileges cartographic over physical space, an idea that should become central to early modern thought. The outlines of the inhabited world are here subject to the dictates of the abstract lines which exert a dual power: they correlate the physical landscape to the framework of the coordinates and they function as a unit of measure themselves, since *stadion*, the standardized unit, is derived from the in-between space connecting two (abstract) meridians.

The abstract metric system, advanced by Strabo and later by Ptolemy, was confidently appropriated by early modern cartographic humanists. It bore the potential to move away from center-enhancing models of mapping which privileged religious and otherwise symbolic centers and instead measure the world through a more secular and objective lens. But the concomitant interpolation of other ancient geographic texts thwarted such enterprise and furthered, instead, an increasingly privileged vision of Europe among the three parts of the *oikoumene*. In his *Naturalis Historia (Natural History)*, Pliny (born in the same year Strabo died) characterizes Europe as the "nurse of the race that has conquered all the nations, and by far the loveliest portion of the earth, which most authorities, not without reason, have reckoned to be not a third part but a half of the world, dividing the whole circle into two portions by a line drawn from the river Don to the Straits of Gibraltar."[89] A towering authority throughout the Middle Ages and well into the early modern period, Pliny's focus on the special status of Europe, morphed into the idea of Europe's geographic grandeur, stood in stark contrast with the Ptolemaic cartographic model.

Far from being dismissed with the rise of a more "objective" cartography, Pliny's idea of Europe's outstanding size was echoed, in the second half of the sixteenth century, in a cartographic model which today is the most authoritative mode of drawing world maps: Gerhard Mercator's cylindrical model of projection, at once the alleged apogee of cartographic standardization and concomitant privileging of Europe, which unabashedly aggrandizes Europe's size. The Flemish cartographer Mercator first introduced his consequential projection method in his world map, published in 1569 with the lengthy title *Nova et aucta orbis terrae descriptio ad usum navigantium emendate accommodata* (*A New and Enlarged Description of the Earth with Corrections for Use in Navigation*).[90] Commonly hailed as a milestone of rational and progressive cartography and the foundation of GoogleMaps and GIS, Mercator's projection "offered the first consistent method of projecting loxodromes [rhumb lines] as straight lines which, at least in theory, enabled navigators to accurately plot a straight line across the globe that took into account the curvature of the earth's surface."[91] Yet by having to stretch out the regions closer to the two poles in order to fill the map's upper and lower edges, Mercator's method necessarily distorted those regions (such as Europe) by disproportionately enhancing their relative size. As Marshall Hodgson provocatively contended:

> Some say the Mercator world map is so popular because it shows the correct angles essential for navigation. . . . But if you use a map not for navigating but for placing and comparing at a glance different parts of the world, shapes and areas are more important than angles. Moreover, areas are more important than shapes, because they have cultural implications. What is objectionable about the Mercator world map in fact is not that it distorts the shape of North America, nor even that it shows Greenland so large. . . . Rather, it is that it shows India so small, and Indonesia, and all Africa. (I call such a world map the "Jim Crow projection" because it shows Europe as larger than Africa.)[92]

The Mercator map serves as a reminder that maps are not objective depictions of space, although they are often imagined and interpreted as such. Maps have the power to—and routinely do—represent territories out of proportion, ranking and evaluating them not according to their actual size but instead according to perceived cultural currency and ecopolitical value. Mark Monmonier points out that "even folks who are routinely suspicious of written texts equate maps with fact and fail to realize that no map is capable of including all information or telling all possible stories. In fact, the process of mapmaking requires cartographers to limit content in order to create a readable map and so allow them to manipulate their audience with the information they choose to include."[93] Maps have the capacity to strategically influence and shape their onlookers' perception of the world, guiding them across the surface of the represented world from the mapmaker's commanding yet invisible position.

CARTOGRAPHIC HUMANISTS

To conceive of the map as an oblique artifact blending science, poetics, and visual arts and capable of manipulating the onlooker with its potent means, was precisely the way Renaissance humanists related to maps all along. Maps, scholars insist, are not only "one of the oldest forms of human communication" but also "fundamental tools" of human knowledge[94]—and "the bedrock of most professions and disciplines."[95] This timeless utility was vitally evident to the early modern humanists who set out to explore geographic texts and produce maps as both epistemological tools and poetic artifacts. In a parallel movement that at once exposed and fostered the affinity between poetic and spatial plotting, humanists designed maps with poetics in mind while incorporating cartographic imagery into poetic and philological writing. With an eye to Renaissance France, Tom Conley has termed "cartographic writing"[96] the fertile tension between discourse and space. Revisiting, translating, editing, and imitating ancient texts under the aegis of cartography became, then, an opportunity to experience and design Europe anew. European Renaissance humanism, the following chapters argue, takes on a wholly new turn if explored through the lens of cartography.

From its inception, humanism unfolded, as this study argues, in tandem with cartography. Educational theory throughout the European Middle Ages divided the *studia humanitatis* into the linguistic *trivium* of grammar, rhetoric, and logic, and the mathematical *quadrivium* of arithmetic, geometry, astronomy, and music.[97] But, by around 1500, the ascendant discipline of cartography turned spatial thinking into an overarching approach that connected and shaped the various domains of learning. Albrecht Dürer's allegorical woodcut of *Philosophia* (fig. 2), included in Conrad Celtis's *Quatuor Libri Amorum secundum Quatuor Latera Germanie* (*Four Books of Love according to the Four Sides of Germany*, 1502), illustrates the rising importance of cartography as a means to introduce spatial thinking into the *studia humanitatis*: here, cartography, represented by Ptolemy, towers over the seven liberal arts, suggesting a heretofore unthinkable spatial outlook of the *studia humanitatis* informed by new and pressing concerns about location and positionality in the pursuit and design of knowledge.

It is quite fitting in this regard that Petrarch, commonly hailed as the first humanist *par excellence*, conveys the coincidence between cartography and humanism across his *oeuvre*: not only do we find his diligent annotations to a twelfth-century copy of Pomponius Mela's *De chorographia* (which Petrarch acquired as early as the mid-1330s) among his first philological endeavors, but also his *Vita solitaria* (*The Life of Solitude*) and

Figure 2. "Philosophia" by Albrecht Dürer, in Conrad Celtis, *Quatuor Libri Amorum secundum Quatuor Latera Germanie* (Nuremberg: Sodalitas Celtica, 1502), n.p. [Typ. 520 02. 269]. Photograph: Houghton Library, Harvard University.

Familiares (*Letters on Familiar Matters*) showcase his unrelenting geographic curiosity about the distant limits of Europe (such as the Fortunate Islands and Thule).[98] The humanists' groundbreaking cartographic desire to merge diverse (one is tempted to say, interdisciplinary) modes of expression led them to a novel understanding of the world surrounding them and to a new approach to ancient and medieval poets, historians, and philosophers. Suddenly, Homer, Cicero, Virgil, Pliny the Elder, Isidore of Seville, and Albertus Magnus became cartographic luminaries, who, when read in this radically different light, indicated hitherto untrodden paths across a continent in the making.

These cartographic humanists (poets, writers, geographers, and historians whose work is driven by geographic and cartographic preoccupations) analyzed here are brought together through their common goal: they all embarked on an unprecedented endeavor to grasp a substantially changing world by calibrating the creation of continental borders against the backdrop of an expanding Europe. While recent studies contend that "it is now a commonplace to argue that the rhetoric of the universal and the global merely masks the imposition of European norms onto 'the rest of the world', often through the violent assimilation of the colonizing process,"[99] *Cartographic Humanism* is the first to venture into the very suture lines that initially set Europe apart from "the rest of the world," those crucial borders where the differentiation between Europe and its continental counterparts started to originate with the rise of cartography as a new humanistic discipline.

Centered on five case studies, *Cartographic Humanism* covers a lot of fresh ground without claiming to be comprehensive. By bringing together unexplored and often untranslated material with texts that have had the good fortune of gaining more currency (among other ways, through available translations into English), this book is the first to engage with the rise of Europe as a continent from a profoundly interdisciplinary and inclusive perspective and in languages that span Europe's cardinal directions. By carefully gauging a geographically balanced view and by adopting, in particular, a dual lens that brings together cartography and comparative literary studies, it provides a long overdue corrective of our understanding of the manifold articulations of Europe's poetic, philological, and translational production, always mindful of humanism's cartographic underpinnings. The texts discussed in *Cartographic Humanism* speak to the astonishing complexity and deep transformation in the imagining of Europe on the brink of modernity: Conrad Celtis's neo-Latin cartographic books of elegies outline a "new Europe" while aspiring to position Germany as Europe's navel; the influential geographic work by Maciej Miechowita (on the complex border between Europe and Asia), which critically challenges Ptolemy's description of Europe's east, constitutes what I call the first early modern border studies manifesto; Geoffroy Tory's manual on the standardization of the French alphabet is revealed as perhaps the most complex—and, at the same time, the least explored—theoretical inquiry into the fundamental question of what constitutes a continent and what role languages play in it; in Girolamo Fracastoro's first new world poem—a unique Neo-Latin blend of cartography, philology, and poetics—the continental boundaries between Europe and the New World are imagined in constant flux through changing sea levels; and, finally, Luís de Camões's epic poem, written in a liminal moment in Portuguese history, when the kingdom's unquestioned dominance and colonial power was fading, heralds the dark side of cartography by epitomizing the conflation, all too easy, between globalization and Europeanization.

CHAPTER ONE

Gridding Europe's Navel: Conrad Celtis's *Quatuor Libri Amorum secundum Quatuor Latera Germanie* (1502)

In 1493, Conrad Celtis (1459–1508), the first poet laureate of the Holy Roman Empire, signed a contract with the influential Nuremberg politician and merchant Sebald Schreyer, who that year had just sponsored the publication of Hartmann Schedel's hexameral *Liber Chronicarum*, widely known as the *Nuremberg Chronicle*,[1] a lavishly illustrated work described as "the most complex printing project before 1500."[2] With the immense success of the *Chronicle* in mind, Schreyer was eager to invest in a second edition of the volume, the contract for which specified the addition of a hitherto unprecedented emphasis: a "New Europe" (*Newen Europa*).[3] The choice of "the poet Conrad Celtis" to "correct anew" and "recast in a different form"[4] the *Chronicle* is hardly surprising: Celtis would rise to become the German lands' first and foremost cartographic writer, whose poetic work displays profound knowledge of Ptolemy's *Geography* and a keen interest in the cartography of his time. He was the first early modern humanist to introduce the word "topography" as a critique of the traditional Ptolemaic dichotomy between chorography and cosmography, which he found insufficient to reflect Europe's rapidly changing contours.

Celtis's project of revising Schedel's *Chronicle* marks the beginning of a momentous shift in the making of Europe. If the tradition of hexameral writing is often seen as offering "poignant responses to a crisis of world order,"[5] Celtis's own work grapples with the delineation of a smaller, newly emerging entity: the continent of Europe. In the end, Celtis did not uphold his end of the contract with Schedel and no second edition of the *Chronicle* was ever published. Yet Celtis's interest in defining a "new Europe," a project executed by Sebastian Münster in 1544, took shape in another large-scale project: the *Germania illustrata*,[6] with an eye to Flavio Biondo's *Italia illustrata*, whose core—four books of love elegies, a description of Nuremberg (*Norimberga*), and a treatise on Germany (*Germania generalis*), among others—appeared in Nuremberg under the title *Quatuor Libri Amorum secundum Quatuor Latera Germanie* in 1502.[7] Prefaced by a woodcut by a young Albrecht Dürer and accompanied by fanciful regional maps of Germany by different artists, it is Germany's most original example of early modern cartographic humanism. The *Quatuor Libri Amorum*, at the center of this chapter, is a unique realization of Celtis's desire both to capture poetically and cartographically the outlines of a "New Europe" in a dramatically changing world and to strengthen Germany's position within it. As he avoids facile explanatory models, Celtis proposes the question of defining Europe, in all its breadth and complexity.

TRANSLATIO CARTOGRAPHIAE

The *Quatuor Libri Amorum*, written in neo-Latin hexameters, has been defined as a work of "erotic topography."[8] Celtis himself referred to the four books of poems as "lasciva quaedam nostra carmina" ("those lascivious songs of mine").[9] In the Latin elegiac tradition of Ovid, Propertius, and Tibullus, the four books of elegies proper—the core of the *Quatuor Libri Amorum*—follow the story of a lovesick protagonist, a secular pilgrim in search of erotic encounters, from his youth to his old age, along Germany's borders, represented by the four cities of Kraków, Regensburg, Mainz, and Lübeck. With an eye to both Francesco Petrarca's *Canzoniere* (revised until the poet's death in 1374) and Matteo Boiardo's *Amorum libri tres* (composed between 1469 and 1476), the *Quatuor Libri Amorum* is a groundbreaking attempt to embed love poetry within a cartographic framework. As scholars have pointed out, the genre of the *poema elegiacum* was "integrative";[10] it allowed the elegiac poet to experiment across disciplines. Yet Celtis's *Quatuor Libri Amorum* is unique in its imaginative formulation of a tight nexus between poetry and cartography, combined and redirected here in a hitherto unprecedented manner: to design the contours of Europe.

The *Quatuor Libri Amorum* opens with a powerful iconographic statement: a woodcut by Albrecht Dürer, probably Celtis's most famous mentee, on one of the volume's first pages (fig. 2). Commissioned specifically for Celtis's volume, the woodcut is one of Dürer's earliest artworks. Here, Philosophy is represented as an allegorized female figure, whose heart coincides with the tip of a pyramid-shaped *scala artium*, at the base of which appear Dürer's initials. Inscribed in the pyramid are the seven liberal arts: grammar, logic, rhetoric, arithmetic, geometry, astronomy, and music. Two mottos, located at the top and the bottom of the woodcut, reproduce Philosophy's words and trace the movement of *translatio studii* from antiquity (with an allusion to

Afranius's racy comedies) to the present (Celtis's own Germany): "Sophiam me Greci vocant Latini Sapientiam / Egipcii & Chaldei me invenere Greci scripsere / Latini transtulere Germani ampliavere" (The Greeks call me "Sophia," the Latins "Sapientia," / The Egyptians and Chaldeans invented me, the Greeks wrote me down, / The Latins translated me, the Germans amplified me).[11] Arranged into three lines, this motto denotes the *trivium*, while the motto on the bottom of the woodcut, composed in four lines, is a nod to the four elements and the *quadrivium*: "Quicquid habet Coelum quid Terra quid Aer & aequor / Quicquid in humanis Rebus & esse potest / Et deus in toto quicquid facit igneus orbe / Philosophia meo pectore cuncta gero" (Whatever holds the Sky, whatever the Earth, whatever the Air and the sea / Whatever in human Things there is and can be / And whatever the Fire god makes in the whole world / I, Philosophy, bear all in my breast).[12] Raymond Klibansky, Erwin Panofsky, and Fritz Saxl remark that in this woodcut Dürer's figure is deeply rooted in the "medieval tradition"[13] of allegorical representations of "Lady Philosophy" and contend that the allegory, lacking novelty, simply perpetuates a millennial tradition of depicting Philosophy as a female allegory, initiated by Boethius: Renaissance humanism, these scholars seem to suggest, had not yet penetrated north of the Alps. Celtis scholars have largely followed suit.[14]

But a more careful examination reveals that in the first medallion atop the garland Dürer inserts none other than Ptolemy holding an armillary sphere, suggesting that "Philosophia" and the *translatio studii* originate with the second-century-CE Roman geographer from Alexandria. Plato, who stands for the "Grecorum Philosophi" (Greek Philosophers), and Cicero and Virgil, representatives of the "Latinorum Poetae et Rhetores" (Latin Poets and Rhetoricians), are all anachronistically displaced: despite having preceded him in time, they appear on Dürer's woodcut as *followers* of Ptolemy. In the last medallion, Albertus Magnus, the medieval German natural philosopher and cartographic thinker,[15] stands to represent the "Germanorum Sapientes" (German Philosophers). It is with Albertus Magnus that the movement of *translatio*—an itinerary leading from Ptolemy's Egypt to Greece, Rome, and finally Germany—comes to a close. The chronological framework is here disrupted and supplanted by spatial reasoning. In its embrace of the *translatio studii* and the seven liberal arts, Philosophy is here cast under the aegis of geography and cartography as a new epistemic tool and unique key to Celtis's work. The woodcut emerges as what may indeed be termed a *translatio cartographiae*, a critical reflection on the rise of cartography and its impact on the organization of knowledge and learning from Ptolemy's Alexandria to Celtis and Dürer's own Germany. In an epochal shift, cartography becomes a strategic intervention, a critical enabler of philosophy and poetry alike, allowing Albertus (the author of, among other books, *De natura locorum*, a widely acclaimed geographic treastise)[16] to complete the arc of *translatio cartographiae* by vigorously redirecting cartographic thought to the German lands.[17]

A catalyst for cartographic writing, Philosophy in Dürer's woodcut becomes something new, a dynamic interplay of poetics and visual arts under the overarching umbrella of a newly emerging discipline. As he combines text and image throughout *Quatuor Libri Amorum* (by including creative woodcut illustrations and imaginative regional maps of Germany in all four books of elegies), Celtis strengthens Germany's position as a producer of cartographic humanism. Celtis powerfully invigorates the tradition of cartographic writing,[18] even with regard to Italy, whose printed geographies were commonly more "iconophobic": from Biondo's *Italia illustrata* to Leandro Alberti's 1550 *Descrittione di tutta Italia*, cartographic Italian works were consistently published without maps even, as Theodore Cachey observes, "at a time when printed maps were in wide circulation and quite common in geographical literature, *isolarii*, and atlases."[19] By casting Philosophy within a framework that binds the cartographic word and image to one another, Celtis and Dürer effect an unparalleled change in perspective and positionality—from Italy to the German lands.

NUREMBERG: GRIDDING THE NAVEL OF EUROPE

Celtis's cartographic desire to conceive of a "New Europe" was perhaps propelled by Schedel's strategic decision to position at the exact center of his *Chronicle* a visual and narrative description of Nuremberg, which, by around 1500, had reached the peak of its wealth, eminence, and cultural and scientific achievement. It was considered the "secret capital" of the Holy Roman Empire, where the imperial regalia, the so-called "Nürnberger Kleinodien,"[20] were kept from 1424 to 1796. It was also the epicenter of the production of maps and globes and of artistic, cartographic, astronomical, and mathematical innovation. Around 1493, Georg Stuchs printed here the so-called "German Ptolemy,"[21] a treatise containing thirty-five sheets and the first map using a globular projection; here, Martin Beheim and Johann Schöner created the first extant terrestrial globes in 1492 and 1515, respectively;[22] the century's most revolutionary and courageous book, Nicolaus Copernicus's *De revolutionibus orbium coelestium* (*On the Revolutions of the Heavenly Spheres*), was originally printed in Nuremberg in 1543 by Hans Peterlein (Johannes Petreius). Nuremburg was host to many artists and scientists:

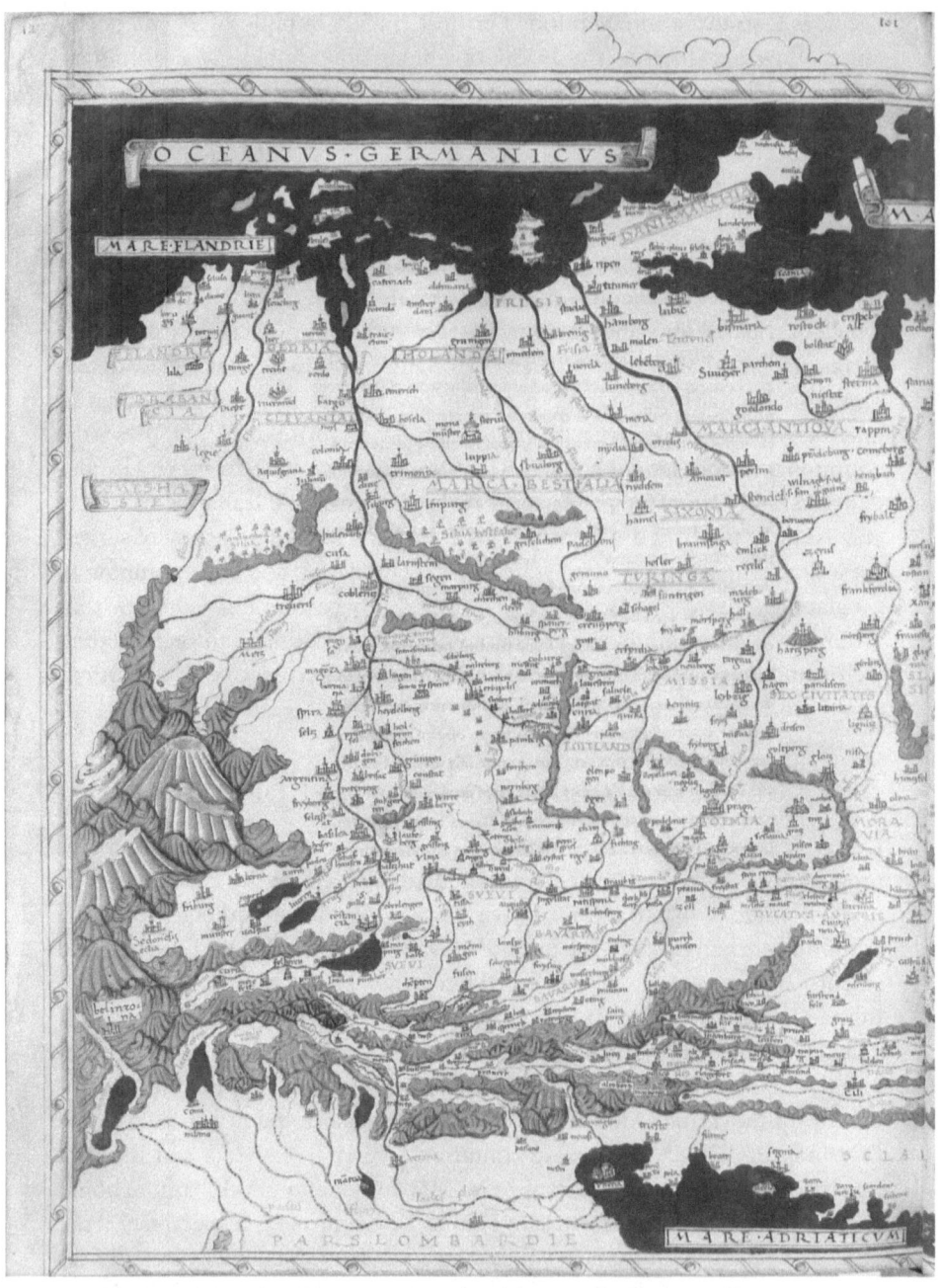

Figure 3. "Insularium illustratum" (Cusanus map), Central Europe by Henricus Martellus (Northern Italy, ca. 1480) [Ms698-folio65recto and Ms698-folio65verso].
Photograph: © RMN-Grand Palais / Art Resource, New York.

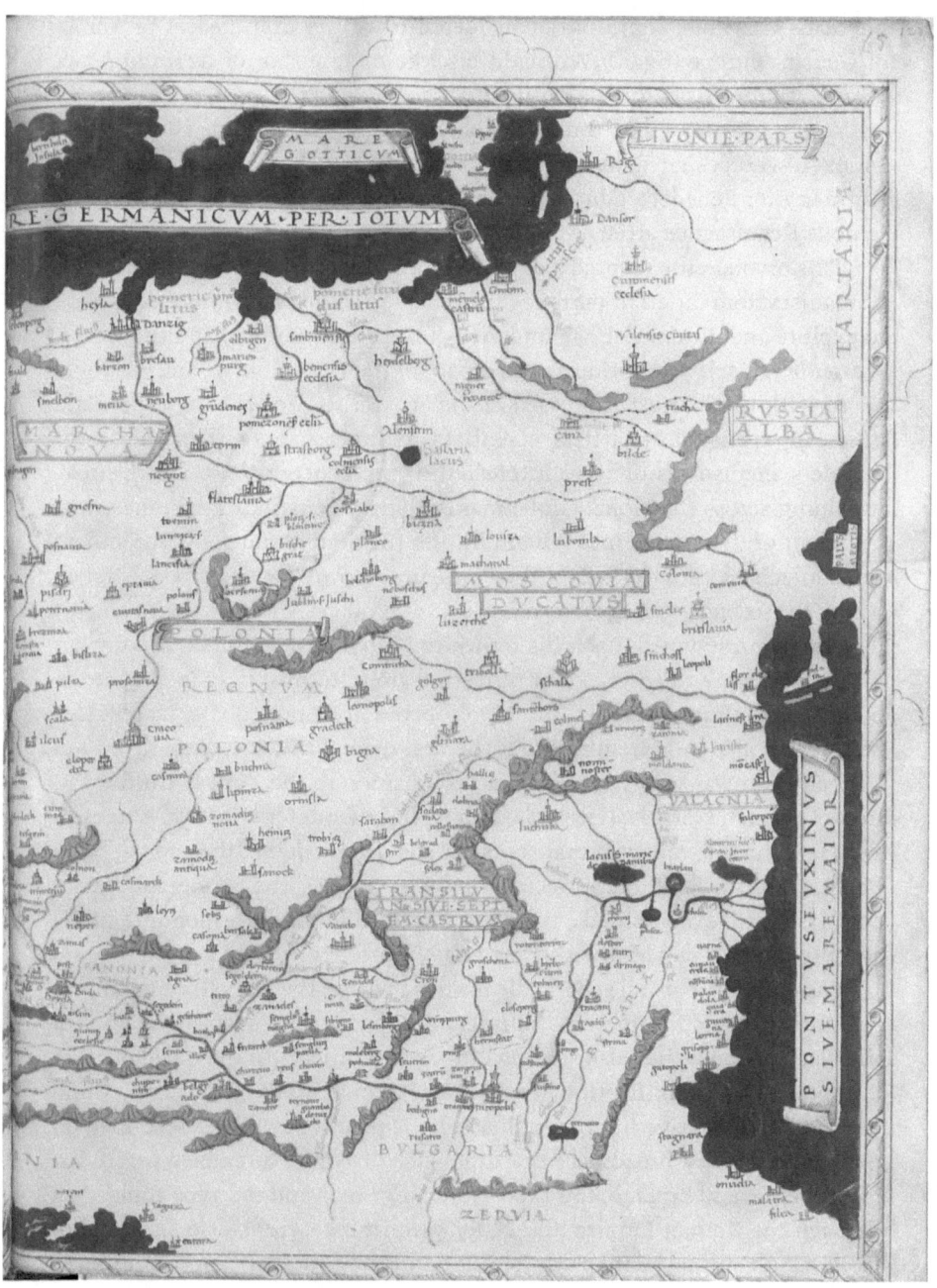

Figure 3. (continued)

Nicolaus Cusanus, commonly considered to be the first creator of a map of Central Europe (fig. 3); Willibald Pirckheimer, author of *Germaniae ex variis scriptoribus perbrevis explicatio* (*Short Exposition of Germany based on Various Writers*) and translator of Ptolemy's *Geography*; Hieronymus Münzer, who designed, among many other things, the first printed map of Germany for Schedel's *Chronicle*; and, last but not least, Germany's most famous Renaissance artist, Albrecht Dürer.

This pivotal city adopted a dual mission: to attract scholars, artists, and humanists from different parts of Europe and to serve as a point of departure to explore and measure the boundaries of Germany, Europe, and the world. Nuremberg's self-promotion as the center of the Holy Roman Empire dovetailed with the interest expressed in the city for territorial exploration and overseas expansion. Its artists, scientists, and merchants either were eager travelers themselves or were interested in the concomitant "discoveries" unfolding across the globe. Globe maker Martin Behaim, for instance, was a student of the astronomer Johann Müller (Regiomontanus) in Nuremberg and a merchant who worked for the Portuguese crown. Having probably taken part in the second voyage of Diogo Cão (1485–86) along Africa's western coast,[23] Behaim commissioned, upon his return to Nuremberg, the creation of what is now known as the first extant terrestrial globe, termed in official contemporary documents issued by the Nuremberg City Council the "apfel" (apple)."[24] It is also in Nuremberg that the first map of the Aztec capital city of Tenochtitlan was printed, as a visual companion to the 1524 Latin translation of Hernán Cortés's letters describing his encounter with the Aztecs in 1519 and the subsequent brutal conquest and destruction of the capital.[25] In a Venetian edition of Ptolemy's *Geography* from 1548, the cartographer Giacomo Gastaldi even named a region in North America corresponding to the coast of modern-day Maine as *Tierra de Nurumberg*.[26]

Nuremberg's central geographic position within the Holy Roman Empire prompted German humanists to imagine the city as a navel—not only of the German lands, but also "totius Europae umbilico" (as the navel of all of Europe).[27] The "umbilication" of the city,[28] which marks its centrality and function as a creative fulcrum and symbolic place of origin, is inscribed on two contemporary broadsheet maps produced by the Nuremberg-based cartographer Erhard Etzlaub, the 1500 "Romweg" map and the 1501 route map through the Roman Empire (fig. 4). By performing a *translatio cartographiae*, Etzlaub shifts the map's emphasis from the spiritual center of Europe, Rome, to Europe's new cartographic center, Nuremberg. Reminiscent of late medieval portolan charts for sea navigation and their use of the compass and rhumb lines, but already endowed with a scale which measures the distance

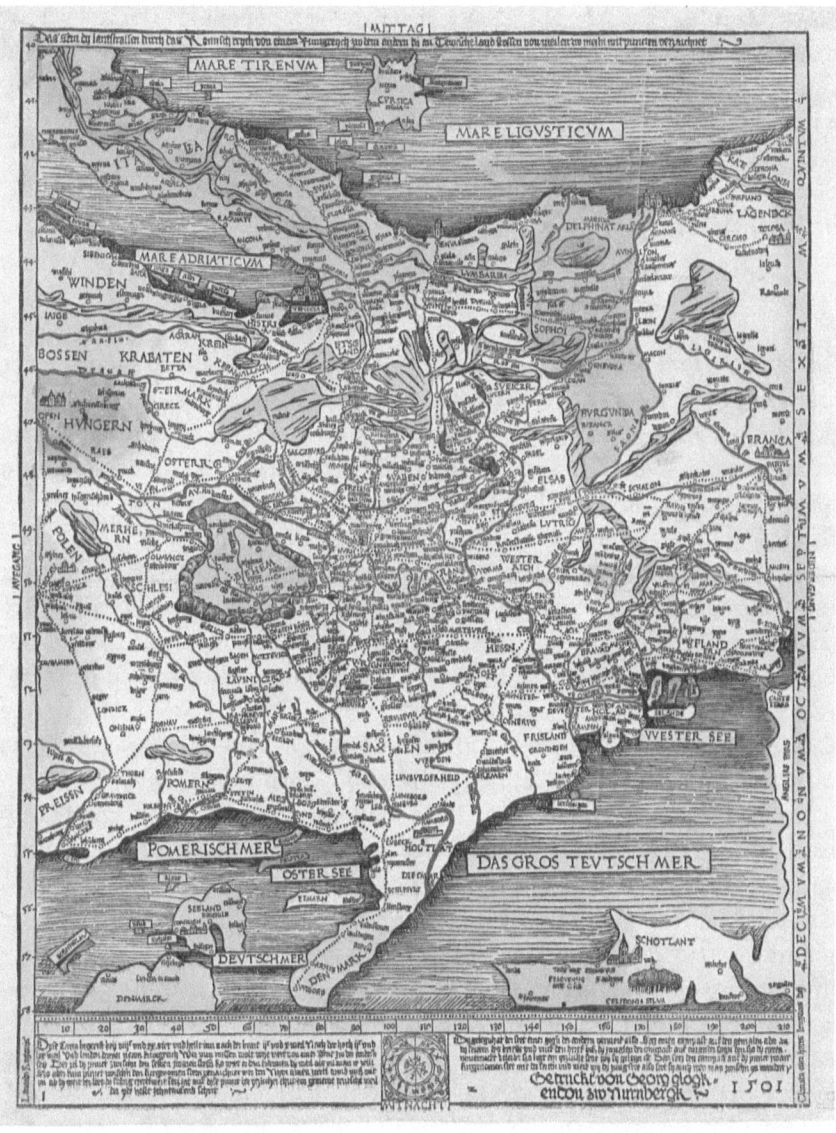

Figure 4. "Das sein dy lantstrassen durch das Romisch reych," colored woodcut map by Erhard Etzlaub (Nuremberg: Georg Glockendon, 1501) [51-2478]. Photograph: Houghton Library, Harvard University.

between cities,[29] the maps enact a tension between Bohemia,[30] the traditional historic heart of the German lands, and the new, cartographically gridded center of Nuremberg. Here Bohemia, enclosed by the Hercynian Forest and emphasized by its exaggerated size and strong chromatic effects, competes with the imperial city of Nuremberg, unmistakably designed as Germany's new center,[31] from which routes across Europe unfold like rhumb lines on a nautical chart.

DISLODGING DIRECTIONS

Schedel's *Chronicle* was critical to Nuremberg's strategic self-aggrandizement and went hand in hand with the city's outstanding record in map and globe production. The work's 1,809 woodcuts, made out of 645 woodblocks,[32] range from hexameral illustrations to stylized cityscapes to a Ptolemaic world map.[33] Its visual and textual program was so influential that it became, shortly after its publication, a blueprint for humanists like Celtis to project their vision of a "New Europe."[34] The *Chronicle*'s ambition was to perform its own impactful *translatio*, a translation of the arts that would emphasize (and visualize) their trajectory from Rome to a Germany whose role within Europe, yet to be established, Schedel's commanding work was co-defining. A careful examination of the volume shows that, from its opening pages, the *Chronicle* ought to be read not only as a work of hexameral writing but also as a cartographic quest to define Europe and Germany within it.

A large woodcut occupies the lower half of the *Chronicle*'s second numbered page: two concentric, empty circles framed by a rectangle suggest a highly abstracted illustration of the first day of divine creation in the tradition of a Christianized Aristotelian cosmos. The presence of God is made visible in the image's upper left corner through God's hand, indexing the outer circle in a gesture of creation and definition alike. Suggestive of a manicule in a contemporary manuscript, the metonymic reduction of God to his hand is perhaps influenced, as Denis Cosgrove surmises, by the idea of God as "a supreme artist manifested in the mathematical harmonies of his creation."[35] Indeed, Georg Alt's German translation of "deus" as "werckmeister"[36] (literally a "master craftsman") in the opening paragraphs of the *Chronicle* further suggests that the work of God and that of a supreme artist move within close proximity. Suddenly, the two concentric circles of the first day of creation become legible as the contours of a *mappa mundi*, framed by the endless circular movement of the *okeanos*, upon which the craftsman's hand sets out to inscribe the four cardinal directions.

The narrative text accompanying the illustration explicitly underscores this epistemic shift. While the text opens with a reference to Genesis 1:1, it ends with a lengthy quotation from Lactantius's early fourth-century *Divinae institutiones* (*Divine Institutes*), from a chapter dedicated to "the World, and its Parts, the Elements and Seasons." The conflation of Genesis and the *Divine Institutes* within one single paragraph performs a subtle, but significant slippage from cosmology to cartography:

> Deinde alteras partes eadem ratione dimensus est, meridiem ac septentrionem, quae partes illis duabus societate iunguntur. Ea enim quae est solis calore flagrantior proxima est, et coheret orienti. At illa quae frigoribus et perpetuo gelu torpet, eiusdem est cuius extremus occasus. Nam sicut contrarie sunt lumini tenebre, ita frigus calori. Ut igitur calor lumini est proximus, sic meridies orienti, ut frigus tenebris, ita plaga septentrionis occasui.

> (Then [God] measured out the other parts on the same basis, south and north, which are related to the first two parts. The one that is more ablaze with the heat of the sun is close to the east, while the one that is stiff with cold and perpetual frost belongs to the furthest west. Cold is the contrary of heat just as light is of darkness. As heat is close to light, then, so south is to east, and as cold is close to dark, so the north zone is to the west.)[37]

Lactantius's passage,[38] here quoted by Schedel, describes the first division of the earth into different parts, which subsequent mapmakers would define as continents. This passage transforms God from the creator of the universe into a geometer and mapmaker,[39] a "werckmeister" measuring and delineating the earth. As the *Chronicle* shifts from Genesis to the *Divine Institutes*, it dislodges a cosmological model and institutes a cartographic paradigm instead. Viewers of the woodcut witness the sudden shift from the diagrammatic representation of the creation of the universe to the cartographic organization of earthly space.

The *Chronicle* rewrites the cosmologic origins of the world in a cartographic key, with a tentative continental division in mind. By strategically placing Genesis alongside Lactantius's work, Schedel merges the epistemological question of the origin of the world with that of spatial division, orientation, and positionality. In a powerful discursive and visual blend, the *Chronicle* presents the cartographic drive of territorial measurement and spatial division as coextensive with ontology and the act of creation itself. The joining of west and north into one hemisphere (labeled in the *Chronicle* as "emisperia," in the Latin edition, and "halbhimel," in the German edition) and of east and south into another emerges as a powerful statement, both spatial and ethical: the Latin terms "oriens" and "occidens," employed in the *Chronicle* in the sense of birth-giving and death-bringing, denote not only the directions east and west. Historically fraught, the cardinal directions formulate here judgments about life and death. Spatial division turns in Lactantius's metaphysical geography into a comment on ethics, so much so that what lies in the (glowing) southeast is associated with life, heat, and divine enlightenment, while the (torpid) west, lacking (God's) light, is correlated with cold and death. Three centuries later, Friedrich Hölderlin would use a cunning wordplay, "Urt(h)eilung," for the joint act of judgment and original division, taking advantage of the fact that the German word for "judgment" (Urteil) can equally mean primordial part (Urteil) and original separation (Ur-Teilung).[40] In the same vein, the etymon of the Greek verb "krinein" (κρίνειν) blends the meanings of "division" and "judgment," framing critique and judgment as a question of positionality and orientation and, vice versa, instilling ethics in the act of division and separation. But the *Chronicle* is also anchored in Old Testament's traditional vision of the cardinal directions, which associates the north with evil, as in Jeremiah 1:14: "Out of the north disaster shall break out on all the inhabitants of the land."[41] Conversely, the east, imagined as the site of the Garden of Eden, has a particularly elevated status. While Schedel reverts (via Lactantius) to the biblical tradition in characterizing the cardinal directions, Celtis dislodges the all-too-traditional dichotomy between life and east/south, on the one hand, and death and west/north, on the other. In *Quatuor Libri Amorum*, the association of specific directions with metaphysical attributes loosens and begins to fluctuate.

Celtis's cartographic work is a unique, strategic attempt to destabilize the historically accepted and metaphysically charged attributes attached to the four cardinal directions, as it reimagines Germany's geography within a "New Europe." Operating within the dynamics of this cartographic turmoil, *Quatuor Libri Amorum* falls into a transitional period, when the northern and western cardinal directions, long associated with darkness, irrationality, and death, slowly crystalized as "the West"—with attributes, that is, that the later Enlightenment tradition would link with light, life, progress, and civilization. One of Celtis's major preoccupations was to resemantize the north in order to underscore the importance of Germany as a key player in the formation of humanism and in the definition of Europe. By tracing his protagonist's travels from east to north (via south and west), and by reckoning his increasing age, the four books of elegies, each dedicated to one cardinal direction, seemingly adopt a traditional framework, in which east is associated with youth and west and north with age and death. Yet read together with the book's iconographic program, the protagonist's linear trajectory and the traditional association of cardinal directions with virtues are disrupted and subverted.

The multilingual woodcut frontispiece of the *Quatuor Libri Amorum* (fig. 5) encapsulates the volume's dynamic spatial stratifications and showcases Celtis's desire to rethink the cardinal directions anew. The frontispiece features an adorned triangle—its tip pointing toward the bottom of the page—in which the book's title is inscribed. Two concentric circular laurel wreaths, perhaps a nod to the representation of the first day of creation in Schedel's *Chronicle*, occupy the lower half of the page. On the outer rim of the wreaths, the four cardinal directions are inscribed in Greek letters: Mesembria (Μεσήμβρια, south), Dysmè (Δυσμή, west), Arktos (Άρκτος, north), and Anatolè (Άνατολή, east). The insertion of the cardinal directions transforms the wreaths into a *mappa mundi*, oriented to the south in the manner of medieval Islamic maps,[42] but also like the 1453 Venetian *mappa mundi* by Fra Mauro and Erhard Etzlaub's more recent 1500 and 1501 maps. The first impression of the frontispiece as a *mappa mundi* representing the *oikoumene* soon vanishes, however, once one scrutinizes the four rectangular tags inserted in the wreath, which contain the names of four cities, Ratispona, Moguncia, Lubecum, and Croca—the Latin names for Regensburg, Mainz, Lübeck, and the capital of the Polish kingdom, Kraków. Each tag contains a fourfold set of information: the name of the city, river, cardinal direction, and stage in the human life cycle. Kraków is associated with the Vistula River, east (*oriens*), and adolescence (*adolescencia*); Regensburg with the Danube River, south (*meridies*), and youth (*iuventus*); Mainz with the Rhine River, west (*occasus*), and old age (*senectus*); and Lübeck with the Baltic Sea, north (*nox*), and death (*mors*). Here, the *oikoumene* of the medieval *mappa mundi* has been supplanted by what Celtis considers to be the four sides of Germany, now quite literally spanning the world.

But the frontispiece's particular distribution of places, rivers, directions, and stages of life does not coincide with the order that Celtis follows in the poetic design of *Quatuor Libri Amorum*. In his elegies, the life cycle spans from "pubertas" to "senectus," while the woodcut starts with adolescence and ends with death. The relationship between word and image is disrupted and the cardinal directions are unhinged, suggesting that the circular wreaths, like a globe rotating around its axis, dislodges and redistributes names, places, and phases of life with the contingency of a wheel of fortune: while west

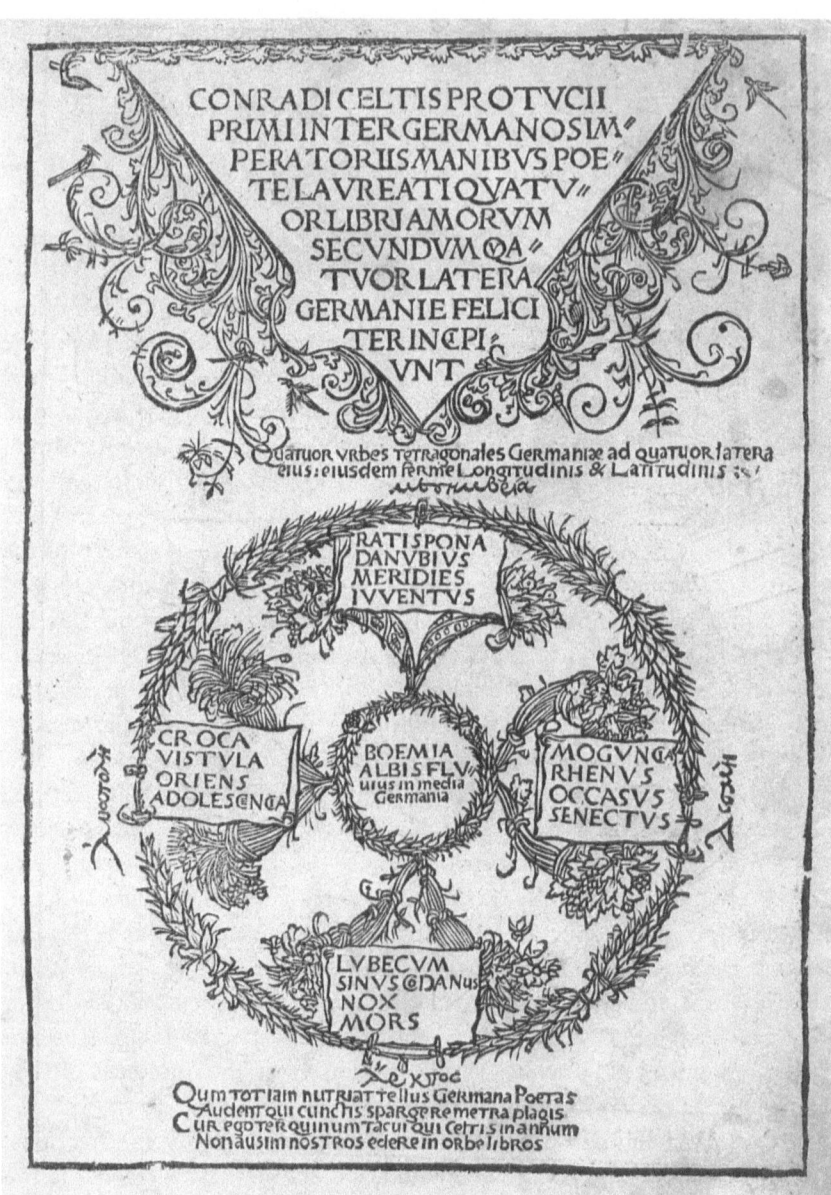

Figure 5. Conrad Celtis, *Quatuor Libri Amorum secundum Quatuor Latera Germanie* (Nuremberg: Sodalitas Celtica, 1502), frontispiece [Typ. 520 02. 269]. Photograph: Houghton Library, Harvard University.

corresponds to "iuventus" in the text, it is linked to "senectus" in the image. Like Dürer's Philosophy woodcut, Celtis's frontispiece disrupts chronology and contiguity. Not only does it disclose the discrepancy and obliqueness between word and image, but it also generates a potent creative and critical framework that forces the reader to slow down and engage more carefully with the manifold spatial layers the work as a whole contains. Rather than illustrating the poetic words of the four books of elegies, the woodcuts "[plot] new and unknown spaces within and about the text,"[43] in the words of Tom Conley on a separate work, Marguerite de Navarre's 1547 illustrated poem *La coche* (*The Coach*), and become a "different idiom."[44] As in many other Renaissance illustrations included in literary texts, in Celtis's *Quatuor Libri Amorum* the image is not, to again use Conley's words, "sensed to be a supplement to the text, but a visual in which language [is] incrusted—and entrusted—with new and unforeseen spatialities."[45]

The work's new spatialities challenge traditional attributes of the cardinal directions. When, in the first book of elegies, Celtis describes his own birth in Europe's northern realms, he strategically dislodges what Schedel had termed the "torpid" traits of the "occidens" and casts the north within a framework of multiple births instead:

Candidus inflexa phoebus tunc stabat in urna,
Proxima cui nitidae stella serena lyrae,
Cumque sagittiferi surgebant sydera signi
Horaeque post medium tercia noctis erat
Tunc mea me genitrix reserata effudit ab alvo
Et dederat vitae stamina prima meae
Illa nocte lyram nemo conspexit olimpo
Phoebus enim roseis hanc sibi iunxit equis
Plectraque pulsabat toto resonantia coelo
Et dixit: phoebo nascere quisquis eris
Ipse meam citharam plectro gestabis eburno
Lesboaque canes carmina blanda cheli,
Seu te germano contingat cardine nasci
Sive Italo, Gallo, Sarmaticove polo
Nam mea sunt toti communia numina mundo
Sim licet arctois languidior radiis.[46]

(Radiant Phoebus [Apollo] stood in the curved Urn [Aquarius], next to him the bright star of the brilliant Lyre, and when the constellation of the Archer was rising, it was three o'clock after midnight. It was then that my mother sent me forth from her opening womb, giving me my life's thread. That night no-one could see the Lyre in the heavens, since Phoebus bound it to his rose-colored horses. Then he plucked the strings, making the whole sky resound, and said: "Be born for Phoebus, whoever you will be! You will take with yourself my lyre with the ivory plectrum, and you will sing charming songs in the style of the lyre of Lesbos, no matter where you are born, under a German sky, or under an Italian, Gallic or Sarmatian; because I have the same power all over the world, even if my rays are weaker in the North."}[47]

The radiant Phoebus Apollo, born on the Aegean island of Delos and intimately linked with sunlight, appears in the northern realms in the company of "the bright star of the brilliant Lyre." As a god of light and poetry, Apollo presides over Celtis's birth, which occurs in the middle of the night. The repeatedly used word "midnight" (*nox*), which also hints at the cardinal direction north (*nox*), is here juxtaposed with Apollo's light-bringing lyre coming from the Mediterranean Sea. Celtis's own birth, wedged into the interplay and co-presence of night and day, recalls Hesiod's description of the beginning of the world, when "from Chasm, Erebos and black Night came to be; and then Aether and Day came forth from Night, who conceived and bore them after mingling in love with Erebos."[48] Celtis's phantasmatic vision of his own birth in the presence of Apollo dovetails with what Conley described as a speculation "about this point of separation that ties the writer or cartographer to the imagination of his or her presence as a godlike form, as a nurturing force." The event with which Celtis chooses to open his cycle of poems is that "point of separation from the mother, Freud's famous mother earth (*Mutter Erde*),"[49] both a belonging to and detachment from a specific locale.

As if haunted by Hesiod's reference to the primordial place, Chasm, from which Erebos and black Night came to be, Celtis obsessively reiterates, later in the poem, that his own birthplace, Würzburg,[50] is the city the ancient Greeks had called Erebos. In the first book of the *Quatuor Libri Amorum*, he establishes a pseudo-etymological explanation with the goal of proving the city's Greek origins. By reverting the German toponym Würzburg to its (then popular) Latinate form, Herbipolis ("the city of herbs or spices"), Celtis hinges his proof on the phonetic similarity between the Greek "Erebos" and "Herbi-," arguing that Würzburg "deque Ερεβου Graeco nomine dicta polis" (takes its name from the Greek City of Erebos).[51] The German Herbipolis (like the Greek "Erebos" and "polis") is coterminous, Celtis contends, with the "City of Erebos," the seat of the God of the underworld.[52]

Celtis undoes the traditional association between north and death by infusing life and light into the darkness of Erebos (Würzburg). Like Petrarch—who in his self-fashioning epistolary (*Familiares* 3.1, "Ad Thomam Messanesem, de Thile insula famosissima sed incerta, opiniones diversorum" [To Tommaso da Messina, the opiniones of various people concerning the very famous but doubtful island of Thule]) narrated his travels "to the north as far as the confines of Ocean" in order to prepare, as Theodore Cachey suggests, "an even more heroic return to the center of the world located in the Petrarchan self"[53]—Celtis inserts into his poem the figure of the god of light. In the guise of a midwife, Apollo is present at the moment of Celtis's birth and accompanies him on his subsequent travels across Germany, Italy, France, and Europe's east ("Sarmatico polo"). The event of Celtis's birth, moreover, deprives not only Greece, but the heavens themselves of "the brilliant Lyre," now deposited, in a movement of *translatio artium*, in Germany. Through his own birth, Celtis's poetic voice triumphantly and self-confidently proclaims, Germany is transformed from a Nordic site of death and darkness into a creative fulcrum productive of a hitherto unknown but now illuminated artistic bounty.

When Celtis set out to reshuffle the connotations clinging to the cardinal directions with the purpose to strengthen Germany's geographic position and the artistic and scientific production within it, he was, together with Schedel, among the first German humanists to do so. Celtis and Schedel both strategically unearthed and manipulated the crisis of directionality and location that would be one of early modernity's central defining features to restore Germany's reputation, heretofore imbued with suspicion from its inauspicious northern location. They did so by highlighting not only the movement of the *translatio artium* from Italy to Germany, but also by reversing the directions and fashioning Germany as the very site where the arts originate. By stressing recent groundbreaking scientific and artistic achievements like the printing press, they framed Germany as the hub of technological innovation and Italy as its beneficiary. Print culture was a particularly strong case in point: when the two German printers Konrad Sweynheym and Arnold Pannartz left Mainz for the Benedictine monastery of Santa Scolastica in Subiaco in 1464 with the purpose of introducing the printing press in Italy, they carried with them instruments never seen before in Italy: "cases of movable type, that marvel of fifteenth-century German ingenuity which had transformed the familiar technology of agricultural extraction (the screw press) into an engine of reproduction."[54]

Lactantius's *Divine Institutes*, passages of which Schedel strategically inserted in his *Chronicle* to illustrate God's first day of creation as an act of cartographic mapping, was among the first books printed in Italy.[55] Known as the *Subiaco Lactantius*, the collected *Opera* of the church father appeared in October 1465, quickly following the first two volumes ever printed in Italy, Donatus's *Ars minor* and Cicero's *De oratore*.[56] To fold into his *Chronicle* Lactantius, whom Giovanni Pico della Mirandola called "the Christian Cicero,"[57] was thus Schedel's remarkably bold and strategic move. It allowed him to incorporate into his *opus magnum*, a masterpiece of German artistry and print culture, an allusion to the very technology that had produced it: the movable type printing press, a groundbreaking German innovation, proudly exported to Italy and across Europe.

Just like Schedel, Celtis fashioned the Germans as a new model to imitate. On August 31, 1492, while Schedel was working on the completion of his *Chronicle*, Celtis delivered an acceptance speech for his new position as lecturer at the University of Ingolstadt, later published under the title *Oratio in Gymnasio in Ingelstadio publice recitata*. Here he explicitly addressed the relationship between Germany and Italy and the rise of the German lands within Europe: "Consider how to add distinction to your dignified office by your virtue, and to increase your fame by learning, so that men may think you worthy of those dignities and may follow after you, instead of your following them, like fowlers after a flock of birds."[58] Referring to Germany as "terram nostram, maximam Europae partem" (our land, the greatest part of Europe),[59] Celtis provocatively positions the Germans, "accepto Italorum imperio" (who have taken over the empire of the Italians),[60] as "Romani reliquias imperii" (the last survivors of the Roman Empire).[61] In his speech, Celtis confidently overrides a millennia-long prejudice against the Germans succinctly encapsulated in Tacitus's rhetorical question: "Quis porro, praeter periculum horridi et ignoti maris, Asia aut Africa aut Italia relicta Germaniam peteret, informem terris, asperam caelo, tristem cultu aspectuque, nisi si patria sit?" (Besides, who, to say nothing about the perils of an awful and unknown sea, would have left Asia or Africa or Italy to look for Germany? With its wild scenery and harsh climate it is pleasant neither to live in nor look upon unless it be one's fatherland).[62] While it is not surprising that Celtis exhorts his students to shed the image of barbarians by embracing the study of philosophy, eloquence, and the ancient languages (Latin, Greek, and Hebrew),[63] he pioneers in programmatically announcing his humanistic program under the aegis of geography and cartography: "Consider it a great disgrace to be ignorant of the histories of the Greeks and Latins, and the height of shame to know nothing about the topography (*situm*), the climate, the rivers, the mountains, the antiquities, and the peoples of our region and our own country, in short, all those facts which foreigners have so cleverly collected concerning us."[64]

Celtis will then deplore that the Germans have so far best been described only by the ancients, who surveyed German customs and the contours of the territory with care and "expressed our customs, our emotional make-up and our spirits as graphically as a painter might delineate our bodies."[65] Just like the contours of a human body, so, too, the contours of a nation have to be brought together into a single geographic and cartographic image. It is now the task of the young Germans whom Celtis is addressing to collect the bodily parts, which lie disjoined and scattered across Europe, into one whole body-nation: "Assume, O men of Germany, that ancient spirit of yours, with which you so often confounded and terrified the Romans, and turn your eyes to the frontiers (*angulos*) of Germany; collect together her torn and broken territories."[66] Celtis's allusion to Germany's frontiers as "angulos" (corners, angles) dovetails with his predilection for the cartographic table as a structuring model for his thought and poetics alike: despite being a topos (such as in the expression "the four corners of the world"), Celtis's "angulos" refer more readily to the cartographic image of a map's stylized rims and margins. They conjure up the image of a Ptolemaic table, which frequently collapsed, as we will see, the map's borders with Germany's rectified (natural) boundaries. This seemingly fleeting word choice signals what Terence Cave terms the "trouble du texte," a small but abrupt change in register indexing a larger epistemological shift.[67] The word "angulos" signals to the reader that Celtis is thinking with a map in mind as he imagines the contours of Germany: such thinking toward "collecting" Germany's "broken territories" like bodily limbs into a new geometric shape is crystallized here as a powerful act of creating a cartographic table productive of a new poetics.

BODIES AND BORDERS

Little attention has been paid to an early passage in the *Geography* where Ptolemy refers to the importance of the *tabula* as cartography's fundamental generative principle. In the first book, Ptolemy contends that, while "the goal of chorography is an impression of a part, as when one makes an image of just an ear or an eye,"[68] cosmography shows "the known world as a single and continuous entity, its nature, and how it is situated."[69] The goal

> Cosmographie vero totum inspicere iuxta proportionem: ut si integrum quis caput designaret. Integris enim imaginibus cum oporteat potiora membra primum adhiberi: deinde ea que imagines picturasque suscipiunt, ita equa dimensione inter sese locari, & ex iusta distantia visu possint discerni, an totum seu pars sint illius quod pingitur.

[of cosmography is a general view, analogous to making a portrait of the whole head. That is, whenever a portrait is to be made, one has to fit in the main parts [of the body] in a determined pattern and an order of priority. Furthermore the [surfaces] that are going to hold the drawings ought to be of a suitable size for the spacing of the visual rays at an appropriate distance [from the spectator], whether the drawing be of whole or part, so that everything will be grasped by the sense [of sight].)[70]

While at first glance the relationship between chorography and cosmography as *pars* and *totum* appears uncomplicated, a closer examination of Ptolemy's seemingly clear-cut differentiation in fact reveals the complexity of the dichotomy between them, a relationship not only of scale, proportion, and measurement but also of quality. Their goals differ. Chorography is an independent discipline, as Ptolemy makes sure to emphasize. In contrast, cosmography glosses over nuances to focus on the connectivity of the bodily members. It is partial in its grasp of details, but total in its capacity to create a pattern. But crucial for the cartographic act, as Ptolemy hastens to add, is the presence of a surface large enough to display and arrange the main parts of the body into a meaningful and orderly pattern.

The cosmographer's arrangement of the "members" on a flat surface is the very moment in which a "body," whose mobile limbs are unmeasurable, is transformed into a motionless, measured two-dimensional alignment. Ungraspable and unmeasurable at first, a living body becomes measurable solely as it extends, with the motionlessness of a dead body, across a flat surface. Stretched out on the physical surface, bodily limbs become not only measurable, but also a unit of measurement themselves, as in the English word "foot" or the Italian "braccio" (arm). The myth of Apollo and his brother Dionysus[71] and the story of Saint George and the dragon[72] are just two out of many stories from antiquity and the Middle Ages which powerfully illustrate this specific transitional moment when the physical table, the necessary material support for the practice of cartography, becomes a *tabula*, the core of cartographic thought and the center of the cartographer's attention.

According to Franco Farinelli, both stories (Apollo and Saint George) have in common the centrality of a body subdued by a stronger force: Dionysus by the Titans—or by grape treaders dismembering the god of wine during the annual rite of the wine harvest[73]—and the dragon by Saint George. In certain iterations of the myth of Apollo, Farinelli recalls, the god of measure and containment and, in fact, "the executor of the first exemplar of what we today call a map,"[74] is tasked with reassembling the dead body of his brother Dionysus. But Apollo's act of displaying Dionysus's body parts upon a surface all too often belies the importance of the surface itself. The crucial material support upon which the bodily members are reassembled is what Farinelli identifies as the cartographic *tabula*,[75] one of cartography's most fundamental constituent parts, the origin of any cartographic activity. In an analogous story, Saint George, in the act of killing the dragon, comes to symbolize the emergence of space.[76] Every iconic rendering of George, Farinelli argues, represents the saint sitting up high on his horse, thrusting his lance vertically into the dragon's unruly, contorted body. Here, the straight lance contrasts the beast's coiled shape. But once the dragon lies dead on the ground, flattened by Georges's blow, its full length can be assessed and it becomes measurable.[77] Both stories, then, brought together by Farinelli in a meditation on the origin of cartography and measurable space, trace the initial moment of cartographic creation, coextensive with the productive force of a *tabula rasa*.[78] The narratives capture the act of assembling and aligning the bodily limbs into measureable units on a flat surface as cartography's elemental gesture. This is the very surface, one is inclined to think, to which Ptolemy refers when he emphasizes the importance of a space great enough "to hold the drawings."

A detail in a woodcut inserted in Celtis's *Quatuor Libri Amorum* illustrates this initial process of cartographic creation precisely through the figure of Apollo, the god with whom Celtis most identified.[79] Here, Celtis, the author, is portrayed at work in his *studiolo*, while two lateral columns feature the gods Minerva, Mercury, Hercules, Venus, Bacchus, and (Phoebus) Apollo—the latter occupying the center of the right column (fig. 6). The woodcut refers to Apollo as "Phoebus," an epithet frequently used to denote his "radiance" and "brightness." Standing firmly on the ground, Apollo's bow and arrow point downward, while his head is framed by a halo-like glow emitting rays. Behind his back are strapped a harp-shaped lyre and a quiver, whose strings and arrows, respectively, seem to emanate, ray-like, from Apollo's body. His pose and the manner in which he holds his bow and arrow are reminiscent of medieval iconographic representations of God as geometer.

Perhaps the most intriguing element of the depiction of Apollo is the presence of the winding body of a python next to Apollo's right leg. Pierced by the Greek god's vertical arrow, the meandering body of the python lies on the ground, transfixed. Apollo is highlighted here as the victor over the python, a representation that recalls Apollo's deed in Ovid's *Metamorphoses*:[80]

> Ergo ubi diluvio tellus lutulenta recenti
> solibus aetheriis altoque recanduit aestu,
> edidit innumeras species; partimque figuras

Figure 6. "Phoebus Apollo" (detail) in Conrad Celtis, *Quatuor Libri Amorum secundum Quatuor Latera Germanie* (Nuremberg: Sodalitas Celtica, 1502), n.p. [Typ. 520 02. 269]. Photograph: Houghton Library, Harvard University.

rettulit antiquas, partim nova monstra creavit.
Illa quidem nollet, sed te quoque, maxime Python,
tum genuit, populisque novis, incognita serpens,
terror eras: tantum spatii de monte tenebas.
hunc deus arcitenens, numquam letalibus armis
ante nisi in dammis capreisque fugacibus usus,
mille gravem telis exhausta paene pharetra
perdidit effuso per vulnera nigra veneno.

(When, therefore, the earth, covered with mud from the recent flood, became heated up by the hot ethereal rays of the sun, she brought forth innumerable forms of life; in part she restored the ancient shapes, and in part she created creatures new and strange. She, indeed, would have wished not so to do, but you also she then bore, you huge Python, you snake unknown before, who were a terror to new-created men; so huge a space of mountain-side did you fill. This monster the god of the bow [Apollo] destroyed with lethal arms never before used except against does and wild she-goats, crushing him with countless darts, well-nigh emptying his quiver, till the creature's poisonous blood flowed from the black wounds.)[81]

Here, the python stands for the yet unmeasured, for the "innumerable forms of life" that Earth can bring forth. In Ovid's lines, the python is endowed with such immensity that it is likened to the "space" (spatium) of a "mountain." The shape of the monstrous beast whose body is rising vertically induces "terror," a noun Ovid chooses to prominently locate at the very beginning of his verse.

Roman land surveyors such as Siculus Flaccus and Frontinus established a correlation between the phonetically similar words "terror" and "territory."[82] In his *De conditionibus agrorum*, Siculus Flaccus argues that "territoria" (territories), understood as empty land, are created when citizens are "territis" (terrified) and have to flee in times of war.[83] In *De limitibus agroribus*, Frontinus succinctly defines a territory as a space abandoned by a terrified and terrorized population during an armed conflict: "Territorium est quicquid hostis terrendi causa constitutum est" (A territory is that which is constituted because of fear of the enemy).[84] Tacitus, whose *Germania* Celtis edited around 1498–1500,[85] defined terror and fear as an effective generator of territorial boundaries. Already in the opening sentence, Tacitus contends that Germany is separated from the Sarmatians and Dacians "mutuo metu aut montibus" (either by mutual fear or mountains).[86] From antiquity, fear, terror, and the desertion of land by migration have been understood as constitutive of potent territorial boundaries.

THE LIMITS OF NORTHERN EUROPE

In his 1525 translation of Ptolemy's *Geography*, Celtis's friend Willibald Pirckheimer defined "Germania" by using bodies of water as its boundaries:

> Germaniae latus occidentale Rhenus terminat. Septentrionale vero Germanicus oceanus; . . . Meridianum autem latus terminatur a parte occidentali fluvii Danubii. . . . Orientale autem latus terminat distantia, que fit a flexu prefato ad superiacentes Sarmatarum montes. . . . Preterea distantia, quae est post montes ad dictum caput Vistule fluvii, & ipse etiam fluvius usque ad mare.

> (The Rhine ends Germany's western side; the Germanic Ocean the northern [side]; . . . the southern side is bound by the western part of the Danube River. . . . The eastern side is bound by the space which stretches from the said bend to the Sarmatian mountains, extending upon it. . . . Thereafter comes the space which is situated behind the mountains by the said source of the Vistula river, & the river itself down to the sea.)[87]

This definition of "Germania," in book two of the *Geography*, is rehearsed again in the eighth book of the *Geography* (in the section containing the tables), albeit in a more political key, in tune with the political circumstances of sixteenth-century Germany: here, Pirckheimer includes geographic information Ptolemy had used to describe adjacent regions of Europe to redefine Germany, subsuming Germany's neighboring countries, regions, and peoples under the heading of "Germania magna": the migrating people of the Iazyges (Iazyges Metanastes), European Sarmatia (Sarmatia Europe), Raetia (Rhetia), the Celtic Noricum (Noricum), the two Pannonias (duae Pannoniac), and Belgic Gaul (Gallia Belgica). With an unmistakable nod to "Magna Graecia" (Great Greece)—a term which came to denote mainland Greece and the ancient Greek colonies along the coast of southern Italy—Pirckheimer strategically inserts here the aggrandizing adjective "magna" to gesture toward a Germany in expansion. "Magna," more akin to the political ambitions of the German lands around 1500 than to the authority of ancient geographers, was rarely used in earlier editions of Ptolemy's *Geography*. It refers to a Germany "cum insulis sibi adiacentibus" (with its adjacent islands) terminated by the "Oceanus Germanicus" (German Ocean).[88] While already Tacitus's *Germania* had suggested that a part of Germany "is surrounded by the ocean, which enfolds wide peninsulas and islands of vast expanse (*insularum immensa spatia*),"[89] the numerous translations and editions of Ptolemy's *Geography* proved a powerful tool to redefine regional borders through subtle adjectival additions, turns of translations, and the commanding tool of cartographic visualization.

Already a quarter of a century before Pirckheimer's translation of the *Geography*, Celtis used Ptolemy's tables as a template for the visual and poetic design of the *Quatuor Libri Amorum*. In his address to Emperor Maximilian I, Celtis explicitly states that in his book of poems the reader will admire "Germaniam nostram in parvae tabulae modum et sua quatuor latera depictam" (our Germany in the manner of a small table, where its four sides are depicted).[90] With the Ptolemaic *tabula* in mind, the generative power of the map is encoded already in the book's title: the four books of love elegies are designed according to cartography; they follow—or are *second* (*secundum*) to—the four geographic sides of Germany. Celtis's poetic work thus emerges out of the creative space provided by the map's surface. While geography had already been crucial for the Latin elegiac poets, Celtis's poems propose something entirely new: by anchoring the topic of love in German topography, Celtis prioritizes cartography over poetry. Here, geography and cartography become the poem's constitutive frame, generating, orienting, and directing the development of the love theme. Each part of Germany is productive of an erotic encounter and a new elegiac plot.

With the astute eye of a cartographer, Celtis manipulates Germany's boundaries in word and image, offering his own, fictionalized, pseudo-regional woodcuts of Germany in the tradition of Ptolemy's *tabulae* as visual companion pieces preceding each book of elegies. The theme of pilgrimage is here secularized, as Celtis dislodges the medieval pilgrim's goal, Jerusalem, by redirecting the pilgrim's pious steps, now under the aegis of a cartographically induced erotic desire, to the German lands represented by four cities. The cities he chooses are not only aligned with Germany's alleged borders but also stand for the specific cartographic (or astronomical) achievements of each urban site. Starting in Germany's east and (somewhat misleadingly) the city of Kraków,[91] whose university, famous for astronomy, counted the young Nicolaus Copernicus among its student body in the late fifteenth century, the first book—consisting of fifteen elegies—narrates the encounter between the protagonist and a young woman, the Sarmatian Hasilina (Hasilina Sarmatica). It is at the banks of the Vistula River that Celtis the protagonist experiences his "pubertas" (youth). In the second book, divided into thirteen elegies, Celtis travels to the southern German city of

Regensburg, where the youthful Alpine Elsula (Elsula Alpina) keeps the traveling protagonist company as he explores the southern border of Germany along the Danube River during his "adolescentia" (adolescence). Regensburg had a long geographic tradition: it is here, at the Benedictine monastery of Saint Emmeran, that a certain Brother Fridericus (who called himself "astronomunculus")[92] compiled, in the first half of the fifteenth century, the so-called Klosterneuburg corpus consisting of astronomical and geographical texts and tables. In the fourteen elegies of the third book, the protagonist is still in his best years ("iuventus"). Accompanied by Gallic Ursula (Ursula Galla), he continues west to Mainz, where he discovers the beauties of the Rhine River and Germany's western boundary. Mainz was famous not only for its wine but, most importantly, for its first workshop for printing with movable lead type, established by Johannes Gutenberg in the middle of the fifteenth century. It is also in Mainz that the first illustrated travelogue (Europe's first printed book to use Arabic letters), Bernhard von Breydenbach's *Peregrinatio in terram sanctam*, was printed in 1486.[93] Lastly, pressed by old age ("senectus"), Celtis voyages in the fifteen elegies that comprise the fourth book to Germany's north, the Baltic Sea, and the city of Lübeck. Here, the *Rudimentum novitiorum* (*Handbook for Beginners*), the earliest printed universal chronicle, which preceded Schedel's *Chronicle* by almost two decades, was printed by Lucas Brandis in 1475. Rooted in the tradition of hexameral writings and containing accounts of pilgrimages, among others, the *Rudimentum* was most famous for its two remarkable maps: the first extant map of Palestine and a world map (fig. 7)—published only three years after the first printed map, the already mentioned T-O map in Isidore of Seville's *Etymologies* (Augsburg 1472).[94] It is at the shores of the Baltic Sea that Celtis concludes his pilgrimage in the company of Baltic Barbara (Barbara Codonea), with whom he heads, a final destination, toward Thule, the Orkney Islands, charting the limits of Germany's northern realms. The more the protagonist advances through unwieldy eternal ice, the more fantastic the landscape becomes. As he reaches the end of his pilgrimage, Celtis steps out of his part as protagonist and into his role as writer, announcing his return to Emperor Maximilian I, to whom he dedicates his elegies.

The complex boundary of Germany's north, which Celtis explores in the fourth book, was depicted on contemporary Ptolemaic world maps such as the one crafted by Johannes Schnitzer von Armsheim (fig. 8), the first printed map signed by an artist. The map featured in the 1482 Ulm edition of the *Geography*—the first edition published outside of Italy and the first to include modern tables of Italy, France, the Iberian Peninsula, northern

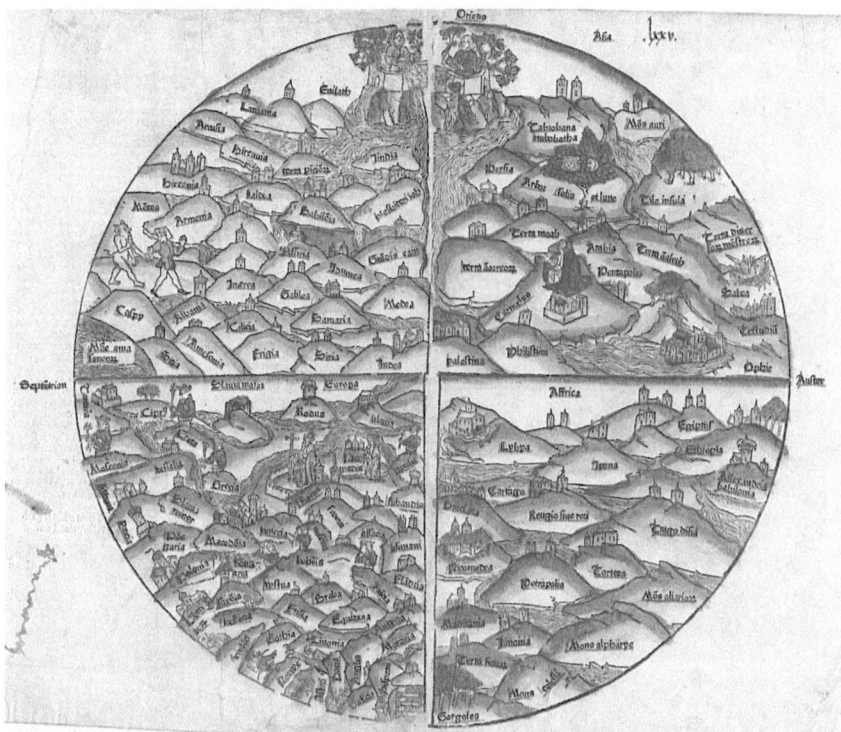

Figure 7. World Map in Lucas Brandis, *Rudimentum novitiorum* (Lübeck, 1475). RBR Milestone of Science 172. Reproduction by Permission of the Buffalo & Erie County Public Library, Buffalo, New York.

Europe, and Palestine.[95] Celtis owned a copy of this edition early on, indeed, it was the first printed book he purchased. On Armsheim's world map, one notices a peculiar protrusion north of Germany: here, the "Mare Glaciale" overflows the map's otherwise well-defined rim. Land becomes hardly distinguishable from water and rivers are easily confused with mountain ranges. The Baltic Sea aggressively disrupts the scale, challenging the reckoning of measured space. Here, north of Germany, a new, incommensurate space that resists the map's framework emerges.

An analogous disruption of metric space and resistance to scale occur in the four woodcuts preceding each book of elegy in the *Quatuor Libri Amorum*. While the woodcuts (whose orientation changes with each city and the moving protagonist) feature scales on three of its sides, the metric system is playfully interrupted on top, perhaps to highlight Germany's potential for expansion.[96] When Nigel Thrift points out that metrication and creativity

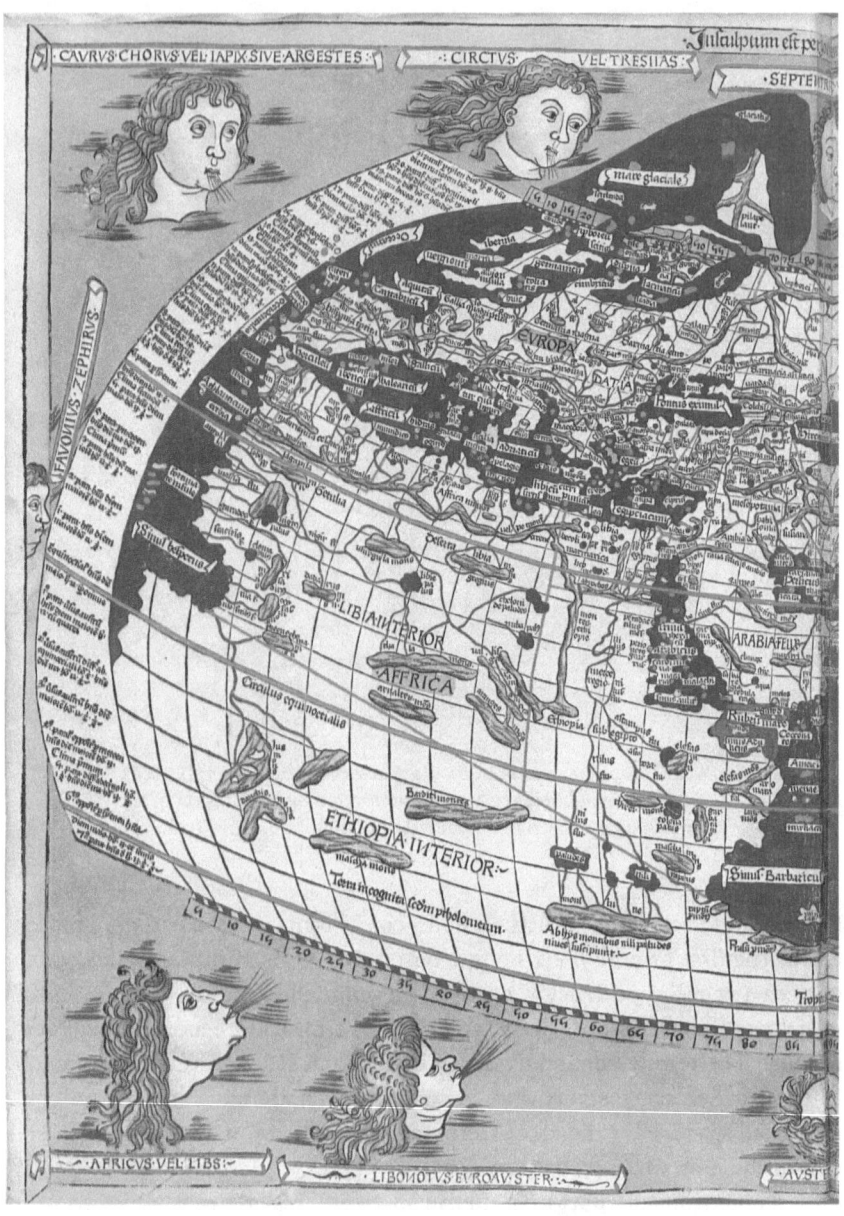

Figure 8. World map (detail of Europe and Africa) by Johannes Schnitzer von Armsheim, in Ptolemy, *Cosmographia* (Ulm: Lienhart Holle, 1482) [Typ. Inc. 2556]. Photograph: Houghton Library, Harvard University.

are not diametrically opposed practices,[97] but instead dovetail with one another, he accurately describes Celtis's poetic and visual program, which at once hinges upon and playfully subverts the Ptolemaic focus on measurement and scale.

A particularly insightful example of Celtis's artistic manipulation of boundaries, in word and image, is the fourth book dedicated to Germany's north and Baltic Barbara. The woodcut is here oriented to the north (the Baltic coastline) and features a disproportionately enlarged depiction of Celtis, who occupies Germany's center together with Barbara (fig. 9). It contains several scales collapsing into a playful optical illusion: while a tag designating "Germaniae Latus Septentrionale" (the northern side of Germany) at first glance coincides with Germany's northern boundary, a closer examination reveals the border as the horizon beyond the Baltic Sea, not the limit of Germany. Germany's territory occupies, in fact, only a modest third of the woodcut, while its space is doubled, or so the woodcut suggests, through the strategic placement of the tag in line with the sea's horizon. The transition from land to sea is almost imperceptible. The woodcut's artistry, its similar carving of land and water, seem to endlessly defer Germany's boundary to the north. Only a careful contemplation of the image reveals the vastness of Germany's territory as an optical illusion and the alleged boundary of the territory, indeed, as the limit of a body of water. Land and sea are here combined into one terraqueous texture that extends limitlessly into what was then the only partially mapped north as the boundaries of both Germany and Europe merge into an ever-expanding horizon.[98]

The poetics of the Baltic Barbara book uniquely render the liminal, uncharted, spaces in Europe's north by expressing them through the disintegration of meter on the level of speech. Traveling along the shores of the Baltic Sea in the company of Barbara, the protagonist notices that his companion's "tongue and feet stagger"[99] (*titubant . . . lingua pedesque*). The words she forms are not German anymore, but resemble, at first glance, "barbarous" infiltrations from the Nordic languages, Danish and Scottish. Embedded in the uncultured condition of the northern climes, or so the text seems to suggest, the sounds (*sonis*) that leave Barbara's mouth (*ore*) become utterly incomprehensible. Playing with the Latin homophone "ore"—denoting both "mouth" (*ore*) and the plural of "border" (*orae* or *ore*)[100]—Celtis ties Barbara's stumbling "tongue and feet" to the liminality of the region. Barbara's decreasing verbal performance, furthermore, contrasts, perhaps in a self-congratulatory move, with Celtis's own regular "pedes," the feet of his masterful hexameter in which his elegy is composed.[101]

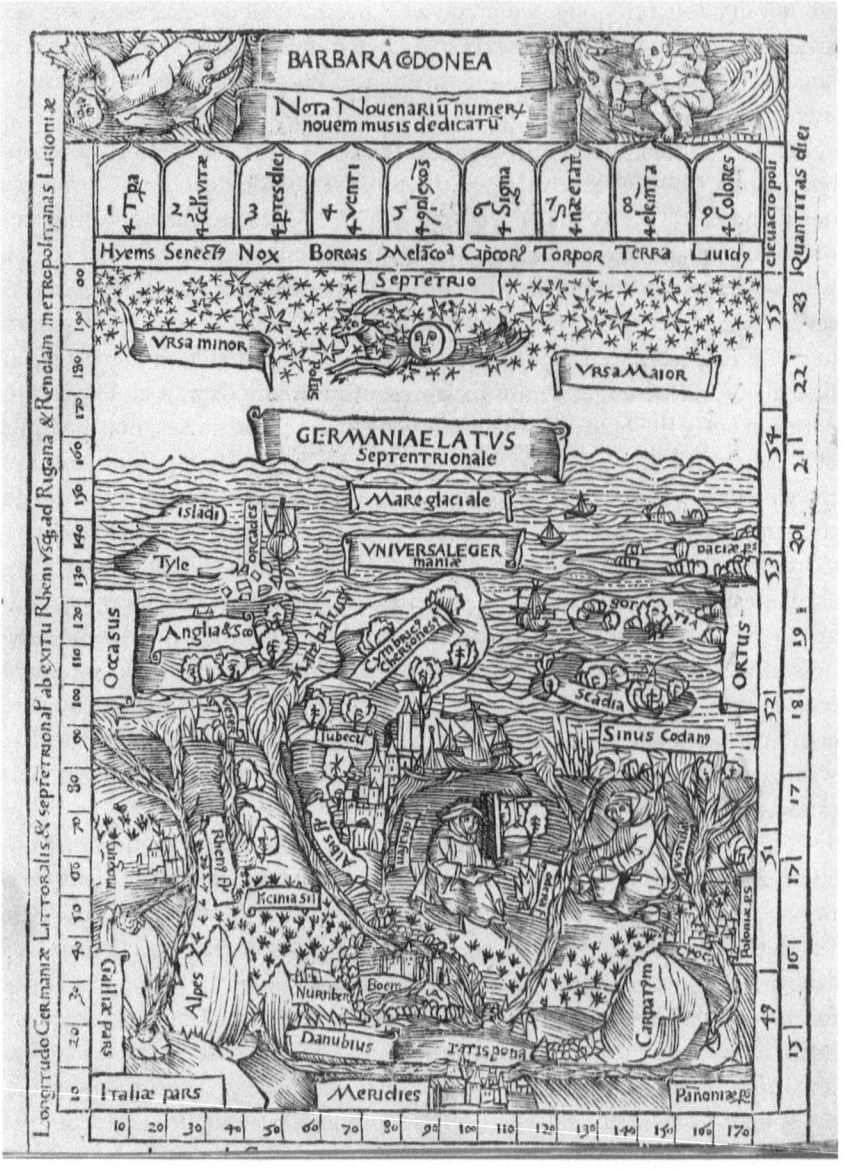

Figure 9. "Barbara Codonea," in Conrad Celtis, *Quatuor Libri Amorum secundum Quatuor Latera Germanie* (Nuremberg: Sodalitas Celtica, 1502), n.p. [Typ. 520 02. 269]. Photograph: Houghton Library, Harvard University.

Celtis imagines Barbara's alleged incapacity to incorporate and assimilate speech as an inability to ingest food: "Dacica cum Scoticis Barbara verba vomens" (She vomits barbarous words by mixing Danish with Scottish),[102] the protagonist exclaims. By using the verb "vomere" (vomit), Celtis marks Barbara, who grammatically merges with the "barbarous words" (Barbara verba) she utters, as someone failing to digest speech. Barbara's failure to properly masticate her "verba" contrasts with the very foundations of humanistic culture and learning, grounded in the capacity to "digest" texts and speeches. According to Quintilian, "Repetamus autem et tractemus et, ut cibos mansos ac prope liquefactos demittimus quo facilius digerantur, ita lectio non cruda sed multa iteratione mollita et velut confecta memoriae imitationique tradatur" (We chew our food and almost liquefy it before we swallow, so as to digest it more easily; similarly, let our reading be made available for memory and imitation, not in an undigested form, but, as it were, softened and reduced to pap by frequent repetition).[103] As the inhabitant of a remote and uncivilized northern shore, Baltic Barbara is seemingly not exposed to the acculturating digestive processes readily available to the refined citizens of the Mediterranean shores.

That the peoples inhabiting the "edges" of the world were "barbarous" and "monstrous," as Surekha Davies has recently demonstrated in a broad comparative study, is a topos that carried over from antiquity to the Middle Ages and to the early modern period in both literature and cartography.[104] In the context of Germany, this topos regained momentum as several new editions of Tacitus's recently rediscovered *Germania* were under way, one of which was prepared by Celtis himself. For Tacitus, the further one progresses to the north, the more one reaches the limits to human speech: where "the earth is girdled and bounded" words are transformed into mere sounds, and all one perceives phonetically is the sound (*sonus*) of the sun's passing:

> Trans Suionas aliud mare, pigrum ac prope inmotum, quo cingi claudique terrarum orbem hinc fides, quod extremus cadentis iam solis fulgor in ortum edurat adeo clarus, ut sidera hebetet; sonum insuper emergentis audiri formasque equorum et radios capitis adspici persuasio adicit.

> (Beyond the Suiones is another sea, sluggish and almost motionless, with which the earth is girdled and bounded; evidence for this is furnished in the brilliance of the last rays of the sun, which remain bright enough from his setting to his rising again to dim the stars: popular belief adds further that the sound of his [the sun's] emergence is audible and the forms of his horses visible, with the spikes of his crown.)[105]

At the limits of the earth, movements become slow and sluggish, while language is replaced by inarticulate sounds. Couched between reality and fiction, liminal spaces are prone to verbal and bodily monstrosities, Tacitus deplores, as he zooms into the life of the Fenni (Finns), who "mira feritas, foeda paupertas: non arma, non equi, non penates; victui herba, vestitui pelles, cubile humus" (live in astonishing barbarism and disgusting misery: no arms, no horses, no household; wild plants for their food, skins for their clothing, the ground for their beds).[106] Beyond the Finns, "cetera iam fabulosa: Hellusios et Oxionas ora hominum voltusque, corpora atque artus ferarum gerere: quod ego ut incompertum in medio relinquam" (all else that is reported is legendary: that the Hellusii and Oxiones have human faces and features, the limbs and bodies of beasts: it has not been so ascertained, and I shall leave it an open question).[107]

In stark contrast to Tacitus, however, Celtis performs an unexpectedly humorous twist in his description of Germany's northern shores. By staging the scene of Barbara's faltering tongue and feet in the vaulted space of a tavern, he facetiously subverts the stereotypes about Europe's north. Against the backdrop of joyful intoxication, Barbara's seemingly stuttering speech is instead revealed to be joyful multilingualism, coming to the fore with the help of "the German vice" (Germano vitio): beer. Further contrasting ancient stereotypes about the northern peoples' sluggishness and slowness, Celtis describes his own bodily members as being revived through Bacchus's libations and Barbara's libido: "Ira, Cupido, furor Veneris, monstrosa voluptas / Ebrietate" (Wrath, Cupid, Venus's frenzy, monstrous lust through drunkenness),[108] the inebriated *senex amans* cries out. Celtis offers a picture of Germany's liminal northern shores as what Ovid termed a "mirifica permixtio" of images that uproots common assumptions conveyed in ancient texts. For Celtis, one prime characteristic of poetic style was to "mix the serious with the playful" (seria mixta iocis) in order to offer his readers a "variety of things" (rerum varietas).[109] By situating his racy scenes at the rim of the *tabula*, Celtis plays with the very tension of the margin—namely, the space between measure and incommensurability, charted and uncharted territories, inclusion and exclusion.

In a discussion of Gilles Deleuze's meditation on the "diagram," Conley has aptly pointed out that what Deleuze means by diagram and "gridding" (quadrillage)[110] is that "what is excluded from a social space is paradoxically enclosed within it."[111] As Ptolemy's map of the "Mare Glaciale" and Celtis's visual and poetic engagement with Germany's northern border make clear, the regions at the margin of the *tabula* are contained within the map by being located outside of it. *Quadrillage*, then, indexes the malleability and ambivalence of the frame and the *tabula*, gesturing toward the aporetic moment that each movement of closure is one of openness—and each movement of openness is equally one of closure. The paradox of Ptolemy's *tabula*—and of every geographic boundary—is that, just like a Moebius strip, it works from the inside while tracing the "line from the outside."[112] The example of Germany's northern border illustrates that what the playful eye of the poet-cartographer delineates from the outside is strategically undone from the inside.

As Celtis's comical tavern scene shows, the delineation of Germany's northern liminal region is an aporia both created and unpacked through the joint workings of poetics and cartography. Here, gridding sets in motion a contradictory poetic discourse—humor—which can reverse directions at will. Indeed, the paradox of gridding and what Deleuze called "sens" ("meaning" and "direction" alike) is that they display linguistic strategies such as humor on the very surface. For Deleuze, then, humor is "the art of the surface."[113] Preferably staged at the rim of the *tabula*, humor becomes the very site of liminal gridding. By using the example of *Alice in Wonderland*, Deleuze contends that "everything happens at the border": "If there is nothing to see behind the curtain, it is because everything is visible, or rather all possible science is along the length of the curtain. It suffices to follow it far enough, precisely enough, and superficially enough, in order to reverse sides and to make the right side become the left or vice versa."[114] By staging the imaginary limitlessness of Germany's northern shores in the enclosed space of a tavern's vaults, Celtis traces his voyage to Germany's and Europe's northern limits with the humor inscribed in the movement of a Moebius strip.

TOPOGRAPHY AND THE FLEETING BOUNDARIES OF EUROPE

Celtis was among the first European humanists to revive the Greek term "topography," introducing it into a cartographically inflected humanistic lexicon. Used sparingly both in ancient rhetorical works such as Quintilian's *Institutio* and in popular medieval renderings of the Alexander romance (most importantly Walter of Châtillon's *Alexandreis*),[115] "topography" was a word Latin writers commonly translated as "loci descriptio" (description of a place).[116] Scholars have compared the word's use in antiquity to a variation of ekphrasis, a means of vividly presenting a landscape in front of the reader's mental eye through the "accumulation of characteristic details."[117] Celtis took up the term "topography" in his encomiastic poem about Nuremberg, *Norimberga*, included in *Quatuor Libri Amorum*, when he notes that a more complete *topographia* of the Hercynian Forest has been included in his *Germania*: "De quibus in Germania nostra illustrata, quae ad topographiam attinere visa sunt, diffusius scripsimus" (As far as topography is concerned, we have written amply about it in our illustrated *Germania*).[118] In Celtis's work, the term enjoyed only a brief appearance: in the second edition of the *Norimberga*, Celtis replaced the term *topographia* with the Ptolemaic "chorographia": "De quibus in Germania quae ad χωρογραφιαν attinere visa sunt diffusius scripsimus" (As far as *chorography* is concerned, we have written amply about it in our *Germania*).[119]

Yet Celtis's initial audacious deviation from the Ptolemaic nomenclature has to be taken seriously. It emerges out of a need to account for a new and more complex spatial and poetic reality for which Ptolemy's binary distinction (between chorography and cosmography) proved insufficient. Celtis's confident introduction of the word "topographia" thus marks a terminological resistance to the Ptolemaic system as it forges a new cartographic framework for early modern times. Used as a wedge to soften what was increasingly perceived as a rigid and inadequate terminology coined by ancient geographers, "topography" launched an unprecedented mobility and poetic dynamics into an early modernity propelled by a novel and complex cartographic reality. The term "topography" thus epitomizes the struggle, not only of Celtis but of his contemporaries as well, to define, delineate, and name new spatial units against the backdrop of the breakdown of the traditional dichotomy between local (chorographic) and global (cosmographic).

That Celtis should use the term "topography" in the specific context of the description of the Hercynian Forest in *Germania generalis* is a powerful statement: in Celtis's hands, this forest, located, as was believed, at the center of the German lands, becomes a place resisting cartographic representation and localization altogether. In Celtis's description, it continuously shifts from one scale to another conflating the local, regional, national, and continental dimensions. Celtis's Hercynian Forest epitomizes the fluidity of early modern boundaries and testifies to the humanists' repeated (and aporetic) attempt to let themselves be guided by and to break out of the dominant Ptolemaic cartographic framework.

The *Germania generalis* appeared, in a second edition, as part of the *Quatuor Libri Amorum*. Celtis uses the Hercynian Forest as a powerful poetic and astute political strategy to blur the boundaries between Germany and Europe. In the central chapter of the *Germania generalis*, the Fichtel Mountain (at the center of the Hercynian Forest) turns into Germany's geographic center and the privileged vantage point to survey the German and European territories:

> Herciniumque iugum medio Germania tractu
> Erigit et multis dispargit cornua terris,
> Ex quibus ingenti dorso stans pinifer atrum
> Tollit in astra caput liquidoque a vertice quatuor
> Quatuor ad mundi fundit vaga flumina partes.
> Menus in occiduum festinat currere Rhenum
> Et Sala qui Drusi gestat praeclara trophea
> Algentes petit Albis aquas. Nabus exit in austrum,
> Egra se Eoos flectit sua cornua in ortus
> Et secum Herciniam trahit alto vertice sylvam.

(In the middle of Germany, the Hercynian Forest rises into the air and spreads its summits across many lands. The Fichtel Mountain, standing on its immense back, raises its dark head to the stars. From its liquid summit, it pours forth four rivers into the four parts of the world. The Main hastens to run into the western Rhine, and the Saale, carrying Drusus's celebrated trophies, seeks the cold waters of the Elbe. The Naab flows to the south, whereas the Eger bends its arms toward the eastern dawn, pulling along the Hercynian Forest with its high peak.)[120]

The Hercynian Forest is here described as a mountain range with the towering Fichtel Mountain (Pinifer Mons) in its midst. From here, four rivers (evoking, also on the woodcut table, the four rivers of Paradise) take their origin and flow down the mountain's "liquid vertex" into Germany's four corners—described here, in a conflation of scales, as the four "parts of the world."[121] Just like its four rivers, the Hercynian Forest stretches out into different directions at once, represented here by three of the four Wind-Gods, the *Anemoi*: Arcton (North Wind), Eurus (East Wind), and Zephyr (West Wind): "Que variis porrecta plagis nunc vasta sub Arcton / Inque Eurum Zephyrumque suis cum saltibus errans / Explicat immensos annoso robore lucos"[122] (This Forest extends across different regions far to the North / Soon it errs with its woodlands and mountainous passes to the East and the West / [And] explicates its immense sacred groves with its aged oaks).

At first, the Hercynian Forest seems to be reasonably confined. Yet once further details come to light, the forest turns into an errant wood reaching significantly beyond central Germany. Indeed, the forest's inclusive "arms" (*brachia*) appear to incorporate all Germanic peoples blurring the center of Germany with its own boundaries:

> Brachia longa iacit, Svevos Francosque Turogos
> Circuit. Obnobios montes, qui particula Alpium, et
> Saxonicisque reducta plagis Phrisiisque bubulcis
> Vissegothis Ostroque Gothis Cimbrisque vetustis,
> Donec Germanorum prope tangat saltibus equor.
> Ast ubi spumiferi fontes contexerit Histri
> Austriacas aditura plagas Ceciumsque comantem,
> Qui dum nubigeras radicem accepit ad Alpes,
> Quas Carnus Stiriusque tenet Slavusque bilinguis,
> Hic ubi Danubius positurus nomina ponto
> Proxima Dalmatice radit confinia terre.

([The Hercynian Forest] stretches its arms wide and encircles Swabians, Franks, and Thuringians. It embraces the Obnobian mountains, a small part of the Alps. It then withdraws to the Saxon lands and the Frisian ploughmen, to the Visigoths, Ostrogoths, and to the old Cimbri, until its lofty forests almost touch the German Sea. As soon as it has joined the sources of the foaming Danube, it seeks the Austrian lands and the wooded Kahlenberg, a last offshoot of the cloud-generating Alps, inhabited by the Carinthians, Styrians, and bilingual Slavs: here, the Danube, which will give its name to the sea, touches the nearby borders of the Dalmatian land).[123]

The Hercynian Forest becomes a geographic paradox that resists cartographic mapping. Firmly seated at the heart of Germany, the forest's all-encompassing arms extend to Germany's border regions, embracing, on their way, all German- and Slavic-speaking peoples; they almost touch the Baltic Sea in the north and follow the flow of the Danube to the shores of the Black Sea in the south. The forest is bounded as well as open, limited and expanding its limits to the boundaries of the Dalmatian lands. With a *double entendre* generated by the astute syntax of Celtis's poetic line, the "nearby borders of the Dalmatian land" turn into the "end of the world" (*confinia terre*).[124] Having covered large portions of Europe, the forest circles back to its center creating a natural wall around Germany's central regions, Franconia, Thuringia, and Bohemia: "Mox Francos Turogosque petit Bemosque feroces, / Tanquam nativo claudit quos undique muro, / Quos medio nutrit bellax Germania tractu" (Soon it rushes toward the Franconians, Thuringians, and the wild Bohemians, / enclosing it from all sides with a seemingly natural wall. / In its midst, it nurtures bellicose Germany).[125] The Franks, Thuringians, and "wild Bohemians" are encircled by the forest, whose arms, like a wall, surround Germany's central regions in a protective gesture, reminiscent of descriptions of the Hercynian Forest enclosing Bohemia, first offered by Cosmas of Prague and later by Enea Silvio Piccolomini.[126]

Celtis's description of Germany's wondrous forest does not stop here. Immediately after returning to its point of departure, Germany's central regions, the forest spreads its arms out again, this time coinciding with the very limits of Europe:

> ... Carpathi culmina spectans
> Marcomanes Gepidasque petit Iasigesque potentes,
> Hic ubi nunc cultis est Transsilvania terris
> Moribus Almanis gentem linguaque creavit.
> Inde iterum vasto diffundit cornua flexu
> ...
> Sarmaticis obducta plagis quosque ultimus orbis
> Angulus abscondit Agatyrsos limite claudit;
> Ad Tanais fontes Rypheosque in vertice late
> Diffusa et fines Europe vasta pererra[t].

> (Looking at the Carpatian summits, it [the Hercynian Forest] seeks the Marcomanni, the Gepids, and the mighty Iazyges. Here is Transylvania with its tilled lands, which created a people with German customs and a Germanic tongue. Then, once again, it [the Hercynian Forest] spreads its arms in a vast bend. ... Surrounded by Sarmatian lands, it sets a boundary with the Agathyrsi, who are hidden in the remotest corner of the world. It reaches the sources of the Don and the Ryphean mountains, where it spreads its broad and vast top wandering to the end of Europe).[127]

Past the Carpathian Mountains, the Hercynian Forest extends to Transylvania, where German settlers have created a German-speaking enclave, and from there to Sarmatia, the Don River, and the Ryphean (Riphean) Mountains, which Ptolemy and other ancient geographers described as Europe's eastern borders. It is at the "fines Europe" (Europe's end), at the continental boundary between Europe and Asia, that the wandering arms of the Hercynian Forest come to a halt.[128]

Celtis's description of the Hercynian Forest serves two purposes. On the one hand, its errant nature allows him to draw attention to the complexity of Germany's "four sides" oscillating between ancient geographic descriptions and the loosely defined boundaries of the Holy Roman Empire. On the other, Celtis turns the forest into a measuring instrument surveying the vast expanse of the German territory. The Latin word "brachia"—just like the Italian "braccio"—denotes a unit of measurement. The "long arms" of the forest thus become a cartographer's tool to convert into a metrical system the yet-unmeasured expanse of the German lands.[129] The oscillation transpiring here between the metrically graspable surface and the serpentine elevations of a sylvan territory is indicative of Celtis's pushback against the dominant Ptolemaic model of gridding, which promotes the flat surface of two-dimensional cartography at the expense of a more complex geography that would encompass the earth's three dimensions. In fact, Celtis recurs to a variety of different cartographic models alongside Ptolemaic *tabulae* such as the Christological *mappa mundi*, with the scope to offer a poetic and cartographic "mirifica permixtio" capable of promoting space and measurement as a multilayered complex endeavor.

Quatuor Libri Amorum explores the possibility of measuring and experiencing space through a productive confluence of different spatial and cartographic models that might reinforce or, contrarily, challenge one another. In Celtis's four books of elegies, each regional description of Germany thus contains the seeds of its own undoing. In the second book, dedicated to the protagonist's adventures with Alpine Elsula, the complexity of spatial reckoning is carefully inscribed in the very texture of the poetic lines:

> Francia germano mihi stirps et origo poetae
> Hercyniae medio cincta beata sinu
> Quam Moenus mediam perlambit vitifer oram
> Cuius quadrifluvio nascitur unda iugo
> Pinifer is: cuius de vertice flumina quatuor
> Quatuor in partes orbis amoena cadunt
> Moenus ad occasum francorum fertur in oras
> Sala sed Arctois perditur Albis aquis
> Egra sed eoi petit ardua regna boemi
> Sed Nabus histrinis insinuatur aquis.

(As a German poet, I have Franconia, happily girded by the Hercynian bend, as my lineage and origin. The wine-bearing Main traverses the region, its waters spring from the four-river mountain, the Fichtel Mountain. From its summit four rivers flow delightfully toward the four parts of the earth. The Main runs westwards, toward the land of the Franks, whereas the Saale throws itself into the northern streams of the Elbe. The Eger seeks the steep kingdoms of the eastern Bohemian, while the Naab empties into the waters of the Danube.)[130]

Scholars have noted that Celtis carefully arranges words within each line to create a poetic cartography. The inextricable link between geographic space and the Latin language is accentuated by Celtis's use of dactylic hexameter, whose lines consists of six "feet." Gernot Michael Müller has shown that Celtis uses long chains of spondees (— —) to describe the immeasurable vastness of mountains and forests,[131] as in the description of the Fichtel Mountain (Pinifer Mons): "Ex quibus ingenti dorso stans pinifer atrum" (— ⌣⌣|— —|— —|— —|⌣⌣|— —).[132] In contrast, Celtis uses dactyls (—⌣⌣, a term with the original Greek meaning of "finger") when imagining the rapid movement of rivers. His description of the flow of the Danube to Austria consists only of dactyls: "Austriacas aditura plagas Ceciumque comantem" (—⌣⌣|—⌣⌣|—⌣⌣|—⌣⌣|—⌣⌣).[133] But readers familiar to metrical notation will notice that Celtis also uses the spondee when rivers take the shape of boundaries: "Solus Germanas Ister sed deserit oras / Lata peregrinis inducens flumina terris" (— —|— —|— —|— —|⌣⌣|— /⌣⌣|— —|— —|—⌣⌣|— —). Here, just as in the line "Austrinos faciunt fines metamque resignant" (— —|—⌣⌣|— —|—⌣⌣|— —), the length of the spondee (— —) morphs into a boundary. Celtis's cartographic poetry, radically new, inscribes the movements and functions of rivers in the very scansion of the poetic line.

Equally poeticized is Celtis's recurrent reference to the Fichtel Mountain (Pinifer Mons), seated at the center of the Hercynian Forest. It serves the poetic purpose to highlight the difficulty of cartographic mapping. From the mountain's "liquid vertex," four rivers flow into the four parts of Germany, from Germany's alleged center to the territory's blurred boundaries. Yet Celtis pushes his metrical aspiration further. It is in the center of a poetic line that he couches the mountain's liquid vertex: "Tollit in astra caput liquidoque a vertice quatuor / Quatuor ad mundi fundit vaga flumina partes" (It raises its head to the stars and from its liquid vertex four rivers / To the four parts of the world it pours forth).[134] In these lines, Celtis's desire to map concentricity and his anxiety about the adequate rendering of an actual landscape, the delicate balance between the horizontality of mapping and the verticality of the mountains, reaches, quite literally, its peak. The mountain's summit is situated in the center of the verse, while Germany's territory is flanked by the word "quatuor," which refers to the four rivers and the four sides of Germany alike. As Müller observes, the rhetorical figure of anadiplosis, the doubling of "quatuor" at the end of one and at the beginning of the next line, maps out the different directions in which the four rivers of the Pinifer Mons flow.

Yet in its conflation of the beginning and the end of a poetic line, the figure of anadiplosis is equally indicative of a moment of hesitation in the metrical establishment of directionality. When the two lines containing "quatuor" are spoken aloud, the directionality of the rivers is disrupted as "quatuor" is turned into an empty repetition, a speech-fluency disorder,[135] unearthing the author's failure to fully grasp the direction of a river's flow and the extent of Germany's territory. As long as the poetic line is kept in place on the printed page, the spatial outlook of the poem and its correspondence with the geographic and cartographic content are upheld. Once the verse is dissolved in speech, the meaning and directionality of "quatuor" are disturbed. The potent position of "quatuor," then, turns into a symptom of a troubled cartography and an echo of the very title of Celtis's work: Quatuor *Libri Amorum secundum* Quatuor *Latera Germanie*. The repetition of "quatuor" allows the reader to reconsider the immediate context of the word's insertion, the very moment when Celtis mentions not only the Pinifer Mons, but, more specifically, its "liquid vertex," an oxymoron embraced by Celtis's powerful cartographic poetics which resists more straightforward cartographic representation. The repetition of "quatuor" expresses something unspeakable and troublesome: the very center of Germany is the origin of all cartographic projections and a vanishing point—a reality impossible to measure and represent.

Celtis's use of the exceedingly rare noun "quadrifluvium," strategically inserted in the context of the Pinifer Mons, likens the mountain's troublesome centrality more to indescribable places such as biblical Paradise than to an actual description of the center of Germany. As Jörg Robert points out, the adjective "quadrifluus" (containing four rivers)—just like its cognate "quadrifluvium"—is "a precious compound,"[136] a *hapax legomenon* found only in one instance in classical Latin, in the third poem of Prudentius's *Cathemerinon* (*Before Taking Food*) in which the Roman Christian poet describes precisely the most topical place which resists mapping—biblical Paradise:

    Tunc per amoena virecta iubet
    Frondicomis habitare locis,
    Ver ubi perpetuum redolet
    Prataque multicolora latex
    Quadrifluo celer amne rigat.

(Then he bids him dwell
in a pleasant green place, in a leafy shade,
where eternal spring is fragrant,
and a swift stream waters meadows
of many colours with its four channels.)[137]

Prudentius explores Paradise as a Christian *locus amoenus* and fantastic place of origin and dwelling. From here, the four rivers of Eden flow into four different directions. The four streams' eternal flow structures the world into four parts and creates a stable framework for living.

Just like Celtis's cartographic center of Germany, imagined as an elusive liquid vertex challenging representation, Prudentius's Eden is empty. Adam and Eve, asked to dwell, do so only fleetingly. What Celtis might have found attractive in Prudentius's description of Paradise is that Eden, a place of expulsion, not attraction, is a place of profound solitude and emptiness which defies representation. Paradise is a *tabula rasa*, the origin of representation from which centrifugal rivers flow like rhumb lines, furrowing the surface of the earth. Briefly touched by the presence of humans, Paradise is remembered as the source of the four rivers (always included on medieval *mappae mundi*) that structure space. The adjective "quadrifluus" is the sole explicit trace that Celtis imports from Prudentius's description of Paradise, nestling it in his own description of the liquid vertex of Pinifer Mons, Germany's ideal cartographic center. But Prudentius's engagement with the complexity of poetic creativity takes a mystical direction that Celtis would have found equally inspiring for his own blend of poetic and cartographic models. In the sixth stanza of the third poem, Prudentius addresses the tension between classical poetry and a Christian framework:

Sperne, camena, leves hederas,
cingere tempora quis solita es,
sertaque mystica dactylico
texere docta liga strofio,
laude dei redimita comas!

(Reject, Muse, the trivial ivy—
You're used to crown your head with it:
You know how to weave sacred garlands—
Tie them with a band of dactyls,
Your hair surrounded with the praise of God!)[138]

Here, Prudentius focuses on the theme of poetry itself. Professing a switch from classical poetry to the "mystical garlands" (*sertaque mystica*) of sacred poetry, Prudentius addresses the Muse, the *camena*, and the wine god, Bacchus.[139] Poised between the Latinate metrical framework of classical poetry and Christological spatial imagery, Prudentius's poem does not propose a transition from one poetic model to the other; instead, it reclaims a "mystical" co-presence of both. Prudentius rejects the poetry of the pagans with the very words of Horace and Propertius—and does so using classical meters. One word in particular, *dactylus*, in its adjectival form "dactylico," epitomizes the impact of classical poetry and meter that subtend Prudentius's lines. Here, classical poetry is ushered in through a reference to the framework underpinning classical poetry itself, its meter.[140]

Viewed against the mystical backdrop of the *Cathemerinon*, Prudentius's reference to feet and fingers evokes the image of Christ's mystical body. The fifth stanza, in particular, has long been interpreted in a Christological framework as a description of the Eucharist:[141]

> Fercula nostra deum sapiant,
> Christus et influat in pateras,
> seria, ludicra, verba, iocos,
> denique quod sumus aut agimus,
> trina superne regat pietas.

> (Our dishes should taste of God,
> and Christ be poured into our bowls;
> things serious, light-hearted, talk, jokes,
> all that we are or do—
> may the threefold holiness from on high guide us!)[142]

Here, Christ's body "streams" and "flows" (*influat*), river-like, into the circular shape of the "patera" (bowl), filling it and forming a marvelous blend that contains everything, the whole world as it were, by combining all (contrary) elements, "seria mixta iocos,"[143] in Celtis's words. In his careful reading of the *Cathemerinon*, Celtis would not have missed Prudentius's use of the Latin "patera," whose connotations encompass not only poetics but also cartography. A shallow, circular bowl, the "patera" was used in sacrificial contexts to pour libations, which the Greeks referred to, quite tellingly, as "spondees." Made predominantly of ceramic or clay, the *patera* was not only created of earth but also *represented* it. Indeed, just like the center of a *mappa mundi* was considered to be the world's navel (Jerusalem), the center of the *patera*, a bulbous protrusion, was named ὀμφαλός (*omphalos*, "navel").[144] In book 18 of the *Iliad*, Homer refers to the center of Achilles's celebrated round shield—which the Romans later called *umbo*—as *omphalos*, as the center of the world.[145]

Prudentius's use of "patera," then, discloses new pathways in the multilayered cartographic imagery folded into Celtis's vision of the liquid vertex.[146] The image of the patera Prudentius offers would have readily been received as one of a *mappa mundi* by Celtis: a map-vessel onto which Christ's body is poured. On a Christological medieval map, the dactyls and feet of Prudentius's poem, previously associated with the dactylic feet of a poetic meter, take on a different meaning: they become Christ's mystical limbs, an important feature on *mappae mundi* (the thirteenth-century Ebsdorf Map is a famous case in point), where not only Christ's head but also his hands and feet visibly embrace the world. While the *mappa mundi* bears a similarity to the form of the *patera*, it also—in its function as a receptacle for Christ's mystical body—represents the Eucharist.[147] When Celtis anthropomorphizes the circular Hercynian Forest by using the word "brachia" ("arms") and by conflating the borders of Germany, Europe, and the *oikoumene*, he allows for a reading of the sylvan limbs as Christ's mystical arms extending to the world's confines, perpetuating a tradition from Prudentius's *Cathemerinon* to the *mappa mundi*.

In Celtis's *Quatuor Libri Amorum*, space is out of joint. Against the backdrop of competing cartographic models and spatial frameworks, ranging from Ptolemy's *tabula* to medieval *mappae mundi*, Celtis seeks to provide a definition of Germany within Europe. Reminiscent of Apollo, who collected Dionysus's disjoint limbs, Celtis embarks on a poetic project to compose "the four sides of Germany" which he finds disparate. Yet what he enthusiastically proclaims in the volume's title is slowly undone as the *Quatuor Libri Amorum* unfolds in both word and image. Despite Celtis's claim that his love elegies are contained within the boundaries of Germany, the assemblage of measured fingers, feet, and arms does not align itself into a smooth rectangular *tabula* with four closed sides. Rather, Celtis's book of poems and series of woodcuts reveals the openness and blurry contours of the map's margins and the complex endeavor of finding Germany's *omphalos*. The aporetic image of the protruding liquid vertex of the Pinifer Mons shows the impossibility of mapping the center of Germany and of defining its contours as they tacitly—and yet so eloquently—merge with those of Europe.

CHAPTER TWO

A Border Studies Manifesto: Maciej Miechowita's *Tractatus de duabus Sarmatiis* (1517)

In 1516, the readers of Ludovico Ariosto's newly published epic poem, *Orlando Furioso*, were drawn into one of Ruggiero's intrepid flights. The protagonist here, "unwilling to forgo the pleasure of discovering the world," charted an entirely "new path" from Asia to Europe: in order to "complete the circle he had started, so as to girdle the earth, like the sun," Ruggiero steered his hippogriff across the world's "bleak northern lands"[1] where Europe and Asia meet:

> Quinci il Cataio, e quindi Mangïana
> sopra il gran Quinsaí vide passando:
> volò sopra l'Imavo, e Sericana
> lasciò a man destra; e sempre declinando
> da l'iperborei Sciti a l'onda ircana,
> giunse in l'ulterïor Sarmazia; e quando
> fu dove Asia da Europa si divide,
> Rossi e Pruteni e la Pomeria vide.

> (On his journey he saw Cathay to one side and to the other Mangiana, as he passed over the great Quinsai. He flew over the Himavian range, and skirted Sericana to his right. From the hyperborean land of the Scythians, he turned in towards the Hyrcanian sea and reached further Sarmatia; then, arriving where Asia divides itself from Europe, he beheld the lands of the Ruthenians and Prussians, and saw Pomerania.)[2]

Ruggiero's *katascopy*, his bird's-eye view of the territories he traverses, is, as scholars have pointed out, akin to a cartographer's gaze upon a map: the Christian paladin meticulously registers the orographic and hydrographic features of the different regions, moving from Cathay to Quinsay, from Scythia to Sarmatia, and from Ruthenia to Prussia and Pomerania.[3] Ariosto discloses the territories and landscapes of the world's northern realms with the mapmaker's power of visualization, as if drawing the reader of the poem's lines into the contours of a new region. Ruggiero's ride on the hippogriff mediates here between the poetic and cartographic registers, allowing the protagonist to explore a landscape halfway between reality and imagination. For Thibaut Maus de Rolley, the flying hippogriff's bird's-eye view offers a "privileged means of access" (*moyen d'accès privilégié*)[4] to an unmapped territory. What scholars have so far failed to grasp, however, is that Ruggiero's gaze upon the northern regions traces the line of an emerging continental divide: it is somewhere in the vicinity of Sarmatia that "Asia separates itself from Europe" (*Asia da Europa si divide*).

Ariosto's poetic lines force the reader to halt before the territorial interstices detaching Europe from Asia. Rather than a catalog of fanciful toponyms randomly cast in the stanza's poetic landscape, the territorial divide emerges here with what Gaston Bachelard has termed (albeit in a different context) the force of the "detail." What Ruggiero passes with alacrity on the back of the hippogriff crystallizes as a territorial suture line which draws an emerging continental divide. The line setting Europe apart from Asia unfolds as a powerful "minuscule," easy to miss in the texture of the poem, but which, in Bachelard's terms, just like "a narrow gate, opens up an entire world."[5] Ariosto's insertion of a list of toponyms for the pleasure of the reader's imaginary eye, then, demands to be investigated within a larger early modern preoccupation: how to plausibly and effectively imagine a continental line across the Eurasian continent whose large territorial body extending from Portugal to China resists a clear-cut divide. The place name "Sarmatia," located at the center of Ariosto's line and in poetic proximity to "Asia" and "Europe," transpires here, Ariosto's poem seems to suggest, as a privileged vantage point to tackle such investigation: it is through this very toponym which in the *Furioso* at once bridges and divides the verse that the early modern formation of the continental divide between Europe and Asia ought to be explored.

In the year that Ariosto published the *editio princeps* of his poem,[6] the Polish historian, physician, and rector of the Universitas Cracoviensis (the future Jagiellonian University), Maciej Miechowita (1457–1523), was putting the final touches to what I call the first early modern border studies manifesto. Printed in Kraków in 1517, the *Tractatus de duabus Sarmatiis Asiana et Europiana et de contentis in eis* (*Treatise on the Two Sarmatias, Asian and European, and What is Contained in Them*)[7] redirected the conceptualization of the emerging continental border between Europe and Asia in an unprecedented way. Miechowita here is the first to revise and reject Ptolemy's description of Europe's east while, at the same time, programmatically mobilizing Ptolemy's distinction between European and Asian Sarmatia. Translated into several languages, the treatise—a "prose cartography"[8] devoid of physical maps—became a template for mapmakers, historians, and travel writers from Muscovy to England: Bernard Wapowski, Willibald Pirckheimer, Sebastian Münster, Sigismund von Herberstein, Gerhard Mercator, Abraham Ortelius, and Richard Hakluyt are among those who drew from Miechowita's *Tractatus*.[9]

The numerous editions and translations of the *Tractatus* and its inclusion in Europe's major travel anthologies cemented the work's reputation as a fundamental and authoritative source of information on the geography, ethnography, and culture of Europe's east. Less than a year after its *editio princeps*, the *Tractatus* was published under the title *Tractat von baiden Sarmatien* in the imperial city of Augsburg in a German translation by Johannes Mair von Eck.[10] In 1532, Simon Grynaeus (with Johann Hutten) included the *Descriptio Sarmatiarum*, as Miechowita titled the *Tractatus* from its 1521 edition on,[11] under the title "De Sarmatia Asiana atque Europea Libri II" in his influential travel anthology *Novus orbis regionum ac insularum veteribus incognitarum* (*New World of Regions and Islands Unknown to the Ancients*),[12] alongside authors such as Amerigo Vespucci, Christopher Columbus, and Sebastian Münster.[13] Grynaeus's volume, later translated into Dutch, served, in turn, as a touchstone for Giovanni Battista Ramusio's three-volume vernacular travel anthology *Navigazioni e viaggi*.[14] In 1561, the versatile Italian geographer, translator, and author of the Island Book *L'isole più famose del mondo* (1572) Tommaso Porcacchi edited and published Miechowita's *Descriptio Sarmatiarum* in an Italian translation by Annibal Maggi:[15] the *Historia delle due Sarmazie* (1561) became the first text by a Polish writer to be translated into Italian,[16] and it was included in the second volume of Ramusio's *Navigazioni e viaggi* (1583), alongside key travel narratives such as Marco Polo's *Historia delle cose de' Tartari* and Paolo Giovio's *Cose della Moscovia*.[17] In the sixteenth century, the *Descriptio Sarmatiarum* was part and parcel of a growing body of travel narratives concerning Europe's east, canonized through its repeated inclusion in anthologies across Europe.

EUROPE'S EAST

In the first years of the sixteenth century, Kraków—a city where Latin, Polish, and German were used on a daily basis—was a hub for scientific, intellectual, and artistic encounters, poised to produce scientists who quite literally revolutionized the world. It was here that Nicolaus Copernicus (1473–1543), undoubtedly the city's most famous student, studied astronomy from 1491 to 1495.[18] Before his central work, *De revolutionibus*, was published in 1543, Copernicus's *Commentariolus* (ca. 1510), the earliest draft which conceives of heliocentrism, soon circulated in Kraków in manuscript form. By 1514, Miechowita already owned a copy of the *Commentariolus*,[19] which he received from their mutual friend Bernard Wapowski, "the founder of Polish cartography" and the creator of the first printed map of Poland in 1526.[20] Kraków was then the capital of the Polish kingdom, and the Polish king, Sigismund I, ruled, in personal union, over the vast and multicultural territories of the Grand Duchy of Lithuania. At its political height in the fifteenth and sixteenth centuries, Poland-Lithuania extended from the Baltic to the Black Sea, and from the Holy Roman Empire to the Grand Duchy of Muscovy. The ambition of the Polish Jagiellonian dynasty—which stemmed from the fourteenth-century pagan grand duke of Lithuania, Jogaila, who, once baptized and renamed Władysław II Jagiełło, became the founder of the Polish Jagiełło (Jagiellonian) dynasty—was to expand east in order to include the Grand Duchy of Lithuania and its vast territories under Polish rule. This occurred in 1569 in the Union of Lublin, when the Polish-Lithuanian Commonwealth (*Rzeczpospolita*, the "res publica") was founded.

The Polish kingdom was intrinsically multilingual and multicultural. It was home to, among others, peoples of Slavic, Baltic, Germanic, Jewish, Hungarian, Turkish, Tatar, Armenian, and Greek descent, who communicated in a sweeping range of languages written in five different alphabets (Latin, Cyrillic, Hebrew, Arabic, and Greek).[21] Rare linguistic phenomena such as a vernacular version of Belorussian written in the Arabic alphabet, used by the Lithuanian Tatars,[22] further enhanced the unique character of a kingdom shaped by and exposed to the experience of a borderland which Miechowita set out to describe in his *Tractatus*. Many inhabitants of the Polish kingdom were "borderlanders," if by that term one is to understand, with Oscar Martínez, a population "liv[ing] and function[ing] in several different worlds: the world of their national culture, the world of the border environment, the world of their ethnic group if members of a minority population, and the world of the foreign culture on the other side of the boundary. Considerable versatility is required to be an active participant in each of these universes, including the ability to be multilingual and multicultural."[23]

Miechowita, the author of the first printed history of Poland (*Chronica Polonorum*, 1519)[24] and the owner of a large map collection,[25] rose to become a transnationally important figure whose insights into the question of which criteria should be adopted to define the border between Europe and Asia not only "drov[e] home the fact that no imposing natural divide separated Europe from Asia,"[26] but also exposed the complex impact and difficult heritage of Ptolemy's *Geography*. Ptolemy had been the first to imagine that the world could be divided into a grid of abstract straight lines, which started to be employed, after the Treaty of Tordesillas in 1494, as a convenient tool to separate regions independently of the physical land they crossed. In the eyes of many geographers and cartographers of the early modern period, these imagined straight lines constituted an attractive alternative to the capricious dictates of unwieldy natural boundaries such as meandering rivers and lofty mountain chains that ancient geographers (including Ptolemy) had used.

In the eyes of Miechowita, however, eastern Europe's borderland, a multicultural crossroad of peoples, religions, and cultures united rather than divided by physical landscape (a flat territory with no significant natural boundaries) resisted the idea of a neat continental borderline. And yet, Miechowita mobilized a Ptolemaic nomenclature—European and Asian Sarmatia—to launch a reflection upon possibilities to distinguish the borderlanders spanning two continents from one another. If early modern "Europe had to be imagined into being,"[27] as Valerie Kivelson aptly put it, Miechowita's *Tractatus* significantly contributed to this process by drawing from Ptolemy while searching for alternative models to trace continental borders. As we will see, the *Tractatus* powerfully negotiates the changing border-making mechanisms, practices, and theories for the complex endeavor to systematically formulate a continental boundary between Europe and Asia.

THE SYMMETRY OF EUROPE'S BORDERS

Scholars of the Renaissance commonly associate Europe's quest for territorial expansion and colonization with western European powers, predominantly Spain, Portugal, France, and later, England and the Netherlands. Only later on, from the Enlightenment period, do eastern Europe and Russia seem to enter the world stage as significant geostrategic players, often through the mediating lens of Western philosophers and writers such as Voltaire, the author of the *History of the Russian Empire and Peter the Great*, and Diderot, in his correspondence with Catherine the Great. All too rarely is the early modern colonizing drive and expansionist desire of eastern European countries acknowledged and discussed as part of a wide-ranging European political, economic, and military push beyond the limits of Europe. A closer examination of eastern Europe's cartographic writing such as Miechowita's *Tractatus* reveals that during the time of world travels and territorial "discoveries" eastern Europe was apprehended as a symmetrical equivalent to western Europe. Cartographic thinking was a powerful catalyst for this perceived symmetry.

Already ancient geographers had pictured Europe's east and west by using a nomenclature that reflected an imagined symmetry of the continent. In his *Description of the World*, Pomponius Mela writes:

> We call the narrows and the entranceway of the incoming water the Strait [Lat. *fretum*], but the Greeks call it the Channel [Grk. *porthmos*]. Wherever that sea extends, it gets different names in different places. Where it is constricted for the first time, it is called the Hellespont [Dardanelles]; then Propontis [Sea of Marmara] where it spreads out; where it compresses itself again, the Thracian Bosphorus . . . ; where it widens again, the Pontus Euxinus [Black Sea]. Where it comes into contact with the swamp, it is called the Cimmerian Bosphorus. . . . The swamp itself is called Maeotis [Sea of Azov].[28]

Here, the length of the Mediterranean Sea, which extends from the Straits of Gibraltar to the Sea of Azov, serves as an axis instilling a sense of ontological permanence through a specular and seemingly inalterable shape. Following the logic of geometry, for Mela's contemporary Pliny the Elder Europe was "a half of the world, dividing the whole circle into two portions by a line drawn from the river Don to the Straits of Gibraltar."[29] Some medieval Islamic mapmakers, as the anonymous author of the twelfth-century Egyptian *Book of Curiosities*,[30] also imagined the Mediterranean symmetrically: as an ellipse in which not only Europe and Africa but also Europe's east and west mirror one another. The ancient geographers' frequent use of identical toponyms for Europe's western and eastern border regions and cities further enhanced the continent's imagined symmetry. In the work of Strabo, Ptolemy, and Plutarch as well as that of Pliny the Elder, the toponym (H)Iberia referred to the Iberian Peninsula as well as to a region in the Caucasus (nowadays Georgia).[31] In his *Rudimentorum cosmographiae*, first printed in Kraków in 1530,[32] the Transylvanian Saxon humanist Johannes Honter declared that "finibus Europae clauduntur Iberica regna" (Europe's limits are bounded by the Iberian reigns).[33] And while in *Ab urbe condita* (*History of Rome*) Livy used the place name Castrum Album for the southern Spanish city Alicante (in the Roman province of *Hispania Tarraconensis*), for Miechowita the same toponym (albeit in a reverse order), Album Castrum (now Bilhorod-Dnistrovskyi in Ukraine), is a city in the vicinity of the Crimean Peninsula.[34] In book 3 of *Pharsalia* (*The Civil War*), the Roman poet Lucan highlights Europe's geographic symmetry by describing its eastern boundaries in analogy to the continent's western border posts:

> ... qua vertice lapsus
> Riphaeo Tanais diversi nomina mundi
> Inposuit ripis Asiaeque et terminus idem
> Europae, mediae dirimens confinia terrae,
> Nunc hunc, nunc illum, qua flectitur, ampliat orbem;
> Quaque, fretum torrens, Maeotidos egerit undas
> Pontus, et Herculeis aufertur gloria metis,
> Oceanumque negant solas admittere Gades.

(The Tanais, falling down from the Riphaean heights, gives the names of two worlds to its two banks, bounding Asia and Europe as well—it keeps the central part of earth from union, and, according to its windings, enlarges now one continent and now the other—and where the Euxine drains the rushing waters of the Maeotian Mere through the strait; and thus men deny that Gades alone lets in the Ocean, and the Pillars of Hercules are robbed of their boast.)[35]

The Tanais River, rushing into the Black Sea and, subsequently, the Mediterranean Sea, performs the function of a continental boundary in symmetry to Europe's western border, Cádiz, next to the Pillars of Hercules, where the Mediterranean and the Atlantic meet. Europe's most prominent symbolical western border post, the Pillars of Hercules in Cádiz (the city's Latin name, "Gades," signifies border), popular already on medieval *mappae mundi*, was mirrored by an imagined eastern equivalent: the Columns (or Altars) of Alexander and the "Sarmatian Gates"—the former typically inserted right in the middle of the Riphean Mountains on maps. Leonid Chekin has long emphasized the symmetrical cartographic layout of Europe, marked by specular border posts:

> The western limit of the world is Gades (modern Cádiz), the ancient city in Spain, which medieval geographers represented as two or more islands. Already in antiquity this city acquired a special significance for theoretical geography, as the westernmost point of the Mediterranean. According to the tradition that Strabo discusses in detail, there the Pillars of

Hercules were erected. Metonymically, those were often designated as *Gades Herculis*. Ancient and medieval geographers tried to designate other extremities of the known world with pillars, columns, or altars, not only because of the characteristic striving for symmetry but also perhaps to stress the fundamental attainability of the "ends of the earth."[36]

Defined by the island city of Cádiz, the limits of Europe's west were imagined with the figure of an island. In a recent study on islands, Marc Shell has highlighted the importance of islands for the act not only of geographic, but also of linguistic and philosophical definition. "Islandness," Shell contends, "informs primordial issues of philosophy: how, conceptually, we connect and disconnect parts and wholes. . . . If there were no islands already, it would be necessary for human beings—the logical and political creatures that we are (or strive to be)—to invent them."[37] One marvels at a temporal coincidence: in the year Miechowita was completing the *Tractatus* with its strong focus on the delineation of Europe's east, Thomas More published *Utopia* (1516), where the question of isolation and definition vigorously comes to the fore with an eye to Europe's (expansion to the) west. Utopia, initially a peninsula, was separated from the mainland by King Utopus, who dug a canal between the peninsula and the mainland. In tune with Shell's concept of "islandness," Franco Farinelli argues that the performed cut across the peninsula in *Utopia*, the first act to give rise to the island, is akin to King Utopus's substitution of a physical landscape with a conceptual reflection thereof; it amounts to the theorization of the originary act of cartographic definition.[38]

Europe's east lacked an equivalent to the western insular border posts—until Miechowita "invented" one. Miechowita's treatise introduces the Crimean Peninsula, a liminal place between Europe and Asia, as an insular boundary. Known to the ancient Greeks as "Thaurica Chersonesus" ("chersos" means "dry land," "nesos" "island"),[39] the topography of the Crimean Peninsula emerges in the *Tractatus* as a *mise en abyme* of continental thinking, foregrounded as a question of (dis)connectivity.[40] Unlike other humanists who refer to Crimea as peninsula, Miechowita mobilizes an insular terminology, calling Crimea "island" (*insula*).[41] His word choice emphasizes Crimea's liminal status and its uncertain connectivity to the mainland by the narrow Isthmus of Perekop (in the Slavic languages, the word "perekop" means "trench," "dugout"):

> Thartari vero Ulani ab ingressu insulae in campestribus eius, ut est innata consuetudo Thartarorum, degunt, et extra insulam idem campestria

Sarmatiae Europianae iuxta Paludes Meotidas et mare Ponti usque ad Album Castrum occupando possident. Ingressumque in insulam seu introitum ad occasum solis fecerunt aggerem de terra longitudinis unius miliaris in modum pontis sternentes, rude tamen et semiplene, ita ut aquae maris in aliquibus locis aggerem pertranseant. Insula ergo antiquitus Thaurica dicebatur, nunc vero Przekop, quod sonat fossatum, quoniam aquae circumdederunt eam et protegunt, tanquam fossata aquis plena civitates.

(From the entrance of the island, the Ulan (= Perekopian) Tatars live on their fields, as is the innate habit of the Tatars, and beyond the island they occupy fields in European Sarmatia next to the Sea of Azov and the Black Sea which stretch to Album Castrum. At the entrance or beginning of the island they made an earthen rampart in the west, one mile long, in the manner of a protruding bridge. But it is poorly made and half-completed, so that in some places seawater flows over the bridge. In antiquity, the island was called Thaurica, now however Perekop, which means trench, because waters surround it and protect it, just like a trench filled with water protects a city.)[42]

An island and peninsula alike, the contours of the Crimean Peninsula depend on the movement of the surrounding sea. In its capacity to change its form, the Thaurica insula is generative of new contours and indexes the process of cartographic definition itself. By isolating Crimea, Miechowita zooms into the delicate and fleeting line between the two continents. If Miechowita is the first to refer to the Crimean Peninsula as "insula" in a geographic treatise, there was a visual precedent which might have informed his word choice: the popular fifteenth- and sixteenth-century cartographic genre of the *isolario*, or Island Book, invented in 1420 by Cristoforo Buondelmonti,[43] who here described and depicted not only the Isles of the Aegean but also cities such as Constantinople as an island (on the city's oldest surviving map). Although Crimea is missing in Buondelmonti, it features in the first printed *isolario* by Benedetto Bordone (1528), who graphically showcases Crimea in an astonishing oscillation between peninsula and island (fig. 10). Here, Crimea's relation to the mainland is surprisingly ambiguous: while detached from the mainland, the "Thaurica Chersoneso" still maintains its connection to it by shallow swamp-like waters ("palus"), symbolized by a myriad of dots.

In his own Island Book, *L'isole più famose del mondo* (1572), Tommaso Porcacchi—the Italian editor of Miechowita's *Historia delle due Sarmazie*—

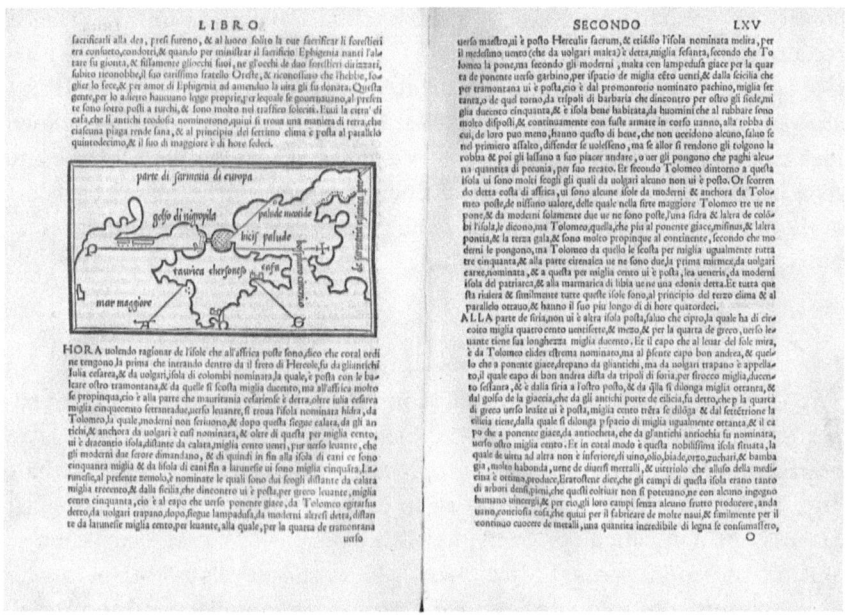

Figure 10. Crimean Peninsula ("Thaurica Chersoneso"), in Benedetto Bordone, *Libro . . . nel qual si ragiona de tutte l'isole del mondo* (Venice: Zoppino, 1528), 64r [G500. B6*]. Photograph: Harvard Map Collection, Harvard University.

imagined the continental line ("linea") between Europe and Asia cutting precisely across the Perekopian Strait, the "Strait of Caffa": "L'Europa confina da Levante co'l fiume Tanai, dalla fonte del quale si tira una linea verso il mar di Tramontana, & cosi ancora dalla foce del detto fiume si tira un'altra linea per la palude Meotide, o mar delle Zabacche fino al Bosforo Cimmerio, o stretto di Caffa" (To the east Europe borders the Tanais River, from whose source one traces a line toward the Western Sea [Black Sea], & from the mouth of said river one traces another line across the Sea of Azov, or the Perekopian Strait, to the Cimmerian Bosphorus, or the Strait of Caffa).[44] If for Porcacchi the continental boundary was a clear-cut "line," for Miechowita the actual border between Europe and Asia was less defined. He imagined it predominantly in flux, changing with the movement of peoples, similar to the connectivity of the Crimean Peninsula to the mainland, which alters its connection status with the potent movement of the waters.

In the centuries that saw the rise of continents, "the fundamental relationship between the world's major land masses was increasingly seen," as Martin W. Lewis and Kären E. Wigen have argued, "as one of separation, not contiguity." What constituted, for the ancient Greeks, "a unitary human terrain" was later "disassembled into its constituent continents, whose relative *isolation* was now ironically converted into their defining feature."[45] An analogy to the island of Cádiz (Gades) at Europe's western limit, Miechowita's nominal creation of an insular space at the border between Europe and Asia is an attempt to delineate a continental boundary through the process of geographic isolation while at the same time drawing attention to the complexity of the process of border-making.

BORDERLINE AND BORDERLAND

The *Tractatus* envisions borders as a constant negotiation between two opposite movements: on the one hand, the definition of the borderline as an abstract and arbitrary cartographic imposition, indebted to the Ptolemaic grid, and, on the other, the exploration of the border as a process of what David Newman called *bordering*, a continuous performative enactment of shifting territorial zones.[46] Miechowita's *Tractatus* develops both pathways as competing, not mutually exclusive, strategies to understand territorial, in particular continental, boundaries, thus uncannily foreshadowing some of the most salient traits in border studies by some five hundred years. Motivated by the at once overlapping and diverging agendas of borderline and bordering, Miechowita formulates an unprecedented agenda to both imagine a line between Europe and Asia and acknowledge the salient similarities of a vibrant borderland spanning the continents. The toponym Sarmatia serves here as a productive pivot that brings both strands of continental thinking together.

The *Tractatus* is divided into two books, book one on Asian Sarmatia and book two on European Sarmatia, which roughly follow the logic of borderline and borderland, respectively. Book one delves into the "Scythian" aggression of European Sarmatia, that is, the Tatar incursions in Poland in the thirteenth century. Book two, in contrast, addresses the potential of European Sarmatia to expand further north and east, into Asian territories. The two books thus negotiate the making of continental boundaries by using the opposed movements of invasion (from Asia to Europe) and colonization (from Europe to Asia) as effective performative acts in the border's crystallization. These movements of performance—understood, with Nigel Thrift, as a "means by which space is produced"[47]—differ not only in directionality, but also in quality. The first movement, pointing from Asia toward Europe, seeks to delineate Europe against invading nomadic peoples from the east, predominantly Tatar and Turkic populations. In a gesture of (en)closure, it follows the logic of a (protective) borderline. The second movement imagines openness as it projects territorial expansion from Europe toward Asia with the goal to extend Europe's continental boundaries, under the aegis of the Polish king Sigismund I, toward the north and the east. It takes advantage of the borderland as an elastic concept, at the core of territorial expansion. Taken together, the two movements turn the border into a permeable site where "originary differentiation"[48] between the two models occurs.

If recent manifestos in border studies such as the *antiAtlas of Borders* push back against the concept of atlases that "express stability, or rather give the illusion of it," and privilege the concept of the *antiAtlas* as an alternative "to reintroduce borders' dynamic nature and complex manifestations, and to provide a critical approach to border representations,"[49] Miechowita's work emerges as even more pertinent: predating the first atlas by half a century,[50] the *Tractatus* addresses, just like the *antiAtlas* manifesto, the cartographic shortcomings of drawing abstract territorial lines independent of the ground they traverse, turning the *Tractatus* into a powerful statement about the difficulty in rendering justice to the multiethnic and multicultural configurations and realities of a borderland difficult to disunite. In its exploration of possible criteria to draw continental borderlines and set Europeans apart from Asians, the *Treatise* is an activation *avant la lettre* of this broad spectrum of possible border conceptions. Miechowita chooses to render the elusive nature of the borderland by deploying a toponym that oscillates between the two poles of borderline and border zone: Sarmatia.

"The idea that the EU has 'borderlands' at its furthest reaches (especially in the east) has become popular in recent times,"[51] theorists of globalization observe. Recent media attention focusing on the territorial conflicts between Ukraine and Russia over Crimea, a peninsula discussed in great detail by Miechowita, has further unearthed the inherent complexity of Europe's eastern borderland. European and Asian Sarmatia, the dual toponym Miechowita chose to define this elusive borderland, was a paradox in itself: while Sarmatia binds the two continents to one another, the added attributes (European and Asian) forcefully divide them. Miechowita's unique continental thinking, developed both within a Ptolemaic framework and without, at once appropriated and explicitly rejected the ancient geographic lexicon. Miechowita's mobilization of the toponym Sarmatia was a powerful and unprecedented lexical and cartographic intervention that sought to elaborate a systematic approach to comprehend an immensely complex multiethnic and multicultural borderland that Ptolemy had equally separated by recurring to insufficient or non-existing natural boundaries: the Don River and the Riphean Mountains.

Situating Sarmatia, then, becomes for Miechowita more than a humanistic act of philological accuracy and translational astuteness: it turns into a test case triggered by the cartographic problem of how and where to interrupt the Eurasian land mass. Scholars have heretofore argued that "the essence of Eastern Europe [even] in the eighteenth century was its resistance to precise geographical location and description"[52] and that "until the early eighteenth century, what we would call eastern Europe fell conceptually under the rubric of the North."[53] Yet the *Tractatus* already manifestly formulates continental thinking as a twofold, albeit paradoxical, proposition: in anchoring the treatise in Ptolemy's *Geography*, Miechowita is the first to privilege the Ptolemaic grid as a powerful tool to trace a borderline separating Europeans from Asians as a distinction between sedentary peoples and nomads (Tatars and other Turkic peoples). Continental differentiation, the *Tractatus* suggests, is motivated not by religion, but by the inclusion in (or exclusion from) the cartographic grid. While the European Sarmatians are cartographically captured, their Asian counterparts are relegated beyond the grid. For the first time, Europeanness is not defined by religion, but by sedentariness. Quite distinct from the first, the second pathway unearths the cultural, linguistic, and geostrategic wealth of Europe's eastern borderland—not only its function as a strategic crossroad that straddles two continents, but also, one might regret, its colonizing potential for further expansion to the east.

MAPPING SARMATIA

Miechowita's contemporary Martin Waldseemüller was one among many cartographers who, in his 1513 edition of Ptolemy's *Geography*, created tables of European and Asian Sarmatia. His "Octava Tabula" (Eighth Table) of "European Sarmatia" (fig. 11) pictures eastern Europe as a disproportionately narrow strait between "Germania" and Asian Sarmatia. Following Ptolemy's description, the Riphean, Hyperborean, and several other mountain chains traverse the territory not so much as a dividing line between Europe and Asia, but as a geographic obstruction of Sarmatia: the mountain chains do not constitute borderlines, but rather extend like an arborescent network across the region. Adjacent to the sources of the Don River ("fons Tanais")

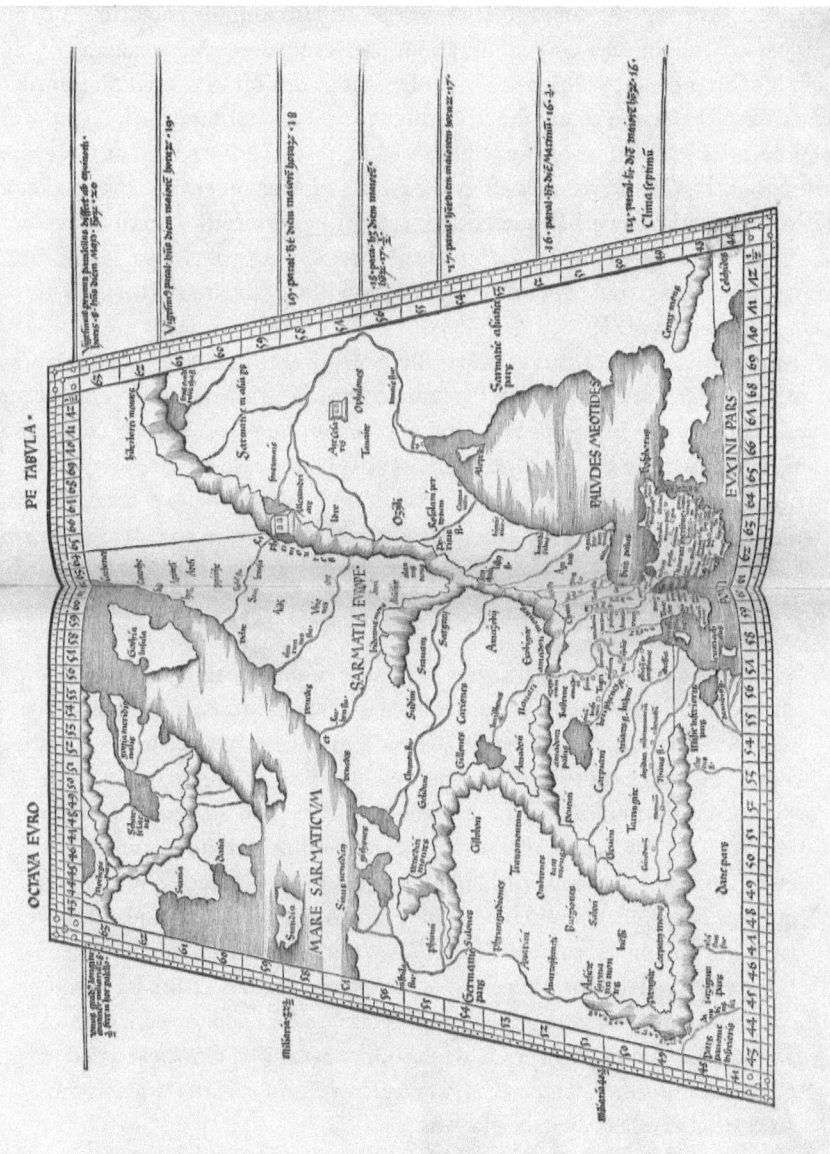

Figure 11. "Octava Europe Tabula," in Martin Waldseemüller, *Geographia* (Strasbourg: Johann Schott, 1513), fol. 178 [GE DD-2043 (RES)]. Photograph: Bibliothèque nationale de France.

and couched in the midst of the Riphean Mountains, the Altars of Alexander ("Arae Alexandri") stake here the limits of Alexander the Great's territorial conquests as a sacralized border between Europe and Asia. Yet Waldseemüller must have felt the insufficiency of the Ptolemaic nomenclature for the purpose of imagining a continental boundary and proceeded, as did all of his cartographic peers, by obsessively multiplying natural and artificial border posts with the scope to artificially enhance and accentuate the character of a border: the very range of the Riphean Mountains harbors here at once the sources of the Don River as well as the Altars of Alexander. The mapmaker's desire to acknowledge his ancient source turns here into a compulsive act of visibly accumulating and strategically displaying tokens of liminality: mountain chains, sources of rivers, and symbolic altars are amassed to mark a continental divide.[54]

Miechowita's *Tractatus* is unparalleled in that it unhinges Europe's east from its heretofore immobile Ptolemaic mold.[55] Following several other Greek and Roman geographers, Ptolemy had argued that Europe is divided from Asia not only by the Don (Tanais) River, but also by the lofty chains of the Riphean and Hyperborean mountains. Miechowita's treatise dramatically revises Ptolemy's *Geography*, arguing that the so-called Riphean and Hyperborean mountains exist exclusively in the imagination of the ancients:

> Sed & montes alanos hiperboreos & ripheos in orbe terrarum nominatissimos, in illis regionibus septemtrionis affirmaverunt. & ex eis non minus famosa erupisse flumina, per cosmographos & vates celebres scripta atque decantata. Tanaim, boristenem maiorem & minorem, volhamque maximum fluminum declararunt. quod cum alienum sit a vero, non abs re (experientia docente que est magistra dicibilium) confutandum & reijciendum est. tanquam prophanum inexperteque provulgatum. Scimus quidem & visu cognoscimus prefata flumina tria (magna siquidem) boristenem, tanaim, & volham ex Moskovia oriri & defluere. minorem vero boristenem (quem aristoteles diaboristenidem appellat) ex Russia superiori cepisse, & in maiorem boristenem decurrere & ei commisceri. Montes autem hiperboreos, ripheos, & alanos nuncupatos illic non existere certo certius scimus & videmus & iam praedictos flumios ex terra plana consurrexisse ac emersisse conspicimus.
>
> ([Cosmographers and famous poets] claim that in the northern regions there are world-famous mountains, called the Alan, Hyperborean, and Riphean mountains, from which flow no less famous rivers such as the

Tanais, the major and minor Borystehenes, and the grand river Volga, described by cosmographers and famous poets in words and songs. Yet this is far from being true, since this information does not originate in a consideration of the thing itself—experience is the teacher of sayable things—and has to be rejected as a sacrilegious declaration lacking experience. We know for certain and have seen that the three mentioned rivers (the great) Borysthenes, Tanais, and Volga originate in and flow from Muscovy. The minor Borysthenes (which Aristotle calls Diaborysthenus), originating in upper Ruthenia, flows into and mingles with the great Borysthenes. Thus, the Hyperborean, Riphean, and Alan mountains do not exist and the above-mentioned rivers originate and have their sources on flat ground.)[56]

The publication of Miechowita's treatise redirected thinking about the topography of eastern Europe and, in particular, about the existence of the mythical Riphean and Hyperborean mountains, all too comfortably used as a dividing line.[57] What is more, in its commanding rejection of Ptolemy, Miechowita's treatise is a bold precursor of Alexander von Humboldt. As Marica Milanesi has argued, "for Renaissance geographers," who followed ancient authorities such as Aristotle, "the great rivers emptying into the northern shore of the Black Sea had to have their origins in great lakes or, preferably, great and distant mountains."[58] Miechowita's observation that a river could originate "in a flat plain" (*ex terra plana*), foreshadows Humboldt's nineteenth-century findings that a great river (such as the Orinoco) "could also emerge from the imperceptible undulations of a plain."[59] Grounding his argument in the Latinized name of the kingdom of Poland, "Polonia," from the Polish noun "pole" ("field" or "plain"), Miechowita claimed that Europe's east was entirely flat, dotted only by a few slight hills.[60]

The toponym Sarmatia circulated widely from the time of Herodotus, who located the Sarmatians northeast of the "Royal Scythians,"[61] the inhabitants of the Black Sea's northern shore. A people of Asian descent, the Sarmatians occupied, at different times, a territory between the Ural Mountains and the Balkans before being defeated by the Goths after the fall of the western Roman Empire. For Ovid, the stereotypical Sarmatian was, just as the neighboring Scythian, a bow-wielding horseman from the Pontic steppes "summon[ing] up the worst type of nomadic marauder" and "driv[ing] his creaking wagon across the Ister" (*per Histrum / stridula Sauromates plaustra bubulcus agit*).[62] Tacitus juxtaposed the Sarmatians with the Germans, arguing that these two peoples were separated "by mutual fear or mountains" (*mutuo metu aut montibus*).[63] If in the sixth century CE the Sarmatians largely, although not entirely, disappear from historic sources,[64] they are resuscitated with the humanists' renewed interest in Ptolemy's *Geography* and, in Poland, with historians such as Jan Długosz and Miechowita.[65]

Ptolemy was the first to divide the elastic toponym Sarmatia into a European and an Asian part. In the *Geography*, European Sarmatia is an impressively vast region, "terminated on the north by the Sarmatian ocean adjoining the Venedicus bay" and "extend[ing] southward through the sources of the Tanais river." It is "terminated in the west by the Vistula river and by that part of Germania lying between its source and the Sarmatian mountains," and bounded, in the east, "by the Tanais river, and by the line which extends from the sources of the Tanais river toward the unknown land as far as the indicated terminus."[66] The contours of Asian Sarmatia, substantially uncertain, are "terminated on the north by unknown land; on the west by European Sarmatia,"[67] and they stretch, in the east, across Caucasia to the Caspian Sea. One of the very first humanists to take up Ptolemy's nomenclature only a few years after the translation of the *Geography* into Latin by Jacopo Angeli da Scarperia was the French cardinal and humanist Guillaume Fillastre, who received, between 1414 and 1418, a copy of the *Geography* at the Council of Constance. In a marginal note where he comments on Ptolemy's map of "European Sarmatia," Fillastre reinterprets Ptolemy's rough staking of the region with an eye to the geographic knowledge of the early fifteenth century and another to competing ancient geographic sources which had suspected the presence of monstrous creatures at the edges of the *oikoumene*:

> Octava Europe tabula continet Sarmatiam Europe, id est illas regiones que sunt ab Germania ad septentrionem versus orientem, in quibus est Polonia, Pruthia, Lituania et alie ample regiones, usque ad terram incognitam ad septentrionem. . . . Item continet, ultra quod ponit Tholomeus, Norvegiam, Suessiam, Rossiam utramque et sinum Codanum dividens Germaniam a Norvegia et Suessia. Item alium sinum ultra ad septentrionem, qui omni anno congelatur in tercia parte anni; et ultra illum sinum est Groenlandia que est versus insulam Tyle magis ad orientem . . . de quibus Tholomeus nullam fecit mencionem, et creditor de illis non habuisse noticiam. . . . In his regionibus septentrionalibus sunt gentes diverse inter quas Unipedes et Pimei; item Griffones sicut in Oriente, velut videtur in tabula.

> (The eighth table of Europe contains European Sarmatia, that is, those regions that extend from Germany to the north and toward the east,

where Poland, Prussia, Lithuania and other vast regions are situated, and to unknown lands in the north. . . . In the same way, it [the table] contains, beyond what Ptolemy mentions, Norway, Sweden, the two Ruthenians and the Baltic Sea which divides Germany from Norway and Sweden. Also, it encompasses another gulf further toward the north which is frozen a third of the year; and beyond that gulf is Greenland located toward the east, toward the island of Thule . . . which Ptolemy does not mention; it is believed that he did not have any knowledge of it. . . . In these northern regions live different peoples such as the Unipedes and the Pygmies; also griffins, just like in the east, as one can see on the map.)[68]

For Fillastre, "Sarmatia" stretches predominantly to the north. It encompasses the Scandinavian countries as it registers the progressive transformation of the Sarmatians into monstrous creatures in the tradition of Pomponius Mela and Pliny the closer they touch the limits of the mapped world. What had been a predominantly geographic framework for Ptolemy concerning a vast area between the Baltic and the Black Sea turned into an ethnonymic concern in the course of the Middle Ages. Several medieval but also the first sources humanists influenced by Ptolemy's *Geography* employed the toponym Sarmatia when referring to the ill-defined peoples north and east of the German lands. In his thirteenth-century *Otia Imperialia*, Gervase of Tilbury, starting his description of "the first region of Europe . . . from the river Don" ([a] Tanay ergo fluvio prima Europe regio), writes that "inter Germaniam et Meotides paludes ab oriente Wandalorum gens ferocissima habitat, inter quam et paludes Meotides pereque Sarmate habitant, a quibus mare Sarmaticum dicitur" (between Germany and Lake Maeotis [Sea of Azov] in the east lives the fierce tribe of the Vandals. Between them [the Vandals] and Lake Maeotis live the Sarmatians. The Sarmatian Sea [Baltic Sea] is named after them).[69] If for Gervase the Sarmatians neighbored the Wends, a western Slavic people often confused with the phonetically close Vandals, for the fifteenth-century Polish historian Jan Długosz, one of the principal sources of Miechowita's *Tractatus*, the region where Slavic peoples originated "[a] veteribus autem scriptoribus et historiographis Sarmatia Europica appellatur, et tam Rutheni, quam Poloni Sarmatae nominantur"[70] (is called European Sarmatia by the ancient writers and historiographers. Both Ruthenians and Poles are named Sarmatians). Yet for Długosz, the capacious quality of the toponym extends markedly further: in a sweeping move, he includes all the peoples living between Germany and the Middle East under this ethnonym:

Ex montibus autem istis Sarmaticis oriuntur tres alii fluvii notabiles etiam et insignes, videlicet Istula alias Wysla, Odra alias Guttalus, et Albia sive iuxta antiquos Albis alias Labya. Item ex Sarmatia praefata egressae sunt gentes hae, quae possederunt Scandiam et novam Daciam, quae dedit ortum Theutonis, Vandalis, Gothis, Longobardis, Rugis et Jepidis, quos vocant aliqui Cimbros, quos hodie vocamus Pomeranos. Illa autem Sarmatia occidentalis ortum habet a Sarmatia orientali, quae est infra mare Caspium, et habet ortum a Saruth, qui fuit quartus ante Abraham de Sem.

(Three other notable and famous rivers, the (V)Istula (or Wisła), Oder (or Guttalus), and Albia (or Albis or Elbe) originate in the Sarmatian Mountains. Peoples coming from said Sarmatia took possession of Scandia [Scandinavia] and new Dacia [Denmark] and gave birth to the Teutons, Vandals, Goths, Langobards, Rugians and Jepids [Gepids], whom some call Cimbri and whom we call today Pomeranians. However, western Sarmatia takes its origins in eastern Sarmatia below the Caspian Sea with its origins in Saruth, the fourth [descendant] of Shem, before Abraham.)[71]

Długosz divides Sarmatia into a western and an eastern part and, grounding his arguments in genealogical precedence, reclaims superiority of the latter. For Długosz, central and eastern European peoples—not only the Slavs[72]—stem from Shem,[73] the ancestor, as was currently accepted, of Asian peoples. Długosz formulates here the twofold idea that the eastern (Asian) part of Sarmatia is superior to its western (European) part and that the Sarmatians are intrinsically heterogeneous. For Długosz, defining Sarmatia amounts to the uneasy task of tracing what Denis Guénoun termed a "movement of tearing away, and carrying off from the paternal lands, from Asia as land . . . toward the strange site without a name."[74] If Długosz retrieved the toponym Sarmatia and its dual geographic division from the work of Ptolemy, Miechowita's intervention proposed a new terminology for what he formulated as a multilayered reflection upon continental contours.

Espousing a critical stance toward his fellow historian, Miechowita's treatise is a determined attempt to substitute cartographic for genealogic thinking, transforming Długosz's description of temporal continuity into a formulation of spatial (dis)continuity. In the *Tractatus*, genealogical and chronological concerns yield to a dominant focus on space. The treatise taps into the very moment in which a rupture in the transcontinental genealogy of the Sarmatians takes place as an "originary differentiation"[75] in space, as a fine suture line setting the stage for continental separation and continuation alike. But Miechowita also used this site of differentiation to reset the toponyms. While nominally embracing the Ptolemaic nomenclature—book one is dedicated to Asian, book two to European Sarmatia—his treatise subverts the terms by strategically introducing a pair of ethnonyms: Miechowita renames the Asian Sarmatians "Scythians"—an ethnonym he equates with the nomadic people of the Tatars—while tacitly enclosing the "Sarmatians" within the bounds of what he designs as a tentatively emerging eastern Europe.

Yet if European Sarmatia—or Sarmatia *tout court*—emerges "by separating itself from and pitting itself against Asia,"[76] a phenomenon Rodolphe Gasché observed for the constitution of Europe at large, the continental line producing difference constantly collapses within the treatise, laying bare the strong and intimate ties that bind Europe and Asia to one another. While the title of the *Tractatus* programmatically announces a binary logic, the body of the work ultimately resist a clear-cut continental differentiation. By continuously shuttling between Asia and Europe in both books, Miechowita ultimately undoes the continental divide he initially sets out to define in the title: the borderline disintegrates within the body of the text, bringing to light the tension not only between ancient toponymy and early modern cartography but also between the theory of a cartographic borderline and the practice of a lived borderland.

CARTOGRAPHIC BELONGING

In *Lines of Power/Limits of Language*, Gunnar Olsson asked a seemingly innocent question: "For what is geography, if it is not the drawing and interpretation of lines?"[77] If and how continental lines ought to be drawn was precisely what Miechowita questioned in his treatise. It is at once paradoxical and unique that the *Tractatus* mobilizes the power of the border region as a thriving microcosm while using Ptolemy as a template to define Europeanness.

Unlike early and contemporary humanists such as Enea Silvio Piccolomini, believed to be among the first to define Europeanness along religious lines, coterminous with the bounds of Christianity, Miechowita developed a concept I call cartographic belonging, in which "the ontological effect of the grid," as Bernhard Siegert put it, "is the modern concept of place"[78]—or, in Miechowita's case, of an imagined and emerging Europe. In its focus on locatability, Miechowita's *Tractatus* uses the dichotomy between sedentary peoples and nomads as a new criterion to establish the boundary between Europe and Asia. Miechowita imagines sedentary peoples as those who can be pointed at, identified, and indexed by the cartographer's taxonomic practice, thus fulfilling what Charles S. Peirce deemed to be the map's dominant function: indexicality.[79] Roaming nomads and itinerant migrants, in contrast, who resist cartographic placement, are relegated beyond the grid. Yet they are not "nomads" in an ontological sense, but rather remain such as long as they resist and are not captured by the grid. Entering Europe, Miechowita's *Tractatus* suggests, is akin to entering the grid—it is an act of being indexicalized and cartographically framed. With Ptolemy's grid in mind, Miechowita develops what Heidegger will much later call a "Gestell" (enframing), a device "aim[ing] at the availability and controllability"[80] of peoples potent enough to convert sedentary peoples into localizable data and bar nonlocalizable peoples from entering the grid system. Europeanness is here a function of cartographic framing and localization.

Scholars have recently raised the question whether "the expansion of Western culture from the sixteenth to the twentieth century [can] be described in terms of a growing totalitarianism of the grid."[81] Miechowita's work addresses precisely the question of cartography's expandability while pondering its possibilities and limits as a marker of continental distinction.

Driven by the question of what a territory "contains," Miechowita popularized the participle "cont(in)ens" for the European continent, after Vespucci's use of the term in the *Mundus Novus*, his letter in which he describes the newly "discovered" territories across the Atlantic as a "New World."[82] Miechowita formulates the question of what a territory "contains" as a question of continental belonging. The treatise powerfully opens with a succinct geographical definition of the two Sarmatias and what they "contain":

> Antiquores duas sarmatias posuerunt, unam in Europa, alteram in Asia, sibi invicem coherentes & contiguas. In europiana sunt regiones russorum seu rutenorum, Lithuanorum, Moskorum et eis adiacentes, ab occidente flumine Visle et ab oriente Tanai incluse. harum regionum gentes olim gete nuncupabant. In asiatica vero sarmatia nunc commorantur et degunt plura genera thartarorum, a flumine Don seu tanai ab occidente usque ad mare Caspium ab oriente *contenta*. Horum imperia, genelogie, ritus et mores latitudoque terrarum, flumina et circum circa adiacentia in subscribendis explanabuntur.

(The Ancients distinguished between two Sarmatias, one in Europe, the other in Asia, which bordered on and were contiguous to one another. In European [Sarmatia] are the regions of the Russians or Ruthenians, Lithuanians, and Muscovites and those adjacent to them. They are enclosed in the west by the Vistula River and in the east by the Tanais. These peoples were previously called Getae. In Asian Sarmatia now dwell and live numerous Tatar peoples, *contained* by the Don (or Tanais) River in the west and by the Caspian Sea in the east. Their empires, genealogies, rites and customs as well as the latitude of their territories, rivers and whatever surrounds them will subsequently be explained.)[83]

The participle "contentus," programmatically folded into one of the first sentences of the *Tractatus*, revives an ancient cosmological and cosmogonic tradition, which can be traced back to Plato's *Timaeus*, via Cicero. In his fragmentary free Latin translation of the *Timaeus*, Cicero introduces the participle "contentus" with the meaning of "self-contained" and "autonomous" in the context of the creation, organization, and division of the world: "Sic enim ratus est ille, qui ista iunxit et condidit, ipsum se contentum esse mundum neque egere altero" (Thus He who brought together and founded everything considered that the world was self-contained and did not need another one).[84] What for Cicero is the self-containment of the world is broken down into smaller, continental, units in Miechowita's treatise.

By spanning Europe and Asia, however, the overarching toponym "Sarmatia" poses, as we have seen, a substantial difficulty for Miechowita's project of territorial autonomy. Both a potent and ambiguous place name operating as a pivot determining inclusion and exclusion alike, Sarmatia gestures to the problem of boundary making itself. The toponym's workings are similar to what Cicero had described as a "tertium aliquid," an intermediary element necessary to set two units apart while making them comparable:

Omnia autem duo ad cohaerendum tertium aliquid anquirunt et quasi nodum vinculumque desiderant. Sed vinculorum id est aptissimum atque pulcherrimum, quod ex se atque de iis, quae stringit, quam maxime unum efficit. Id optime adsequitur, quae Graece αναλογια, Latine (audendum est enim, quoniam haec primum a nobis novantur) comparatio proportiove dici potest.

(Each binary demands a third element for cohesion, it desires something like a knot or a fetter. Of all the fetters the most apt and beautiful is the one which forms the greatest possible oneness out of itself and that which it binds together. The best way to achieve this is what is called in Greek αναλογια and what in Latin (audacity is required, since I am the first to invent this expression) can be called comparison or proportion.)[85]

As a "tertium aliquid," Sarmatia binds, as it were, the two "containers" Europe and Asia together enabling comparative thinking. That Miechowita was grappling with the question of the relationship between the two continents equally transpires in the very structure of the *Tractatus*: a significantly ampler book one dedicated to Asian Sarmatia (Scythia) and the different nomadic Tatar tribes is juxtaposed with a shorter book two centering on European Sarmatia and what Miechowita considers to be sedentary peoples.

In 1535, the *Tractatus* was translated into Polish by Andrzej Glaber,[86] under the title *Polskie wypisanie dwojej krain świata*.[87] In the translation's lengthy title, Glaber took up Isidore of Seville's etymology of the Sarmatians ("Sarmatae") as "armed" (*armatus*) warriors[88] to stylize the encounter between European and Asian Sarmatia as a military confrontation.[89] A woodcut inserted on the frontispiece stages the two parts of Sarmatia as an armed conflict between the Europeans and the Tatars—considered since their incursions in Poland and Lithuania from the thirteenth century on the most formidable threat for eastern Europe coming from Asia. A previous iteration of the woodcut already featured on the frontispiece of the 1521 edition of *Descriptio Sarmatiarum* (fig. 12). It staged the line disconnecting Europe from Asia in an even more dramatic tone. The Europeans, resisting the Tatar aggression, are represented here by the army of the victorious Polish king, riding on horseback in full armor and lifting up high Poland's heraldic banner, a white eagle. The Tatar army, clad in *bashlyks* (cone-shaped hats), brandishes a flag depicting a scorpion, while several Tatars lie defeated on the ground. If Ruggiero's hippogriff, in Ariosto's *Orlando Furioso*, registered the continental border "where Asia separates itself from Europe" as a succession of toponyms, the contiguity between European and Asian Sarmatia is translated here as an armed conflict, further enhanced by the association of the Asian Sarmatians with the scorpion: according to Isidore of Seville's *Etymologies*, a scorpion was, first, "a poisoned arrow, shot by a bow or a catapult, that releases a poison at the spot where it pierces the person whom it strikes."[90] But it was also a geographic marker. During the Middle Ages, the scorpion was associated with the ill-mapped region of Scythia. In his eighth-century *Cosmography*, Aethicus Ister calls Scythia "the mother of dragons, breeding ground of scorpions, native land of snakes, and source of demons."[91] The dichotomy that the illustration spells out in unequivocal

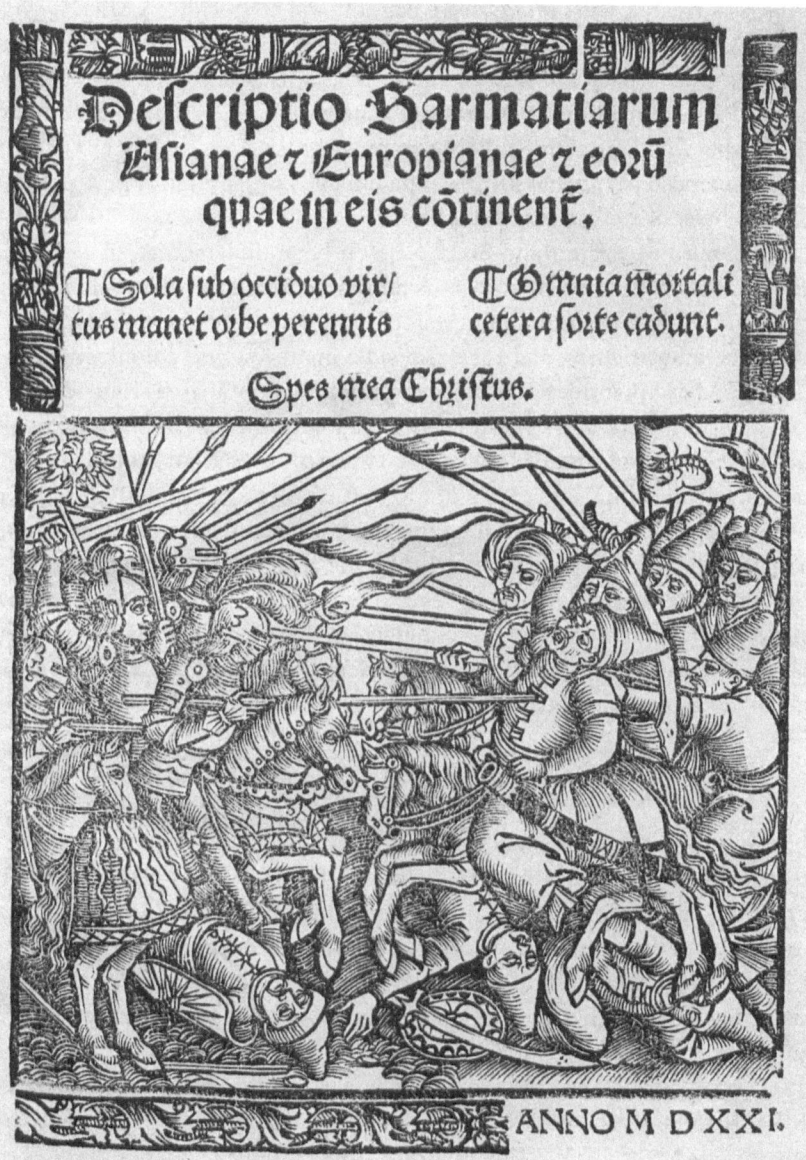

Figure 12. Maciej Miechowita, *Descriptio Sarmatiarum Asianae et Europianae et eorum quae in eis contentis* (Kraków: Jan Haller, 1521), frontispiece [20.Cc.16]. Photograph: © Österreichische Nationalbibliothek, Wien.

terms was treated with much more caution and confusion in Miechowita's treatise. In contrast to the *Tractatus*, which constantly upsets the continental boundary between Europe and Asia, the illustration, a new addition, speaks a different idiom as it translates narrative doubt into pictorial certainty: the woodcut unmistakably frames the two armies as antagonists frozen in a pose of mutual killing. The elusive border zone is here eclipsed at the benefit of a clear-cut borderline.

As Sandro Mezzadra and Brett Neilson have underscored in *Border as Method*, the production of borders is an inherently ambivalent affair. Often romanticized as a desired production of "alternatives to bloc subjectivity," migratory movements—and their flip side: incursion and colonization—are constituted as "practices of mobility" profoundly entangled with "domination, dispossession, and exploitation forged within them as well as the desires for liberty and equality they often express." If migrant mobilities "embod[y] desires, habits, and forms of life that rewrite the normative scripts of national as well as continental belonging,"[92] Miechowita's plan to tap into the performativity of continental belonging through a cartographic lens—with Ptolemy as a guide, but an eye to what Mezzadra and Neilson have termed "cartographic anxiety and metageographic uncertainty"[93]—is a crucial intervention of negotiating early modern continental borders. I call this endeavor cartographic belonging.

SHAPING THE NOMAD

Sigismund I ruled in personal union as the king of Poland and Grand Duke of Lithuania over "terras suas Lithuaniæ et Russiæ Coronæ Regni Poloniæ perpetuo" (his lands of Lithuania and Ruthenia, under the rule of the Polish Crown eternally)[94] and had successfully enlarged Lithuania in the Battle of Orsha in 1514, encroaching on the territory of the Grand Duke of Muscovy, Vasili (Basil) III. Although the victory was short-lived—Muscovy regained Smolensk and other territories soon after—Sigismund I exploited the victory in several encomiastic works in which he was celebrated as the great victor of the Battle of Orsha. One such work in praise of the king was a small volume, titled *Carmina*, published in Rome in 1515, only a few months after the battle. This volume was the most immediate source for Miechowita's *Tractatus*. It contained poems, epigrams, and epistles by Poland's most renowned diplomats, poets, and cartographers such as Johannes Dantiscus, Andreas Krzycki, and Bernard Wapowski (who collaborated with the Italian cartographer Marco Beneventano on the 1507 edition of Ptolemy's *Geography*).[95]

The volume is unique in its attempt to merge a variety of literary forms, contemporary politics, and ancient geography into a cartographically inflected celebration of Sigismund I.

Perhaps the most intriguing feature is the volume's lengthy descriptive title, where the Polish king is celebrated as an heir of the Altars of Alexander the Great: "Carmina | De memorabili cede Scismaticorum | Moscoviorum per Serenis. ac Invictis. | D. Sigismundum Regem Polonie | magnum Ducem Lituanie, | Russie, Prussie, Sarmatieque | Europee dominum et heredem | apud aras Alexandri ma(-) | gni peracta" (*Carmina* completed for the memorable defeat of the schismatic Muscovites by the most Serene and Invincible Sigismund, King of Great Poland, Duke of Lithuania, Ruthenia, Prussia, European Sarmatia, Lord and Heir of the Altars of Alexander the Great).[96]

From the outset, the catalog of Sigismund I's territorial possessions is aligned cartographically according to a double axis—historical and contemporary—that establishes a tension between ancient topography and modern politics. The document defines the Polish king from the vantage point of cartography by blending the actual boundaries of Poland-Lithuania not only with the blurred and ill-mapped Ptolemaic toponym Sarmatia, but also the Polish king's desired eastward expansion, bound solely by the mythical Altars of Alexander the Great.[97] Jan Łaski, archbishop of Gniezno, grand chancellor of the Polish Crown, and one of the contributors to the *Carmina*,[98] further expanded the term "Scythian" to subtend also the orthodox "schismatic" Muscovites, whom he identifies as a "savage and bellicose Scythian people" (*truculenta ac bellicosa gen[s] Scitica*).[99] Like Miechowita, Łaski distinguishes between the nomadic Scythians inhabiting a territory beyond the Tanais and the sedentary Sarmatians who are European dwellers.

While the Asian Sarmatians are referred to as Scythians, both authors consider the Sarmatians either as a sedentary people from Europe or as former Asian migrants now firmly settled (and thus pertaining to Europe). For Łaski, "Poloni namque Slavinorum gens" (the Poles are a Slavic people), who came to "ea Europe loca quibus Maiestas tua iuste feliciter ac victoriosissime imperat (ut grecorum latinorumque monumenta testantur) ex Asia ab hiperboreis montibus" (these European places which are justly, happily, and victoriously ruled by his Majesty [Sigismund I], as Greek and Latin documents show, out of Asia, from the Hyperborean mountains).[100] What separates the (European) Sarmatians from the (Asian) Scythians is the crucial distinction between sedentism and nomadism. The European Sarmatians themselves are an astonishing case in point, as Miechowita lays out in one of his (numerous) myths of origin: once an Asian tribe from the eastern Mediterranean

(Miechowita derives the Slavic name "Jan" from the biblical Javan, Japheth's fourth son, and locates the origin of the Slavs at the Ionian coast in Asia Minor),[101] the Sarmatians have morphed into a sedentary people. Through the commanding power of Ptolemy's cartographic framework, they are captured by the cartographic grid and incorporated into the "continens" named Europe. Localizable, they are now disconnected from their nomadic (Asian) past.

Miechowita sought to formulate cogent criteria to exclude what he considered to be potentially dangerous nomadic tribes while programmatically highlighting the vibrancy of Europe's eastern realms and their potential of territorial expansion. His treatise is as much a vigorous emphasis on Europe's (sedentary) east as it is a strategic attempt at dislodging nomadic tribes such as the Tatars beyond Europe's borders.[102] Book one of the treatise thus strategically opens with a lengthy description of the Tatars, their invasion of Poland under Batu Khan in 1241, and their subsequent conversion to Islam.[103] Unlike his western counterparts, however, preoccupied with a potential new crusade against the Ottoman Turks, Miechowita is mostly concerned with the Muslim Tatars, whose past incursions onto Polish territory he deems a more urgent threat than any potential aggression by the Ottomans. Yet the *Tractatus* focuses not so much on the religion of the Tatars as on their nomadic lifestyle.

Although the Tatars were, in the sixteenth century, an important part of the Polish-Lithuanian Union with their own literary language,[104] Miechowita subsumes them, together with peoples he deems nomadic or non-autochthonous, as "Scythians" and relegates them to "Asian Sarmatia." Alongside the Tatars, book one counts such peoples as the Goths, Alans, Vandals, Swabians, and Hungarians among the nomadic (Asian) Scythians. Already in his 1507 edition of Ptolemy's *Geography*, Beneventano identified the Scythians as nomads who "agrum non exercent nec domus ulla aut tectum, aut sedes est. Armenta & pecora semper pascentibus & per incultas solitudines errantibus" (do not plow their field nor do they have a house or a roof, or settlement. Their domestic animals and sheep graze and err on uncultivated desert land).[105] Valerie Kivelson has pointed to the sharp distinction that the early modern cartographic practice made between sedentary and nomadic peoples: "The elaborate cartouches and ornamental flourishes that adorn early modern printed maps spell out the details of ownership and the boundaries of properties and depict hardworking Europeans tilling neatly fenced fields or tending their livestock. [In contrast,] on maps of 'Tartaria', camels and tents make the same point: the nomads of North Asia enclose no property and invest no labour in the land. Private property and intensive labour conferred rights to landholding that the indigenous population had manifestly failed to earn."[106]

When Piotr Kochanowski translated Ariosto's *Orlando Furioso* into Polish in 1620 (*Orland szalony*), he further enhanced the contrast between the field-tilling Europeans and the nomadic Tatars by adding the adjective "inhuman" (*nieludzkiej*), absent in Ariosto, as a qualifier of the Scythians: Ruggiero flies "Wzdłuż od Hiperborejskiej *nieludzkiej* Scytyej / Do Hirkany, i zajął części Sarmacyej" (From the *inhuman*, Hyperborean Scythia / To Hircania, and occupied a part of Sarmatia).[107] The Scythians are "inhuman," one ought to assume, because they are unable to cultivate the soil for husbandry (also due to the inhospitable land they inhabit).

Contemporaneously with Miechowita's *Tractatus*, Niccolò Machiavelli evoked the Tatars in his *Discourses of the First Decade of Titus Livius* as a people who historically threatened Italy and Christianity. Similar to those European peoples who in the past successfully created a shield for themselves against the nomadic Scythians, Machiavelli suggests, the Hungarians and the Poles have served as a powerful bulwark to protect not only themselves, but nations such as Italy against another nomadic tribe coming from Asia, the Tatars: "E spesse volte occorrono movimenti grandissimi de' Tartari che sono dipoi dagli Ungheri e da quelli di Polonia sostenuti; e spesso si gloriano, che, se non fussono l'armi loro, la Italia e la Chiesa arebbe molte volte sentito il peso degli eserciti tartari" (Nevertheless, movements on a great scale have oftentimes been begun by the Tartars, and been at once withstood by the Hungarians and Poles, whose frequent boast it is, that but for them, Italy and the Church would more than once have felt the weight of the Tartar arms).[108]

Here, Machiavelli employs the popular image of the Poles and Hungarians as a bulwark against invaders from the east. Also, Jan Łaski described the advance of the Tatars as an "infestation," while painting the Poles as the only people capable of resisting this "barbarous force": "Scitice gentes post Gotthos, Vandalos, Hunnos, Alanos, & alios pene infinitos Scitie populos Europam tumultuose infestarint: solique Poloni tantam barbarorum vim sustineant, quam nec Romanum adhuc florens imperium" (Scythian peoples restlessly infested Europe, following the Goths, Vandals, Huns, Alans, and other innumerable peoples from Scythia. Only the Poles withstood the barbarous force that not even the flourishing Roman Empire was able to contain).[109]

Interestingly, the first to use the terms "propugnaculum" (bulwark) and "propugnator" (defender) for the Hungarians and Poles was not a Polish or Hungarian, but an Italian humanist: Francesco Filelfo. Filelfo used the term for the young Ladislas III, king of Poland and Hungary, who died in 1444 in the Battle of Varna while fighting against the Turkish army and the Ottoman Sultan Murad II.[110] In his letter to Ladislas III, Filelfo describes the king as the "propugnaculum" of "the Christian Republic": "Te Christianae Reipublicae propugnaculum nominat" (The Christian Republic calls you a bulwark).[111] For Filelfo, the Polish king's function is twofold: he serves as a firm bulwark against the approaching Ottoman army while exerting his geostrategic power to extend his territories further into Asian land, stepping, as Filelfo explicitly suggests, in the footsteps of Alexander the Great.[112] If the Polish king did not, at first, take up the first function, Filelfo's flattering analogy between the Polish king and Alexander the Great was happily taken up by sixteenth-century Polish humanists, who emphasized King Sigismund I's power to subdue the "Scythian" East.

MOBILIZING EUROPE'S EAST

Miechowita's treatise is at once path-breaking and alarming as it instills unprecedented dynamics into eastern Europe and its borders: the *Tractatus* articulates itself as one appreciative of the region's abundance as well as one plainly exhibiting the Polish king's desire to expand eastward. It strategically mobilizes the *"productive* power"[113] of the border to shape Europe's east with the same complexity and malleability as its western counterpart. In fact, parallel political developments in the first half of the sixteenth century revolving around territorial expansion and colonization unfolded in Europe's east and west alike. If Ivan IV ("the Terrible") was crowned emperor in 1547, his new title, "Tsar," from the Latin "Caesar," referred to his successful establishment of an eastern empire in analogy to the Holy Roman Empire. At the same time when the Spanish emperor Charles V invented his device "Plus Ultra" (1516) with the aim of promoting Spanish expansion beyond the Pillars of Hercules, Miechowita registered and fueled the ambition of King Sigismund I (to whom he was particularly close, also in his capacity as the king's personal physician) to expand the Polish kingdom. In the dedicatory epistle of the *Tractatus* to Stanisław Turzo,[114] Miechowita establishes a symmetry between western and eastern attempts at colonizing new lands by comparing Sigismund I to the Portuguese king Manuel I, who had enlarged his territories to the southern hemisphere:

> Quare ut haec & complura alia in sarmatijs contenta, tue doctissime presul amplitudini vera veraciter enarrarem. subsequentem tractatum de duabus sarmatijs ab antiquoribus minus cognitis nominibus, quibus tem-

poribus nostris nominantur. tibi domino & patrono meo semper colendissimo, scribere breviuscule, ut res expostulabit, ad incitandum alios, qui maiora noverunt, & elegantiori stilo scribere facile poterunt curabo. Utque sicut plaga meridionalis cum gentibus adiacentibus oceano, usque ad Indiam, per regem Portugalie patefacta est. Sic plaga septemtrionalis, cum gentibus oceano septemtrionis imminentibus, & versus orientem spectantibus, per militiam et bella regis Polonie aperta, mundo pateat et clarescat.

(I wrote the subsequent treatise on the two Sarmatias, which the Ancients referred to by less-known names than our contemporaries to tell you, most learned patron, truthfully about these and many other things contained in the Sarmatias. I write to you briefly, my dearest master and patron, as the topic demands, and will make sure to encourage others, who have discovered greater things to write more freely and in a more elegant style. Just as the Southern Hemisphere with peoples adjacent to the ocean as far as India was discovered by the Portuguese king, so the Northern Hemisphere with its peoples [living] closely to the northern ocean and oriented toward the east, discovered through the army and warfare of the Polish king, should be opened up and become known to the world.)[115]

Akin to King Manuel, the Polish king is here imagined as a bellicose conquistador. Miechowita powerfully launches the discussion about Europe's boundaries by establishing an analogy and hemispheric symmetry between two expansionistic powers at Europe's geographic extremes. Miechowita's recurrent use of verbs denoting openness gestures toward a desired disclosure of new territory through invasion and mapping of the Northern Hemisphere,[116] propelled by recent victories of the Polish king alluded to in the text: the Battle of Orsha against Muscovy ("gentes imminentes") in 1514 and the Battle at Lopuszno in 1512, when the combined Polish-Lithuanian forces defeated the Crimean Tatars ("gentes versus orientem spectantes").[117] The king's victories in the east turned into a powerful political tool to transform a scarcely known liminal region into "the field of expansion of the Polish Kingdom" (pole ekspansji Królestwa Polskiego).[118] Miechowita contends that the knowledge of Europe's north and east has hitherto been obscured and obstructed by the ancient geographers' lack of interest in the "two Sarmatias." These territories were further dismissed as "unknown" (incognitas) by later writers who, following in the footsteps of the ancients, described Europe's north "vaguely" (indistincte) and "obscurely" (obscure):

Plures scriptorum orbem terrarum lucubrationibus suis & elucidationibus exararunt. Sarmatias vero, tanquam incognitas praetervecti dimiserunt. Qui autem aliquicpiam de ipsis posteris scriptis carminibusve relinquere curarunt, indistincte & antiquitate premente, tanquam in media nocte, obscure dixerunt. Et quod intollerabilius est, multa ficta et fabulas inextricabiles nusquam adaptandas superaddiderunt.

(Many writers have used their late-night vigils to write about and comment upon the world. But they have dismissed the Sarmatias as something unknown. Those who cared to leave to posterity some information in their writings or poems wrote about it vaguely and obscurely, as if they were writing in the middle of the night. And what is even more intolerable, they added a lot of fictitious and modified fables which are impossible to disentangle.)[119]

As he did in the case of the geostrategic comparison between the Portuguese and the Polish kings, Miechowita privileges here a division of the world into a Northern and Southern Hemisphere. While the image of a Western and Eastern Hemisphere was slowly emerging,[120] the north-south divide, grounded in the Macrobian zonal map which imagines a specular divide of the world along the equator (torrid zone), persisted well into the sixteenth century. In fact, Giovanni Battista Ramusio's *Navigazioni e viaggi* puts an emphasis on a reevaluation of the north as an important hub for international trade. In his *Discorso sopra varii viaggi per li quali sono state condotte fino a' tempi nostri le spezierie e altri nuovi che se potriano usare per condurle* (*Discourse on Various Routes Used until Now to Import Spices and Other New [Routes] that Might be Used to Import Them*), Ramusio discusses how new navigable routes to India via Europe's north could be explored in the wake of the foreclosure of traditional passages to India "per la via del mar Rosso" (*via the Red Sea*). Alluding to the disclosure of possible shorter routes, but also to the rise of the Ottoman Empire and the "mutazioni grandissime e delle religioni e delle signorie" (great changes in religions and dominions) in recent history, Ramusio stresses the potential of Europe's north to guarantee the flow of intercontinental trade. In recent times, he contends,

i mercatanti cominciarono a navigar[l]e pel fiume Indo a contrario dell' acqua, e tanto andarono che giunsero appresso la provincia Battriana, ... e da quello le navigarono a traverso del detto mare [Caspio] insino a un luogo detto Citracham, il qual e dove il grandissimo fiume Rha, ora detto

Herdil o Volga, sbocca in detto mare; poi le condussero a contrario dell' acqua del detto fiume per la Tartaria, e di nuovo con camelli le portarono nel fiume Tanais, che è in capo del mar Maggiore, ora detto la Tana, nel qual luogo non sono ancora centocinquanta anni che andavano le galere e navi veneziane e genovesi a comprar dette spezie e gioie.

(The merchants started to navigate the Indus river upstream, and they proceeded so far as to reach the Bactrian province, . . . and from there they crossed the said [Caspian] Sea until they reached a place called Astrakhan, where the very great Rha river—nowadays called Erdil or Volga—flows into the sea. They pulled them [the ships] upstream along the said river through Tartary, and then again transported them with camels to the Tanais river and Tana which lies at the mouth of the Black Sea. Here, less than hundred fifty years ago, Venetian and Genoese galleys and ships went back and forth to buy spices and jewels.)[121]

The Tanais River, formerly a continental dividing line, acquires here a new, geoeconomic dimension: if navigated upstream, the historic border between Europe and Asia turns into an aquatic highway facilitating international commerce.[122] While the Portuguese have chosen to take "la via del ponente, circondando tutta l'Africa, per la virtu e industria de' gran capitani delli serenissimi re di Portogallo" (the western route, circumnavigating all Africa, by virtue and effort of the great captains of the illustrious Portuguese kings),[123] the heretofore ill-mapped north should be opened up for a navigation that spans continents. With a nod to Miechowita's comparison between the geoeconomic politics of King Manuel I and King Sigismund I, Ramusio praises the advantages of a passage to India via Europe's north:

anco questa tal cosa potria verificarsi ai tempi nostri, quando li principi che confinano sopra quelli mari vi volessero metter qualche poco d'industria e diligenza, e che non sapeva imaginarsi navigazione alcuna di tanta utilità e profitto a tutta la cristianità quanto saria questa, cioè che per questa via si potesse penetrar nell'India, e che si trovasse il paese del Cataio che fu discoperto gia ducento anni per messer Marco Polo. . . . Il viaggio saria molto piu breve di quello che fanno ora li Portoghesi, e anco di quello che si dice che potriano far li Castigliani, all'isole Molucche. E comincio a dire che la citta di Lubecco, ch'e cosi nobile e potente republica posta sopra il mar Germanico, la qual di continuo naviga li mari della Norvega e Gottia, e anco il serenissimo re di Polonia, che vien con li suoi regni di Lituania sin sopra detto mare, sariano atti a far fare questo scoprimento, ma sopra tutti il duca di Moscovia averia la maggior commodità e facilità di ciascun altro principe.

(Another thing could well happen during our lifetime: if the princes who border on those [northern] seas put in a little bit of effort and care, one could not imagine a navigation more useful and profitable for all of Christianity than the one that passes [north] to India and where the land of Cathay is found, discovered two hundred years ago by Marco Polo. . . . The travel would be much shorter than the one by the Portuguese as well as the one that the Castilians could undertake to reach the Moluccas. And I think that the city of Lübeck—a noble and powerful republic on the shores of the German [Baltic] Sea, engaged in navigations across the seas of Norway and Sweden—and the illustrious king of Poland, whose kingdom of Lithuania reaches said sea, are poised to undertake this discovery. But it is above all the Duke of Muscovy whose benefit would surpass that of any other prince.)[124]

For Ramusio, the economic benefits of investing in Europe's north (to reach South Asia) are as important as they are manifest: with only "minimal investment and work" (*poco d'industria e diligenza*) "tanta utilità e profitto" (a lot of utility and profit) can be drawn for "tutta la cristianità" (all of Christianity). Traveling northward to India would take considerably less time and cover a shorter distance than the circumnavigation of Africa undertaken by the Portuguese. When Ramusio emphasized the advantageous geostrategic position of Europe's northern cities and kingdoms, he closely followed Miechowita, who raised, for the first time, a larger awareness of the geostrategic potential of Europe's northeast.

Miechowita's *Tractatus* paints the liminal regions between Europe and Asia as a vibrant ecosystem thriving amidst cultural, linguistic, and religious diversity, pithily referring to these traits as "abundantia" (abundance).[125] The borderland that spans the continental divide is thriving, characterized as it is by a wealth of different cultures living in and traversing it. One is struck, in this regard, by Miechowita's description of Ruthenia ("Russia"):[126]

> In Russia sunt plures sectae: est enim religio christiana, Romano pontifici subiecta, et illa regit et praevalet, quamquam sit exigua numero. Est altera secta Rutenorum amplior, quae ritum Graecorum insectatur, totam Russiam adimplens. Est tercia secta Iudeorum, non usurariorum, velut in terris christianorum, sed laboratorum, agricolarum et mercatorum

magnorum, praesidentque ut saepe theloneis et exactionibus publicis. Est quarta secta Armenorum, praecipue in civitate Camenecensi et Leopoliensi. Hi sunt mercatores peritissimi, ad Caffam, Constantinopolim, ad Alexandriam Aegypti, ad Alkairam et partes Indiae penetrantes et merces afferentes.

(In Ruthenia, there are several religions: the Christian religion, subject to the Roman Pontiff, which rules and dominates, although it is the minority. The religion of the [Orthodox] Ruthenians, who follow the Greek rite, is more significant and stretches across the entire region of Ruthenia. The third religion is Judaism. It is practiced not by usurers, but by workers, farmers, and big merchants in the lands of the Christians. They preside over tolls and public taxes. The fourth religion is practiced by the Armenians, especially in Kmianets-Podilskyi and Lviv. There, one encounters skillful merchants who travel as far as Caffa, Constantinople, Alexandria in Egypt, Cairo, and parts of India to sell their merchandise.)[127]

A major crossroads for exchange and trade, Ruthenia emerges here as a hub for merchants from all three parts of the *oikoumene*. Their circulation and interaction constitute the region's *abundantia*, which articulates itself equally dynamically through the confluence of manifold languages and alphabets. Just as the Ruthenians have "proprias litteras et abecedarium instar et proximum Graecis" (their own letters and alphabet equal in form and close to the Greek one),[128] the Jewish population "similiter Hebreorum litteris utuntur et disciplinis, veru etiam artes liberales, astronomiam et medicinam perscrutantur" (likewise uses Hebrew letters and customs, in their pursuit of the liberal arts, astronomy, and medicine), while "Armeni suo ritu gaudent et litteris" (the Armenians take pleasure in their own rites and letters).[129]

Miechowita's careful attention to the region's co-presence of a variety of languages and alphabets is also a nod to Jakub Parkosz, a professor of law and rector of the University of Kraków, who already in about 1440 designed a new (Latinate) alphabet for the Polish language with the goal of adequately representing its phonetic system. In an explicit reference to John of Mandeville, who included classical alphabets in his travel writings, Parkosz discusses the alphabets of the three classical languages—Hebrew, Greek, and Latin—in conjunction with the alphabet of the Ruthenians in Europe's east, equally touching upon Turkish, Chaldean, and Russian.[130]

In tune with Parkosz, Miechowita commandingly showcases the linguistic *abundantia* as the border regions' "*productive* power," understood as an intrinsic part of a broader system of circulating languages and learning. In the *Tractatus*, Europe's vibrant east does not emerge as a *translatio* or copy of western models, but rather as a unique confluence of cultures, religions, and languages that thrives in analogy and conjunction with the west. The *Tractatus* describes a microcosm whose paths chart itineraries into the different parts of the world as they, conversely, bring merchants, scholars, and artists from the entire *oikoumene* into the region. Mezzadra and Neilson rightly contend that "the modern cartographical representation and institutional arrangement of the border as a line . . . has somehow obscured this complexity [of the border] and led us to consider the border as literally marginal."[131] But borders articulate, the authors continue with an eye to contemporary global and postcolonial capitalism, "global passages of people, money, or objects."[132] Miechowita's *Tractatus* is a unique example showing that borders as thinly drawn, abstract, demarcation lines were only slowly emerging and coexisted alongside boundaries which took the shape of a vibrant borderland. The treatise exposes and operates within this tension. Here, the border slowly morphs into a "method,"[133] the pursuit of a pathway, for continental thinking. From the border as a productive site, processes of translation, circulation, and colonization unfold against the backdrop of the rising practice of cartographic gridding.

BEYOND CARTOGRAPHIC BELONGING: THE POETICS OF ZŁOTA BABA

In book two of the *Tractatus*, formally dedicated to European Sarmatia, Miechowita not only blurs his own criteria between sedentary and nomadic peoples, but also offers, in a concluding section, an insight into the fleeting demarcation of the regions adjacent to the North Pole by mobilizing not only cartographic but also new poetic imagery:

> Post terram viathka nuncupatam in scythiam penetrando iacet magnum idolum *Złota baba*, quod interpretatum sonat *aurea anus* seu *vetula*, quod gentes vicine colunt et venerantur, nec aliquis in proximo gradiens aut feras agitando et in venatione sectando vacuus et sine oblatione pertransit, quinimo si munus nobile deest, pellem aut saltem de veste extractum pilum in offertorium idolo proijcit et inclinando se cum reverentia pertransit.

> (Beyond the land called Viatka, deep inside Scythia, one finds a great idol, *Złota baba*, which means "the Golden Old Woman." Neighboring

peoples worship and venerate it, and no one walking by or out hunting passes it without making an offering. Even if they do not have a precious gift, people leave a fur or even a thread plucked from their jacket. After which, bowing reverently to the idol, they resume their walk.)[134]

The Golden Old Woman—or *Złota Baba* in Polish—is Miechowita's unique poetic in(ter)vention that marks not only the end of the known territory but also the end of the *Tractatus* itself. Venerated as a pagan idol to whom one offers gifts, the *złota baba* became a "popular figure in Western mythology about the Russian North,"[135] challenging the impact of religion as the sole adequate marker of borders. As Miechowita specifies in his dedicatory letter to the Kraków-based printer Johannes Haller:

In die nativitatis Christi descendit grando glacialis & confecit imaginem virginis tenentis infantem coronatum, & imago infantis pedetentim defluebat & minorabatur, donec die circumcisionis Jesu Christi tota resoluta est. Imago vero virginis, quia glacialis videbatur, ignibus adunatis per accolas resolvi procurabatur, sed ignis penitus nihil de ea resolvit. Hanc imaginem est videre in metis regionis dictae Viatka post Moskoviam & vocatur usque in hanc diem *aurea vetula*, alias *Złota baba*, quemadmodum in Sarmatiarum descriptione olim dixi.

(On Christmas day, icy hail descended and created the image of a virgin holding a crowned child in her arms. The image of the child diminished and got smaller, and on the day of Christ's circumcision it entirely dissolved. The image of the virgin, however, which seemed to be made of ice, remained untouched, although the local population sought to dissolve it by putting fire under it. One can see this image at the border of the region called Viatka, beyond Muscovy. Until this day, it is called *aurea vetula*, or *Złota baba*, as I already mentioned in the description of the Sarmatias.)[136]

The *aurea vetula* discloses Miechowita's preference for pagan symbols to mark border posts. It is Miechowita's unique poetic response to the traction of religious belonging and to the limits of cartographic gridding. Miechowita's *aurea vetula* is as much an idol to be worshiped as she is a territorial demarcation line staking out the borders of the world's northern realms under the auspices of a newly invented Slavic mythology.

The Golden Old Woman (also called "Golden Wife" in sixteenth-century English texts) had a surprisingly long-lasting cartographic career in the wake of the publication of Miechowita's treatise. Anton Wied depicted the *aurea vetula* on his 1542 map of Muscovy—used by Sebastian Münster for his 1544 edition of the *Cosmography*[137]—while both Sigismund von Herberstein and Richard Chancellor were among those who took up Miechowita's poetic invention in their prose cartography of Muscovy.[138] The travel narrative of Chancellor, included in Richard Hakluyt's *The Principal Navigations*,[139] offers an account of his voyage to Muscovy in which he writes about the "Golden Old Wife": "There is a certain part of Muscovy bordering upon the countries of the Tatars wherein those Muscovites that dwell are very great idolaters. They have one famous idol amongst them which they call the Golden Old Wife; and they have a custom that whensoever any plague or any calamity doth afflict the country, as hunger, war, or such like, then they go by to consult with their idol."[140] Miechowita's poetic invention fueled cartographic imagination: the image of the *aurea vetula* was activated throughout the sixteenth century, in both travel narratives and on maps.

In their design of New World maps, both Augustin Hirschvogel and Gerhard Mercator inserted the *Złota baba* to mark the world's northern realms close to the North Pole.[141] In 1551, Hirschvogel created a map of Muscovy as an illustration of Sigismund von Herberstein's travel account (fig. 13), *De rerum moscoviticarum* (1549), where the *Aurea anus* (*Slata Baba*) is located just beyond the "Montes dicti Cingulus terrae" (the Mountains called the girdle/zone of the earth). Here, the *Slata baba* reinforces those orographic border posts, which allegedly gird the end of the earth. In 1569, Mercator inserted the *Zolotaia baba* (in a transliteration of the Russian phonetics) on the first map of the North Pole. Stylized as a statue, the Golden Woman is here positioned at the very rim of the Arctic Circle, in close proximity to a mountain range symbolically enhancing the limits of the North.

In Abraham Ortelius's first atlas, *Theatrum Orbis Terrarum* of 1570,[142] the *Złota baba* features (as the map's cartouche explicitly states) on the regional map of Russia based on a wall map originally created by the English traveler and cartographer Anthony Jenkinson in 1562.[143] Here, the *aurea vetula* is seated[144] on a stone altar with a child in her arms. Several individuals surround her, kneeling, in a position of worship, while hunters killing stags spread animal furs as gifts in front of the *vetula*. A river separates the Golden Woman from a cartouche, which describes the *Złota baba* in the words of Herberstein's account of the "Golden Woman" in his *Rerum moscoviticarum*. Yet an additional, oracular, quality is bestowed upon the Golden Woman: "Zlata Baba, id est aurea vetula ab Obdorianis, & Iougorianis religiose colitur. Idolum hoc sacerdos consulit. quid ipsis faciendum, quove sit

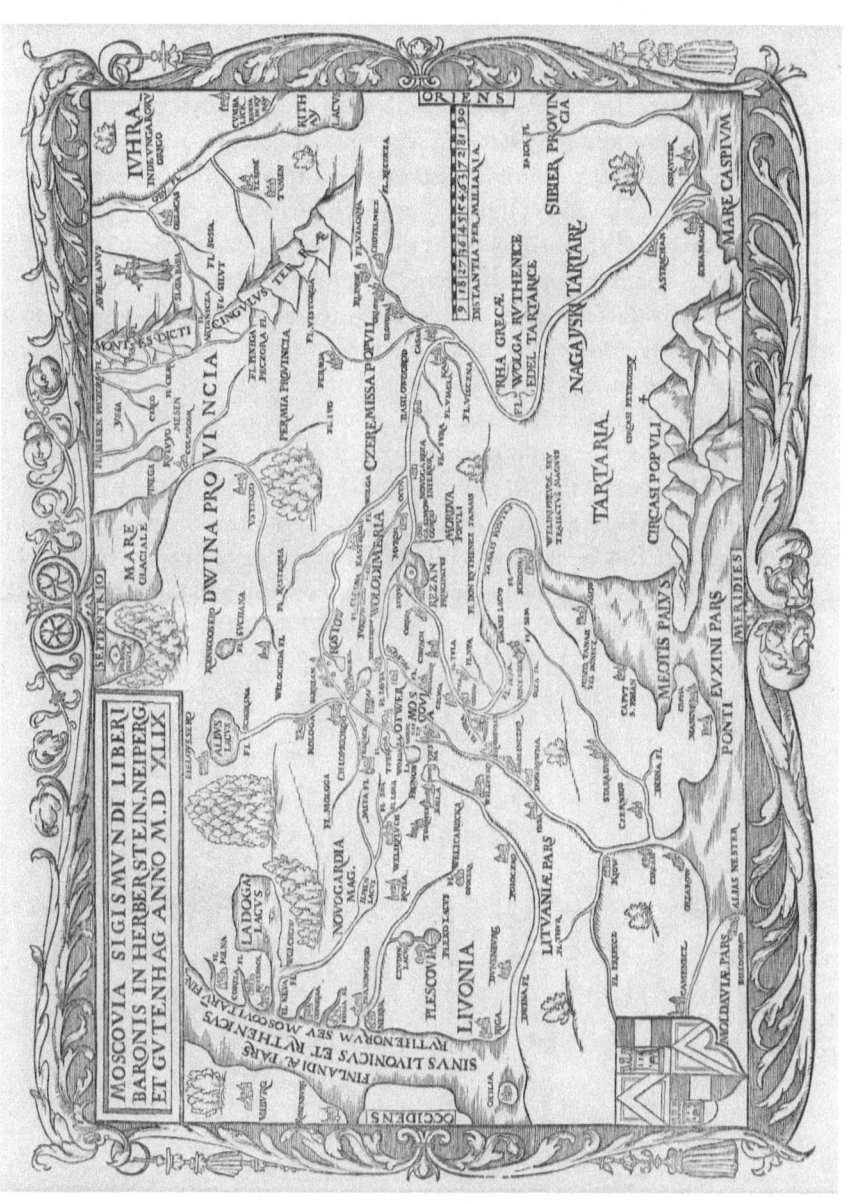

Figure 13. "Moscovia" by Augustin Hirschvogel, in Sigismund von Herberstein, *Rerum moscoviticarum commentarii* (Basel: Johannes Oporinus, 1551), n.p. [Typ. 565 51.450]. Photograph: Houghton Library, Harvard University.

migrandum, ipsumque (dictu mirum) certa consulentibus dat responsa, certique eventus consequuntur" (Zlata Baba, that is the Golden Old Woman, is religiously worshiped by the Obdorians and the Iougorians. The priest consults this idol about what to do and where to go, and—it is a wonder to relate—she gives clear answers to those consulting her, leading to certain outcomes).[145] Here, the *aurea vetula* embodies the power to apprehend and direct the itinerant and migratory movements of peoples. As a border post (or cartographic oracle), the Golden Woman performs that "constitutive relationship between exile and mapping" in which cartography emerges as what Karen E. Bishop has recently called "the muted history of exile, a history of lines drawn and redrawn, interpreted and reinterpreted."[146] While scholars have investigated the presence of the Golden Woman on Jenkinson's map as one of its "decorative elements,"[147] the *złota baba* performs, instead, a Janus-like function as a territorial boundary: she delineates a territory while also manipulating its limits. A new Alexander, she serves as a cartographic figure that structures the itineraries of peoples by restricting or impelling their movements. From her position at the suture line separating the known and unknown world, she performs—from the perspective of the northern territory as well as with the oracular eye of her inventor, Miechowita—the act of cartographic belonging as an artful and arbitrary process in the making of continents.

CHAPTER THREE

The *Alpha* and the *Alif*: Continental Ambivalence in Geoffroy Tory's *Champ fleury* (1529)

Blois, 1518. A newly printed catalog of the Royal Library lists a precious and richly illustrated manuscript of Ptolemy's *Geography* (here titled *Cosmographia*), presumably seized by the French king Charles VIII during his military campaign in Naples in 1495.[1] Prepared in 1490 for the Neapolitan aristocrat and bibliophile Andrea Matteo III Acquaviva and illustrated by illuminators in the orbit of Cristoforo Majorana, the manuscript is unique among fifteenth-century Latin manuscripts of the *Geography*. Unlike other contemporaneous editions of Ptolemy, it contains full-page illustrations of the three parts of the world known to the ancients. Alongside the traditional Ptolemaic maps (a world map and twenty-six regional maps), Europe, Asia, and Africa are here represented as allegorical female figures. For the first time in the history of cartographic editions, the continent is added as an entirely new geographic entity. Destabilizing the Ptolemaic binary between chorography and cosmography, the manuscript presents Europe as the mythical maiden Europa, daughter of the Phoenician king Agenor, abducted by Jupiter in the form of a white bull (fig. 14). Yet, unlike the other allegories of Asia and Africa, standing firmly on the territory they represent, Europa touches no foot on the ground she is made to stand for. Seated on the bull with her eyes cast back to the fertile Asian shore that extends into the horizon, Europa is led away by a figure reminiscent of Poseidon, across the sea and to a continent not only unmapped, but also unseen: Europe is entirely missing from the picture, while ample space is given to the Phoenician shore and the sea Europa traverses.[2]

Basel, 1561. The French cartographer, translator, and scholar of Semitic languages Guillaume Postel dedicates his *Cosmographiae Disciplinae Compendium* to the Holy Roman Emperor, Ferdinand I. In this heterodox geographic work, Postel advances the argument that the three parts of the

Figure 14. "Europa," in Ptolemy, *Cosmographia* (Naples, 1490), fols. 292v-293r [MS, Lat. 10764]. Photograph: Bibliothèque nationale de France.

oikoumene have wrongfully been named Asia, Africa, and Europe: Postel contends that, since all the peoples of the earth stem "ex unius Noachi domo" (from the house of one and the same Noah),[3] the "parts" (*partes*) of the *oikoumene* owe more to the migration and colonization of Noah's three sons—Shem, Cham, and Japheth—than to a fanciful Greek myth. Postel suggests that Asia should be called "Semia" and Africa "Chamesia." As for Europe, he writes, it is Noah's son "Japetus, whom the true Holy Scripture calls Japhet or Jephet . . . who should give—or give back—the name Japetia to our part of the world [that] is one with that part of Asia that extends from the Amanus mountain to the West."[4] Employing what he terms an "etymological method" (*ratio etymi*) as a central tool for his arguments, Postel promotes the Hebrew language as a powerful explanatory model for the understanding of the migratory movements that made up the origins of the continent of Europe. Anchoring his geographic arguments in the roots of words allows Postel to connect the Europeans, in particular the French, to what he considers to be the original biblical language and, moreover, to proclaim France's undisputed supremacy by linking his nation's beginnings to prediluvian times: "For it is from [the Hebrew word] Gal, which designates the flood waves, that the Gauls are named: they are the only ones who preserved for the entire world numerous miracles never obliterated from memory, which otherwise would belong to all the inhabitants of the world."[5] Postel thus admonishes, with a nod to the myth of Europa, that it is wrong to name the continent "from the encounter of a wicked scoundrel with a cow."[6]

From Ptolemy's 1490 manuscript to Postel's 1561 *Compendium*, the question of Europe's contours was at the center of humanistic debates, and its intimate relationship with Asia was increasingly broached using the allegorized body as a cartographic, linguistic, and graphic sign. Early modern editions of ancient geographies mobilized the image of the abducted maiden Europa gazing away from Phoenicia's shores[7] to imagine a corporeal Europe holding the continent as at once derivative of Asia, highly indebted to its regions, cultures, and languages, and being slowly disconnected from it, "'severed into' the world,"[8] as Tom Conley would remark in a different context. Postel's *Compendium* proves that this dual movement still pulled strongly in opposite directions in the mid-sixteenth century. While the etymological method of Postel's prose cartography entangles the two continents by revealing their common linguistic and geographic proximity, a small woodcut (fig. 15), which explicitly references Johannes Putsch's first allegorical female representation of Europe of 1537, instead emphasizes Europe's separation from Asia. A crude

Figure 15. "Iapetia" (Europa), in Guillaume Postel, *Cosmographiae Disciplinae Compendium* (Basel: Johannes Oporinus, 1561), 5 [*FC5 P8458 B561c]. Photograph: Houghton Library, Harvard University.

prototype for Sebastian Münster's later depiction of "Europa Virgo,"[9] Postel's woodcut imagines "Japetia" as an autonomous continent, detached from any Asian ties. The illustration pushes against or even undoes the narrative work of the *Compendium* (and vice versa), thus generating a semiotic tension between word and image that would stimulate the vigorous rise of cartographic language in the sixteenth century. In binding language to geography, the cartographic conceptualization of continents seen in such humanist works is crystallized as a semiotic ambivalence, as a liminal space carved out between the fantasy of a holistic world with an attendant antediluvian universal language and the division of the *oikoumene* into continents and vernaculars as new geographic and linguistic entities.

Geoffroy Tory, the writer, translator, editor, and the first printer to serve French royalty, offered what is perhaps the most complex and, at the same time, the least explored investigation of the nascent delineation of Europe. His foremost work, the *Champ fleury* (1529), at the center of this chapter, is a manual on the standardization of the French alphabet. It not only prescribes the exact measures of French letters, both printed and handwritten, but also imagines the alphabet as spatialized and even cartographically oriented body parts. Some scholarship has been devoted to the cartographic framework cleverly conceived for the design of the letters in the *Champ fleury*, as well as to Tory's obsessively reiterated drive to bind together word, body, and number. But the continental ambitions of the work have yet to receive attention. The work's lengthy title, *Champ fleury. Au quel est contenu Lart & Science de la deue & vraye Proportion des Lettres Attiques, quon dit autrement Lettres Antiques, & vulgairement Lettres Romaines proportionnees selon le Corps & Visage humain* (*Flowery Field. Wherein is Contained the Art & Science of the Proper & True Proportions of Attic Letters, Otherwise Called Antique Letters, and in Common Speech Roman Letters, Proportioned According to the Human Body and Face*), already invites the readers on a wondrous journey to the limits of Europe and disciplines alike, a journey in which the boundaries between lands, bodies, and the liberal arts are artfully merged. The *Champ fleury*, organized in three books, constitutes an astounding cartographic surface, a vibrant map in its own right upon which letters, as graphic, somatic, and numeric signs are deployed as a new cartographic language, in constant transformation and translation. In Tory's hand, the "flowery field" of the title becomes a platform for the generation of complex cartographic signs that I will call "cartographemes," Ján Pravda's term for the "most elementary and visually perceivable graphic formation[s] with a mainly graphic function."[10] In the *Champ fleury* and in Tory's other humanist projects, cartographemes operate toward a dual inquiry: into the nature of the smallest graphic units of a map and into the exploration of the creative power of these units as they align themselves to form the new entity of Europe.

In equating words displayed on a page with flowers arranged on a field—the "grans Champs Poetiques et Rhetoriques plains de belles, bonnes, & odoriferentes fleurs de parler" (vast fields of poesy and rhetoric, full of fair and wholesome & sweet-smelling flowers of speech)[11]—Tory plays with the popular etymological proximity between the Latin words *pagus* (field) and *pagina* (page).[12] Yet in its careful design and spatialization of the Latinate alphabet, the *Champ fleury*—the "flowery field" or "field in bloom"—accomplishes much more. Here, *gramma*, the Greek word for "letter," is transformed into *grammen*, the dividing line:

> Avant que ie commence a enseigner nostre p[re]miere lettre A, devoir estre faicte de le I, Ie veulx cy prier le bon estudient quil sache premierement que cest que le Point, que cest que la Ligne.... La Ligne est une longueur sans largeur, de la quelle les extremites sont deux points ... A————B. Aulus Gellius au XX. Chapitre de son premier livre, pareillement dit, Linea autem a nostris dicitur, quam ΓΡΑΜΜΗΝ, Graeci vocant.
>
> (As I am about to begin to teach how our first letter, A, is to be made from the I, I would beg the good student to learn, first of all, what the Point is; and what the Line is.... The line is length without breadth, of which the ends are two points ... A————B. Aulus Gellius, in chapter XX of his first book [*Attic Nights*], says to the like purpose: *Linea autem a nostris dicitur, quam Graeci* ΓΡΑΜΜΗΝ *vocant*.)[13]

Disassembled on the page into their constituent lines and circles, Tory's letters are suddenly translated into mathematical signs or human features, in accordance with Vitruvius's precepts of *De architectura* and with the science of melothesia that imagined the human body as a small-scale projection of planetary and zodiacal movements.[14] At the core of the *Champ fleury* are the very vibrancy and generative force of the smallest graphic, numeric, and biometrical digits, the basis, for Tory, of the cartographer's interdisciplinary task. By adopting a spatial lens, but equally sensitive to the articulations of typography, philology, and translation, the "prototypical cartographer"[15] Tory blends *grammen* and *gramma* into a powerful cartographic tool used to investigate Europe's nascent contours as an intersection of linguistic, graphic, somatic, and numeric lines.

HUMANISM AS EUROPEANIZATION

When Tory left his native Bourges to settle in Paris around 1507,[16] he would become part of a vibrant circle of French and international humanists, such as Guillaume Budé, Desiderius Erasmus, Jean Lemaire de Belges, Charles de Bovelles, and Jacques Lefèvre d'Etaples, who were working to radically transform the traditional scholastic division of the *trivium* and the *quadriv-*

ium and establish new disciplines, including geography and architecture. The humanists' research was driven by the growing impact of the *quadrivium*—in particular of mathematics—upon the *trivium*, the study of languages. Timothy Reiss has long identified a substantial shift, in the course of the sixteenth century, "from natural language to the 'measurable' language of mathematics."[17] Though humanists in France and beyond, as Reiss contends, "were increasingly distinguishing between language as a means of rational communication (grammar and rhetoric) and language as a means of discovery (logic), of 'productive knowledge'," they ultimately "turn[ed] to mathematics and a new idea of 'method' for actual validation" of their inquiries into the "relations of language and non-linguistic phenomena (*res*)."[18]

Anticipating later mathematical works by Gerolamo Cardano, Gottfried Leibniz, and Isaac Newton, these early humanists transformed the study of mathematics, both arithmetic and geometry, into "a central part of the humanist programme."[19] Yet what commanded their advocacy of mathematics, this chapter argues, was the growing impact that cartography exerted upon the entire liberal arts curriculum. The cartographic *table*, which began to figure regularly in geographic publications in the mid-fifteenth century following the resurfacing in Europe of Ptolemy's *Geography*, emerged as a crucial pivot that helped shape early modern thought. "Without the table or a flat surface which resembles it," Franco Farinelli forcefully argues, early modern "geometry, which for the Ancients was equivalent to civilization, would not exist."[20] Farinelli surmises that early modernity (and, more specifically, the discipline of cartography) was born when "a line in the abstract geographic grid became material, [and] the Earth was modelled in the form of its own drawing."[21] The map thus became a foundational support driving the production of new sign systems, from vernacular languages to calculus.

Encouraged by the French king François I's own interest in the Kabbalah,[22] humanists sought out new (hermetic) pathways to inquire into the nature of words through ancient or non-European languages such as Hebrew, Arabic, and Chaldean as well as alternative sign systems such as the riddle, rebus, and the Egyptian hieroglyph. Tory's fascination with rebuses and hieroglyphs alike recurs throughout the *Champ fleury* as an "obsession for the hidden structures of the alphabet."[23] In contrast to J. L. Austin's formula "how to do things with words,"[24] hieroglyphs create words with things—as the original meaning of the Latin word "rebus" (with things), in which words are represented by a combination of pictures and letters, illustrates. Marsilio Ficino, one of the first translators of the *Hieroglyphica* by Horapollo, aptly described the Egyptians' use of hieroglyphs as a pragmatic linguistic system: "The Egyptian priests, when they wished to signify divine things, did not use letters, but whole figures of plants, trees, and animals; for God doubtless has a knowledge of things which is not complex discursive thought about its subject, but is, as it were, the simple and steadfast form of it."²⁵ Just like Ficino, Tory believed that he had found in Horapollo²⁶ a possible key to the hidden meaning of words.²⁷

The humanists' hermetic pursuits were coterminous with a renewed interest in the spatialization of words. What propelled their quest for a redefinition of disciplines was their profound belief in the unlimited transformative power of what Ricardo Padrón has recently termed the "spacious word."²⁸ Through their spatial lens, canonical ancient Greek and Roman authors such as Plato, Aristotle, Horace, Tacitus, and Ovid suddenly became cartographic luminaries and the so-called hermetic writers from Antiquity and early modernity—Pythagoras, Ramon Llull, Ficino, Pico della Mirandola, Nicolaus Cusanus, and Johann Reuchlin—were all read spatially, as harbingers of new modes of conceiving and delineating the world. To redefine the spatiality of word and world alike, French humanists pored over texts as diverse as Cicero's *Republic* and the *Corpus Hermeticum*, Plato's *Timaeus* and Francesco Colonna's *Hypnerotomachia Polyphili*, Aristotle's *Poetics* and Ficino's translation of Horapollo. When Tory translated pseudo-Lucian's *Tabula Cebetis* into French and prepared an edition of the *Antiquitates* by pseudo-Berosus, he did so in the firm belief that he was providing access to foundational texts that were no less significant in reshaping spatial thinking than his later editions of Enea Silvio Piccolomini's *Cosmographia* and Leon Battista Alberti's *De re aedificatoria* (1512). It is not a coincidence, then, that at the same time Tory's Spanish counterpart, Elio Antonio de Nebrija, most famous for having authored the first grammar of the Castilian language whose preface makes explicit the link between language and empire, also edited the spurious *Antiquitates*, which emphasize the role of Egypt as the source of civilization disclosing, as Anthony Grafton shows, "the connections between the biblical Orient, Troy, and the states of fifteenth-century Europe."²⁹

A profound spatial concern can be seen to motivate the humanists' study of the power of the word. Be it Pythagoras's inquiry into the letter Y and the perfect number ten, Johann Reuchlin's treatise on the wondrous workings of words in *De mirifico verbo*, or the revelation of genealogic and linguistic origins embedded in the *Corpus Hermeticum*—all texts explicitly referenced in the *Champ fleury*—investigations like these were driven by an unprecedented spatial approach to the *trivium* and *quadrivium* as world-structuring tools. Conversely, humanists in early modern Europe unearthed cartographic potential in late ancient and medieval texts on the seven liberal arts, such as Isidore of Seville's *Etymologies*, Martianus Capella's *De nuptiis Philologiae et Mercurii*, or Boethius's *Consolatio Philosophiae*. The publication of the first printed world map—a schematic T-O map included in the first printed edition of Isidore's *Etymologies*—is a powerful case in point: published in 1472 in Augsburg by Günther Zainer, the map featured Noah's three sons as the legitimate heirs to the three parts of the world. With the rise of cartography, the encyclopedist Isidore was now a "cosmographer,"[30] as he is described by the French cardinal Pierre D'Ailly,[31] an early reader of Ptolemy's *Geography* and the author of the geographic work *Ymago mundi* (circa 1410).

As a prolific editor of ancient and modern geographic work alike, ranging from Pomponius Mela to Piccolomini, Tory was actively propelling the process of continental creation. What at first glance resembles dry editorial philology is revealed upon closer examination as a restructuring of a carefully chosen cartographic corpus. Tory's spatial lens produces coherence in an otherwise eclectic body of works that his careful editorial hand relates to the delineation of a continent. While his earliest work, a 1507 edition of Mela's *De totius orbis descriptione*,[32] privileges his preference for geography as an epistemic tool for comprehending the world, Tory's subsequent project pushes his cartographic intervention even further toward continental delineation. His edition of Piccolomini's two geographic treatises, *De Asia* and *De Europa*, is the first to bring the two together in a single volume. By joining them into one work, *Cosmographia in Asiae & Europae eleganti descriptione* (1509),[33] Tory forces the reader to halt before the toponyms displayed in the title and therefore contemplate the fine continental line emerging within the Eurasian territory.

Tory designed editions of eclectic collections under the unifying topic of spatiality. His 1510 edition of *De interpretandis Romanorum literis opusculum*,[34] by the Roman grammarian Valerius Probus, is a collection of treatises that appears, at first glance, highly divergent. Ranging from Roman antiquity to sixteenth-century France, it includes texts by authors "very worth knowing" (*scitu dignissimi*)[35] addressing topics as diverse as weights and measurements, descriptions of cities and monuments, and the art of agriculture.[36] Yet subtly binding the volume together is Tory's geographic vision that correlates spatial and linguistic measurement. In the anthology's perspective, Valerius Probus's treatise on letters can be read productively as an inquiry into spatial measures. Conversely, Columella's book on agriculture—which had heretofore been edited alongside thematically comparable Roman treatises, typically under the generic title of *Scriptores rei rusticae* (*Writers on Agriculture*)—is here dislodged and placed under the aegis of grammar and philology. By choosing to unite in one volume Probus (an author originally from the eastern Mediterranean colony of Berytus, modern Beirut), the Roman Sergius from Pula in modern Croatia, and Columella (a writer from the Iberian city of Cádiz), Tory not merely reproduces a mental map of the Mediterranean at the time of the greatest extent of the Roman Empire. Rather, he actively draws the contours of the newly emerging continent of Europe—from its borders with Asia to its boundaries in the west.

Renaissance humanism is commonly tied to the new discipline and method of philology by which early modern scholars, as Denis Robichaud recalls, "spent countless hours comparing Greek manuscripts, considering scholia and textual variants"[37] to prepare critical editions and more reliable translations. "Humanists," Jill Kraye contends, "studied philosophical works in the same manner that they dealt with literary or historical texts, that is, as philologists," as "devotees (philoi) of the study of words (logoi)," who drew, in particular, on "expert knowledge of the language, culture and history of Greece and Rome to determine the precise meaning of an ancient author's words in a specific context."[38] Within the traditional framework of Renaissance studies, Greco-Roman philology came to signify the epitome of European humanism. Yet the ancient Greeks' relationship with Europe was far more convoluted and ambiguous than it might appear at first glance. Aristotle shows in the seventh book of the *Politics* that Greece itself was not exclusively situated in Europe but, rather, occupied a privileged liminal position in the continental interstices between Europe and Asia:[39]

> The nations inhabiting the cold places and those of Europe are full of spirit but somewhat deficient in intelligence and skill, so that they continue comparatively free, but lacking in political organization and capacity to rule their neighbours. The peoples of Asia on the other hand are intelligent and skillful in temperament, but lack spirit, so that they are in continuous subjection and slavery. But the Greek race [Ἑλλήνων γένος] participates in both characters, just as it occupies the middle position geographically, for it is both spirited and intelligent; hence it continues to be free and to have very good political institutions, and to be capable of ruling all mankind if it attains constitutional unity.[40]

Aristotle's Greece features the best qualities of each continent precisely because it draws from both. As the inhabitants of an in-between space, the Greeks experience no disadvantage but are instead geographic beneficiaries of a truly intercontinental dynamics that allows for spirit, intelligence, and skill to manifest in a most advantageous manner. For Aristotle—unlike his commentators and interpreters from the Renaissance onward—qualities such as "intelligence" (*ingenium*) and "skill" (*artificium*) are attributes characteristic of Asians, not Europeans.[41] When Edmund Husserl, in his Vienna lecture entitled "Philosophy and the Crisis of European Humanity," conjured up Europe's "Greek heritage in the idea of a universal rational science . . . intimately tied to the very idea and promise of reason and rationality,"[42] his conception of Europe could not stand in starker contrast to the sentiment Aristotle expressed in his *Politics* more than two millennia prior.[43]

With the rise of cartography and philology, Renaissance humanists set out to strategically dispel continental ambivalence contained in ancient and medieval authors and render them unambiguously European. The object of their philological study (the documents imported, translated, adapted, rewritten, and transcoded), all too often branded and promoted as the hallmark of European Renaissance humanism, is not so much an intrinsically European phenomenon, as a powerful process of Europeanization, by which authors of Greek, Hebrew, and Arabic texts were spatially disambiguated and incorporated into the bounds of a newly emerging Europe (or excluded from it).

Tory's *Champ fleury* speaks to this process of Europeanization. It is unique—and perhaps hyperphilological—in that his study of the French alphabet is driven by a profound understanding of the manifold complex processes of Europeanization at the root of an emerging French humanism. A lengthy catalog of European, non-European, and imaginary alphabets appended to the three books of the *Champ fleury* testifies to Tory's ambition to investigate French letters within a comparative framework, highly sensitive to continental ambivalence. His sustained focus on the complex geographic provenance underpinning each alphabet reveals an ambition to envision alphabetic tables as a function of cartographic *tabulae*. For instance, the uneven numbers of letters of each tabulated alphabet do not determine the layout of the page, but, instead, Tory's strategic manipulation of the alphabets, through the addition and subtraction of single letters, disciplines the letters to conform to the table. Suddenly, the letters morph into "equipollent" cartographemes,[44] deployed on the table like carefully planted flowers on a field in realization of Tory's own metaphor. With Ptolemy's equipollent map in mind, Tory insists that alphabets are not hierarchically structured, but unfold as horizontal cartographic networks which can only be contemplated in relation to one another. By adding "les Lettres Phantastiques, Les Utopiques, quon peut dire Voluntaires" ("the Fantastic Letters, the Utopian, or, as one may say, Voluntary")[45] to his tables of European and non-European alphabets, moreover, Tory unabashedly points to the poetic underpinnings that drive each linguistic and cartographic endeavor, especially crucial in the context of the construction of Europe's contours.

HERCULES: THE SURVEYOR OF EUROPE'S CONTOURS

Tory's approach to texts is that of a cartographic philologist, sensitive to a complex cartography of thought and to questions of liminal spaces. His philological method is seen as early as the work that would launch his career as an editor, his 1507 edition of Mela's *De totius orbis descriptione*.[46] "Josephus, Pomponius Mela, & the poet-historian Lucan," Tory contends, "are of the opinion that the Phoenicians, who are in Syria, invented the said [French] letters" ("Iosephus, Pomponius Mela & le Poete historien Lucain sont dopinion que les Pheniciens qui sont en Syrie, ont invente les dictes lettres [Françoises]").[47] Tory turns Mela, strategically inserted between the historian Josephus and the poet-historian Lucan, into a cartographic passe-partout, highlighting the potent mediation geography exerts among disciplines of the liberal arts. Tory captures Mela not only as an authority on the origins of the Roman alphabet but on the very problem of continental division:

> Dicte lettre Attique a este inventee en ung pais de Grece nomme Ionie, qui est comme dict Pompone Mela en lextremite Dasie la mineur, entre Carie, & Eolie. Ioniens lont premierement inventee, figuree, & proportionnnee. Mais les Atheniens qui ont este seigneurs & dominateurs de toute Grece, lont mise en usage & honneur, si bien quelle en a & retient encores le nom.
>
> (The Attic letter was invented in a country of Greece called Ionia, which is, as Pomponius Mela says, at the edge of Asia Minor, between Caria and Aeolia. Ionians first invented, drew, & proportioned it. But the Athenians, who were the sovereign lords & masters of all Greece, brought it into use and credit, so that it has and still retains their name.)[48]

Here, Tory frames the invention of the letters as a problem of continental in-betweenness: the origin of the "Attic letter" is found at the edges of Asia Minor, a liminal space between Europe and Asia that emerges as the very site where *gramma* and *grammen* collapse.

For his illustration of continental ambivalence through a blend of cartography and philology, Tory mobilizes, in a euhemeristic gesture, the mythological figure of Hercules, in particular of the "Gallic Hercules," to explain how on one of his voyages from Asia Minor, Hercules imports the alphabet to France, more specifically, to Paris:

> Avant que . . . Cesar vint icy, & y traynast sa dicte langue Latine, les lettres Grecques y pouvoient estre, & de faict elles y estoient en cours, considere que long espace de temps, & grant nombre dans par avant, . . . quant Hercules alla oultre Espaigne aux iardins des Hesperides, passa par ceste contree, & quant il fut en lisle de ceste cite de Paris, il print si grant plaisir a veoir le pais & la riviere de Seyne, quil y commenca a edifier, puis sen volant aller oultre a ses entreprinses, y laissa une bande & compaignie de ses gens darmes qui estoient appellez Parrhasians selon le nom de leur pays en Grece du coste Dasie, qui est nommee Parrhasia. Iceulx Parrhasians laisserent leur nom icy et en mutation de A en I les habitans de ceste dicte Cite ont este & sont encores dicts & appellez Parrhisiens.

> (Before . . . Caesar came hither & brought in his train his Latin language, the Greek letters may have been, and in fact were, current here; inasmuch as, a long period of time and a great number of years earlier, . . . when Hercules journeyed beyond Spain, to the isles of the Hesperides, he passed through this country; and when he was on the island of this city of Paris, he took so great delight in viewing the country & the river Seine, that he began to build there; then, desiring to proceed with his enterprises, he left a company of his men-at-arms, who were called *Parrhasians* from the name of their province of Greece, on the coast of Asia, the name of which is Parrhasia. These Parrhasians left their names here, and by the change of A to I, the inhabitants of this city, were, and are to this day, called Parrhisiens.)[49]

The *Champ fleury* casts continental ambivalence as a truly Herculean deed: the exchange of one letter for another achieves no less than the substitution of a continent. The transformation of the Parrhasians—the inhabitants of a liminal space on the coast of Asia—into Parrh*i*sians hinges upon the change of a single letter, from an A to an I. Not merely the founder of a new city, Hercules is also the bearer of a jointly philological and cartographic knowledge that binds the *grammata*, the letters I and A, to the *grammen*, the continental divide.

The "Attic letter" "I" here takes the shape of a signifying frontier, an originary cartographeme that precedes the differentiation between *gramma* and *grammen*. Conversely, the letter A is an expanded, multiplied I. No longer a frontier-letter, the letter A is imagined in the *Champ fleury* in the shape of a compass,[50] a spatialized surface underlying the principles of triangulation: "Iay dict que la lettre A qui est la premiere de Lalphabet, quon dit aultrement Le A.B.C. est faicte de la lettre I & est chose vraye, en le figurant en triangle, qui est nombre imper. Les deux pattes dudit A & la teste sont le dict triangle, mais ce triangle vault estre assis en ung quarre" (I have said that the letter A, which is the first letter of the alphabet—otherwise called the A B C,—is made from the letter I; and this is true, representing it in a triangle, which is an odd number. The two feet of the A and the head make the said triangle; but it must be placed within a square).[51] At the same time, the letter A, "made from the letter 'I'," incessantly recalls the permanence of the boundary and spatial ambivalence within each delimited surface.

Hercules has long been associated with cartography. He is the earth's "pacifier and surveyor," as Gilles Deleuze notes: if Dionysus belongs to the underworld and Apollo to the skies, Hercules is "of the surface, in his dual battle against both depth and height" and, consequently, effects a "reorientation of the entire thought and a new geography."[52] But he is also a surveyor of continental contours. By journeying beyond the Garden of the Hesperides, situated according to Hesiod "beyond glorious Ocean," Hercules stakes out the western boundaries of the European continent, described in the *Theogony* as "the limits of the earth."[53] Mela's *Description of the World*, one of Tory's foremost geographic sources, credits Hercules with the separation of the two gigantic rocks Abyla (Jabal Musa) and Calpe (Gibraltar) at the mouth of the Mediterranean and reveals that what are commonly called the Pillars of Hercules is in fact the demigod's creation of a continental divide:

> Deinde est mons praealtus, ei quem ex adverso hispania attollit obiectus: hunc abilam, illum calpen vocant. columnas Herculis utrumque. addit fama nominis fabulam herculem ipsum iunctos olim perpetuo iugo diremisse colles, atque ita exclusum mole montium oceanum, ad quem nunc inundat admissum. Hinc iam mare latius funditur, summotasque vastius terras magno impetu inflectit.

> Next comes a very high mountain, facing the one that Spain raises up on the opposite shore. The one on this side they call Abila [Jabal Musa], the one on the far side Calpe [Gibraltar]; they call them together the Pillars of Hercules. (Oral tradition goes on to give the story of the name: Hercules himself separated the mountains, which had once been joined in a continuous ridge, and Ocean, previously shut out by the mole of the mountains, was let into those places that it now inundates [the Medi-

terranean Sea]. On this side of the Strait, the sea already pours in over a rather broad area, and with its great rush it bends back rather far the lands it has cleared from its path.)[54]

In the guise of the Roman Hercules Gaditanus, intimately tied to the western Spanish city of Cádiz (Gades, in Latin), Hercules became synonymous with boundary making: the "pioneer of world dominion" was so closely associated with expanding territories and boundaries that the Roman emperor Hadrian minted coins showing Hercules together with the earth-girding Ocean.[55] In Tacitus's *Germania*, Hercules is credited for having marked Germany's northern boundaries,[56] while several ancient traditions associate him with the fragile border separating Europe and Asia. For many Romans, as Fabio Barry recalls, Hercules Gaditanus "was none other than a reincarnation of the Phoenician Melqart,"[57] the tutelary god of the city of Tyre, where, following ancient myths, the Phoenician alphabet was invented and where Europa, king Agenor's daughter, was abducted.

In sixteenth-century France, the popular figure of the Gallic Hercules came to convey the dual idea of rhetorical astuteness and spatial knowledge. Commonly associated with the rise of the French language that would be officially sanctioned as a national language, for the purpose of legal documents, by the 1539 Treaty of Villier-Cotterêt, the Gallic Hercules is used in the *Champ fleury* in a much more sophisticated context—namely, as a tool to contemplate the productivity of linguistic and spatial systems. When Tory inserts a woodcut of the Gallic Hercules as an illustration of his own French translation of Lucian's myth of Hercules Ogmios, he intimately binds rhetoric to geography through the overarching theme of measurement. In this woodcut, Hercules leads a crowd whom he pulls with the help of a long chain stretching from his tongue to his followers' ears. While Tory contends that the pierced tongue represents Hercules's powerful rhetoric, it equally stands for the human's most mobile body part. "There is no other part of the body," Erasmus contended in *De recta Latini Graecique sermonis pronuntiatione* (*The Right Way of Speaking Latin and Greek*), first published in 1528, when Tory was putting the final touches to the *Champ fleury*, "so quick and so pliable and so ready to take up different shapes, nor any other on which a man's acceptability and success so much depends. In short it is the tongue which distinguishes human from animal."[58] An attempt to discipline what is otherwise subject to continuous uncontrollable movements, the chain straightens Hercules's tongue, forcing it to follow the correct usage of language. Katie Chenoweth has pointed to the paradox of the tongue as a metonymy of the human being, oscillating between hypermobility and immobility, that in its attempt to perform the superhuman task of correct speech runs the risk of becoming inhuman:

> Humanist education for Erasmus would be above all a *techne* of this overly susceptible organ: a fashioning or crafting—above all, a straightening—of the (maternal, vernacular) animal out of the tongue. The Latin title of the treatise gives us a more palpable sense of this straightening: correct pronunciation is *recta*, right, ruled, straight. It is up to the tongue to draw a straight and hard line between human and animal. . . . In this sense, the tongue becomes the ultimate transhuman instrument: that body part capable of overcoming embodiment, as long it remains *recta*.[59]

The correct usage of language subjects the human to a straightening of the tongue—or of the entire body, as Tory's other images of bodies tortured as they are stretched into letter shapes painfully illustrate.

Tory reiterates on various occasions that only in its straightened state, thus as a dehumanized and lifeless instrument, can the human body be employed as a *techne*, a tool of mensuration. Unsurprisingly, Tory's image of the straightened body takes the shape of the straight letter I: "De le I, toutes les autres lettres, comme iay dit, prennent & ont comancement a estre faictes & escriptes. Cest ascavoir, ou en estant garde en sa droitte ligne, ou en estant reflecte & courbe, ou en estant brise. Et luy seul entre toutes les lettres garde sa droicte ligne perpendiculaire, a limitation du corps humain" (As I have said, 'I' is the starting point for making all the other letters: whether it is kept as a straight line, or is bent and curved, or broken. And it alone of all the letters keeps its straight perpendicular line, in imitation of the human body).[60] Letter and body discipline one another in a dyadic movement of straightening and conforming to the scope to better approximate themselves to the *grammen*, the straight line of cartographic demarcation.

The lettered and embodied "I," the minimal cartographic signifying unit, acquires in the *Champ fleury* the value of a cartographeme, as further underscored by Tory's source for his Hercules vignette: the frontispiece of Mela's 1522 edition of *De orbis situ*, which prominently features a woodcut of "Hercules Gallicus" (fig. 16), who, as in Tory's adapted illustration, leads a crowd to which he is attached with a chain.[61] Yet unlike with the *Champ fleury*, Mela's Hercules is framed by a landscape and cast in an angled posture: his upper body is straightened—in consonance with the domineering, I-shaped columns beside him, perhaps an allusion to the eponymous Pillars, and an unambiguous call for straight, perpendicular alignment—while his legs are slightly bent into an undulation that resists the ultimate push

Figure 16. Pomponius Mela, *De orbis situ libri tres* (Basel: Andreas Cratander, 1522), frontispiece [UBH CF III 12:1]. Photograph: Universitätsbibliothek Basel, Switzerland.

toward rectification. Oscillating between regulator and regulated, the figure of Hercules is here caught in the very process of taking shape, in the very moment of turning into cartographer and cartographeme alike.

IO: THE CONTINENTAL IMPRINT

With the unfolding of the *Champ fleury*, Tory rehearses the continental and linguistic tension (already performed through the myth of Hercules with a focus on the frontier-letter I) from a still more theoretical angle. With an eye to cartography and philology alike, Tory mobilizes the Ovidian myth of Io, whose very name is composed of the frontier-letter I and the globe-shaped letter O, which, Tory explains, is identical to the I, only bent into a circle: "I & O sont les deux lettres, desquelles toutes les aultres Attiques sont faictes & formees. . . . On peult aussi dire que le O est faict de le I mais nous pouuons estimer que le O est modele pour les panses & arrondissemens de aulcunes aultres lettres que de luy" (These two letters, I and O, are those from which all the other Attic letters are made and fashioned. . . . We may say also that the O is made from the I; but we may well consider that the O is a model for the handles and curves of other letters than itself).[62]

To illustrate the productive and transformational power of the I and the O and their capacity to illustrate the complex problem of cartographic measuring, Tory strategically manipulates Ovid's myth by setting the story of Io not in the central Greek province of Thessaly,[63] but in Ionia, at the edge of Asia Minor. It is worth recounting the myth through Tory's lens: Jupiter, enamored of King Inachus's daughter Io, invented a subterfuge to seduce her and escape Juno's ire by changing Io into a cow. Juno, noticing Jupiter's ruse, asked her husband to give her Io as a gift. Once her demand was granted, Juno exposed Io to continuous mistreatments by her shepherd Argus, "qui auoit au visage, & par toute la teste cent yeulx qui ne dormoient iamais tous ensemble, mais deux a deux tandisque les aultres veilloient" (who had in his face and all over his head a hundred eyes, which never slept all at the same time, but two by two, while the others kept watch). Unable to speak, Io suffered in silence until Jupiter finally took pity on her and sent his messenger Mercury, in the guise of a shepherd, to Argus with the goal to make him asleep through his sweet melodies and decapitate him. Moved by Mercury's music, Argus's "yeulx, qui estoient, comme est ia dict, cent en nombre, se vont tous endormir tres parfondement, & tantost Mercure prent son Bracquemard et luy trenche la teste" (eyes, which were, as has been said, a hundred in number, fell asleep at once; whereupon Mercury took his shepherd's knife and cut off his head).⁶⁴ Io, finally liberated, set out on her journey home,

> tant quelle vient en ung endroit ou son pere Inachus estoit mue en Dieu de fleuve, quon dict aultrement, en Dieu Marin. Iceluy Inachus ne cognoissant linfortune cas de sa fille, mais pensant que ce fuste un [sic] vraye vache, luy tendoit plaines mains de doulces & odoriferentes herbcs, & la sadeyoit amyablement en luy touchant & la pariant de ses divines mains par le front, par le dos, et par les costez, iusques a ce que en allant & venant entour elle, il veit le nom de sa fille escript au pas & en la place ou auoit presse le pied de celle belle vache qui est de deux lettres seullement I. & Ω. au nom de la quelle le pais a este nomme Ionia, & les habitans Ioniens.

> (until she came to a place where her father, Inachus, had been changed into a river-god, otherwise called a sea-god. This Inachus, knowing naught of his daughter's unfortunate plight, and thinking that she was in truth a cow, offered her handfuls of tender and sweet-smelling grasses, & patted her affectionately, touching her with his divine hands on brow and back and flanks, until, as he went and came about her, he saw his daughter's name written in the place where the beautiful cow's foot had pressed the earth: a name of two letters only—I and Ω, from which name the country was called Ionia, and its people Ionians.)⁶⁵

Tory's story of Io is much more than a narrative in which the emergence of a new alphabet is coterminous with the survey and naming of a new (heretofore unbounded and unnamed) land, Ionia. What Tory exposes here is the careful elaboration of a new sign system in which linguistic, somatic, graphic, and numeric signs are formed into interdisciplinary cartographemes, used as an epistemic tool to formulate the slowly emerging method of triangulation, crucial for the development of linguistic and cartographic models alike. Here, Io's act of etching the two letters IO into the ground and the first naming and localization of Ionia is the result of the confluence of two different trajectories: Io's own strenuous physical journey across the countryside and the messenger Mercury's swift descent to earth.

The two itineraries that ultimately give rise to the alphabet as an imprint of a toponym upon a surface belong to two different sign systems and stand for two contrasting cartographic models. Io's itinerary focuses on the physical travel in which the foot (or hoof) is turned into a tool of linguistic and cartographic measurement. In this sense, it is bio-metrical: the imprint of her step is essential for the naming and localization of Ionia. Mercury's descent on earth is equally crucial for the process of naming in that the emergence of the toponym is propelled by his act of severing Argus's globe-shaped head. Yet in contrast to Io, Mercury's contribution to naming and locating is dissociated from the body in the moment he violently eliminates the head of Argus, nicknamed the Panoptes, "all eyes."[66] While merged into one story, the step and the eye become metonyms standing for two different models of measuring space: one which relies on the physical presence of the measurer, and another, no longer grounded in the body, in which the step is substituted by the volumetric eye and abstract metrics. For Farinelli, the former model is characteristic of the medieval world and exemplified, for instance, by the travels of Marco Polo, for whom "neither space nor time exists," but "things endure" revealing to him "their proper duration . . . , and at the same time measure the duration of his life."[67] Conversely, Mercury's itinerary is akin to what Farinelli considers to be "the first of the modern travelers":[68] Columbus, who traveled with a map in mind as he transformed the earth "into a gigantic table"[69] of uniform and continuous space. In the *Champ fleury*, this new type of cartographic space, with the letter I standing for uniformity and the O for continuity, is indebted to Mercury, who by severing Argus's head symbolically emphasizes the increasing rise of the map at the expense of the physical body.

For Farinelli, the new method of measuring space, which dispenses of the body for the benefit of abstract calculation, announces the birth of triangulation as a new epistemic tool which "functions through the substitution of step by sight."[70] With triangulation, the act of measuring shifts from symbol to sign:

> For the ancient Greeks, the symbol was an object split in two halves and containing two different people who, meeting each other and returning to match the separate pieces, guaranteed, reciprocally, the identity of the other. It was, then, a system of recognition. . . . The sight of the symbol was thus something which stood for the total of steps effectively made from the beginning of the affair, and only because it represented them (that is, it returned to make them present) was it capable of substituting them.[71]

The sign, on the other hand, is quite different:

> The connection between expression and meaning is wholly arbitrary in the sign, in the sense that it is external and formal. That is to say, re-

turning to our example, that between the sight of the sign and the steps, there is no relation whatsoever, and the sign substitutes steps without any longer representing them. The sign is no longer the product of the voyage; on the contrary, it dispenses with the journey, it renders it superfluous.[72]

When Tory describes Io's imprint as the result of two confluent trajectories, he has in mind, alongside a geometric triangle at the basis of cartographic triangulation, also a semiotic triangle *avant la lettre*, formulated centuries later by Gottlob Frege. Modeled on the image of a geometric triangle, Frege's illustration of the semiotic triangle's distinction between reference (Sinn) and sense (Bedeutung) uncannily dovetails with Tory's much earlier description of the origin of a toponym through the confluence of two separate trajectories. In a triangle, Frege contends, a point (A) has one reference (Sinn), but two senses (Bedeutungen) in that it is created through the intersection of two converging lines: A-B and A-C. The sense is the dual route to the reference.[73] If Tory's Ionia has one reference, it is created and pointed at ("bedeuten"—the German verb "to mean"—is a cognate of "deuten," "to point at") at once by Io and Mercury in their respective itineraries. Tory's description of the dual journey and the confluence of two movements in the site in which Io imprints her hoof makes the emergence of language and place coterminous and unearths the inextricable tie that binds language and geographic space to one another.[74] By bringing Io's and Mercury's trajectories together Tory dramatizes the juxtaposition of foot and eye as metonymic placeholders standing in for two distinct, and yet overlapping, sign systems.

The *Champ fleury* forcefully captures the tension between symbol and sign in a lengthy elaboration on the substitution of step by sight as it vigorously illustrates the rise of triangulation not only as a new cartographic system, but also as a new semiotic system whose tentative first beginnings it formulates. Not only Io's myth, but also Tory's address to his readers, "all true and devoted Lovers of well-formed Letters," with which the manual opens, is a miniature exposition of sign systems which uncannily dovetails with Farinelli's description of the ancients' use of the symbol:

> Linvension descripre en Raouleaux . . . est venue des Anciens Lacedemoniens qui en temps de Guerre avoient deux bastons faictx p[re]cisement dune mesme longueur & grosseur, & en bailloient lun au Prince qui alloit en Guerre puis gardoient lautre iusques a ce quilz luy vouloient mander quelque segret. Et quant ilz luy en mandoient ilz prenoient ung Parchemain ou Cuyr, ou autre chose semblable, Long et estroit comme

une sainctures & lenvyronnoient bourt a bourt au tour & le Long de leur baston quilz avoient retenu, puis escripvoient sus leur Parchemain le long & tout au tour de leur dict baston en sorte que la plusgrande partie des L[ett]res se trouvoit ou a demy, ou a tiers, ou a bien peu sus les bours & assembleurs de leur dict Parchemain puis le desployent & lenvoyent tout desploye a deur dict Prince qui incontinent qui lavoit receu le mettoit au tour de son baston, & tantost pour la grace de la mesme Mesure des deux bastons semblables, toutes les Lettres se rencontroient iustement en leur entier, comme quant on les escripvoit.

(The device of writing in scrolls . . . came from the ancient Lacedemonians, who in time of war had two truncheons made, of exactly the same length & thickness, & gave one to the Prince who was going to war, and kept the other until they should have occasion to write to him in secret. And when they wrote to him they took a strip of parchment, or leather, or some similar thing, long and narrow like a girdle, & wrapped it edge to edge around the length of the truncheon which they had kept; then wrote upon their parchment along and around their said truncheon in such wise that the greater part of the letters were a half or a third or very little way over either edge and place of jointure of their said parchment; then they unrolled it & sent it all unrolled to their said Prince, who, as soon as he received it, placed it around his truncheon, and thereupon, because the two truncheons were of the same size, all the letters fitted together exactly as they were written.)[75]

Tory exhibits an acute awareness of the deficiencies which the cumbersome system of coding and decoding advanced by the symbol entails. He acknowledges that if one should "desire to write but three or four verses lengthwise, the scrolls must needs be longer than from here to the Isles of Molucca, & especially if he would write in large letters. The fashion of writing in scrolls is a great abuse in many ways."[76] Unfolding and displaying a lengthy scroll, Tory argues, is as inconvenient or impossible as attempting to physically track the distance (from Paris) to the Moluccas. With the new tools of cartography in mind, then, Tory proposes the substitution of the scroll by the *tabula*: "Laissez doncques la ces Raouleaulx, & escrivez en belles & patentes Tavletes & autres choses semblables, afin que vostre lettre soit veue toute dung front Et Notez que Lespace dentre les Lignes vault tousiours estre aussi large que la Lettre I, est haulte" (Put aside these scrolls and write on good open tablets and other like things, to the end that your letter may be seen face to face. And note that the space between the lines should be always as wide as the letter I is high).⁷⁷ This transition from scroll to table succinctly summarizes what Farinelli had identified as the dramatic retreat of the human body for the purpose of spatial measurement: "In the five centuries elapsing between Alberti's first triangulation and the first semiotic triangle, the visual relationship, that is, the cartographic gaze, becomes the prototype for direct relation, to the detriment of the relation which involves the entire body."⁷⁸ With cartographic and linguistic triangulation on the rise, the human body is slowly removed from the picture.

Tory's blend of linguistic and geographic thought, enhanced by the explicit reference to the Moluccas, discloses his keen interest in contemporary cartographic developments. In 1529, the year the *Champ fleury* was published, the question of the exact geographic location of the Isles of the Moluccas came into sharp focus. Considered to be a precious asset by the Spanish and Portuguese alike, the Moluccas (known as the spice islands) were notoriously (and consciously) mislocated on contemporary maps: the longitude determining their exact location tended to shift according to who commissioned the map. The Treaty of Saragossa of 1529 officially settled the dispute.⁷⁹ Just like the Treaty of Tordesillas of 1494, the fight over the Moluccas showed what immense political and economic power was accumulated in a single cartographic line. While previous forms of spatial measuring relied on the physical itinerary and involved the body of one single person, Tory's address to his readers is a vivid demonstration of the accomplishments of triangulation and the rise of the equipollent map, grounded in abstract longitudes and latitudes—associated in the *Champ fleury* with the letter I.

INTERCONTINENTAL CARTOGRAPHEMES

Woodward's "equipollent-coordinate" mode of mapping, taken up by Tory, is "entirely geometrical and abstract, independent of the geographical content beneath."⁸⁰ It heralds the retreat of the human body from the sphere of cartographic measurement and forcefully demonstrates that the somatic yields to the numeric register, replacing the human body by a mathematical corpus. In consonance with Vitruvius, Tory upholds the number 10 as the standard for alphabetical, bodily, and cartographic measurement alike:

> Ie treuuve que noz bons Peres Anciens ont volu entendre consommee et entiere perfection au nombre dixiesme entendu quil est nombre Per, compose de nombre Per & Imper. Martianus Capella en son VII livre ou il parle De Decade, nous en est bon tesmoing quant il dict: "Decas vero

ultra omnes habenda quae omnes numeros diversae virtutis ac perfectionis intra se habet." La dixene dict il, voirement est de passe & dexcellence, en tant quelle contient & a en soy tous les nombres de Per & Imper. Cest a dire de vertus & perfection.

(I find that our excellent Ancient Fathers intended to attribute consummate and absolute perfection to the tenth number, inasmuch as it is an even number, composed of both even and odd numbers. Martianus Capella, in his Book VII, where he speaks of the *Decade*, is a good witness when he says: *Decas vero ultra omnes habenda quae omnes numeros diversae virtutis ac perfectionis intra se habet*. The tenth, he says, "in truth is of surpassing excellence, seeing that it contains and has in itself all the numbers both odd and even, that is to say, all good qualities and perfections.")[81]

The Pythagorean number 10 is a pivotal number from which also disciplinary perfection unfolds. Not only does 10 testify to "all good qualities and perfections," but it is also the very principle driving cartographic and linguistic thinking. The standard measure of a plane surface, Tory contends, is a square (*quarreau* or *quarré*), ten units long and ten units broad, "so that the large square will contain a hundred small squares, which I shall call 'units' (*corps*), because the length of the I, which will be of the same proportions as all the other letters, will be contained in one of these small squares."[82] Tory's cartographic inflection of the geometric square turns the plane into a gridded map upon which the letter I, in its function as a standard measure, is geographically oriented. A woodcut illustration suggesting that one end of the letter I points to the "orient," while the other is directed toward the "occident,"[83] reveals that the letter I is malleable enough to equally become a longitudinal line. As a cartographic surface, Tory's grid rests on a decadic principle itself productive of different letters and alphabets in which 10 is graphically identical with the letters I and O, Tory's frontier-letters. Binding numeric perfection, border-thinking, and alphabetic order to one another, 10 (IO) turns into a cartographeme holding the keys to the very principles of the alphabet's hidden secrets, cartographic thought, and the linguistic and geographic lines that separate European from non-European alphabets.

The complex and interdisciplinary function performed by the number 10 is perhaps most vividly showcased in the alphabetic table, here accompanied by an explanatory description, featuring the Arabic alphabet (fig. 17):

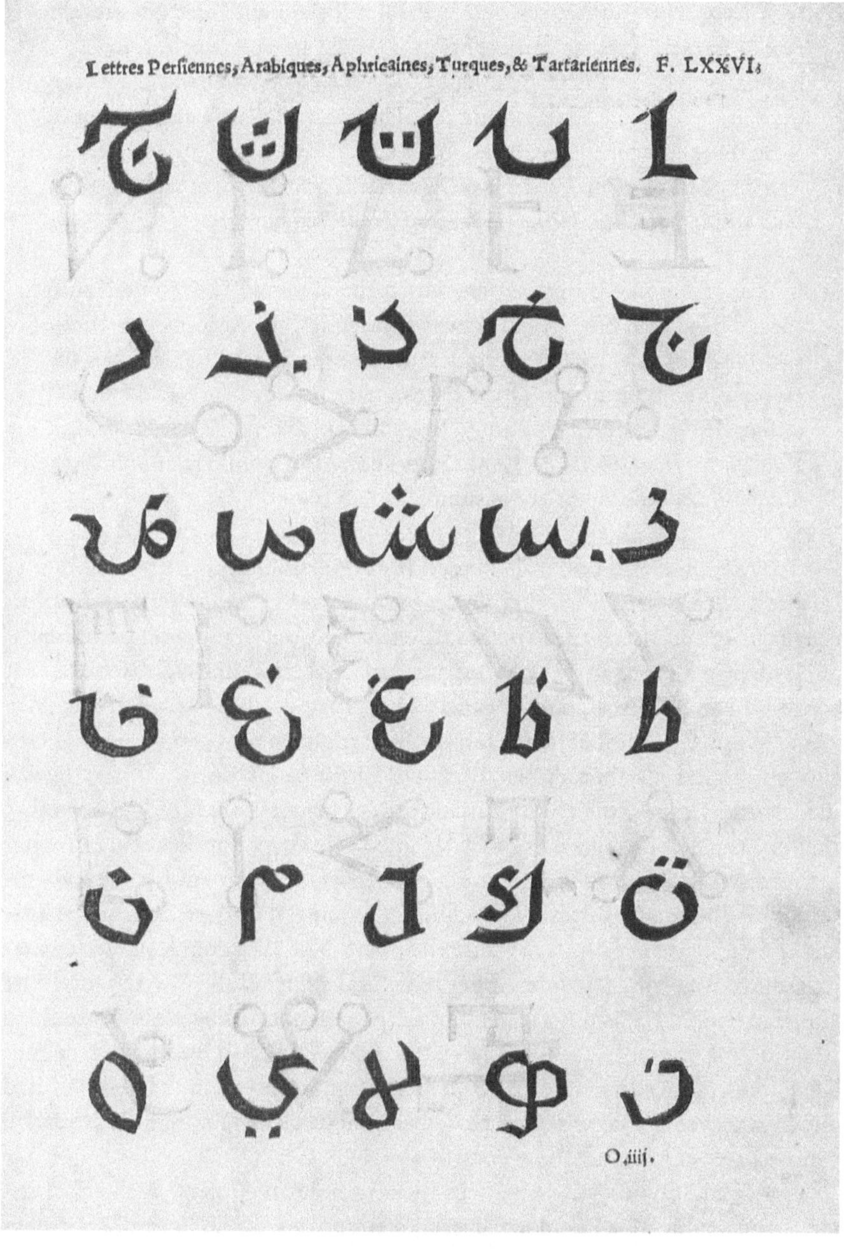

Figure 17. "Lettres Persiennes, Arabiques, Aphricaines, Turques, & Tartariennes" (Arabic Alphabet), in Geoffroy Tory, *Champ fleury* (Paris: Gilles de Gourmont, 1529), fol. 76r [Typ. 515.29.844]. Photograph: Houghton Library, Harvard University.

Ie vous ay faict les Lettres que le susdict Sigismunde dict qui servent aux Perses, aux Arabes, aux Aphricains, aux Turchs & aux Tartares. . . . Icelles l[ett]res veullent estre leuues a gaucge [*sic*] comme les Hebraiques, & leurs noms sont comme il sensuit en commanceant tousiours a la fin dune chascune ligne. Aliph, Be, Te, The, Zim, Che, Chi, Dal, Zil, Iz, Xe, Sin, SSin, Sat, Zat, Ty, Zi, Hain, Gain, Fe, Caph, Eiep, Lam, Mim, Nim, Vau, Eliph, Lam, Ge, Nulla. Elles sont Trente en nombre.

(I have drawn for you the letters which the aforesaid Sigismund [Fanti] says to have been used by the Persians, the Arabs, the Africans, the Turks, and the Tartars. . . . These letters must be read toward the left like the Hebrew, and their names are as follows, beginning always at the end of a line: *Aliph, Be, Te, The, Zim, Che, Chi, Dal, Zil, Iz, Xi, Sin, SSin, Sat, Zat, Ty, Zi, Hain, Gain, Fe, Caph, Eiep, Lam, Mim, Nim, Vau, Eliph, Lam, Ge, Nulla.* These are thirty in number.)[84]

Tory's alphabetic design is copied from Sigismondo Fanti's *Theorica et practica . . . in artem mathematice professoris de modo scribendi fabricandique omnes literarum species*, the first printed manual on calligraphy, published in Venice in 1514.[85] But instead of the twenty-eight letters that compose the Arabic alphabet, Fanti's (and Tory's) table is enlarged by two signs, featuring a total of thirty letters. The addition of two new signs is crucial: the round number allows Tory (via Fanti) to lay out the thirty letters in a spatially more conscious way, akin to a cartographic *tabula*, subjecting them to the dictates of a carefully designed geometric framework. But what is more striking still is the meaning of the two supplementary signs themselves: the popular ligature "Lam-Alif" (لا), a special form of *Lam* and *Aliph*, and the final cipher "Nulla." While the "Lam-Alif" is a composed letter that in itself means "no" ("la") in Arabic, the final sign, "Nulla," with which the alphabet concludes, is not a letter, but a cipher denoting negation. The double negation inserted into the last row of the Arabic alphabet forces the reader to slow down the reading pace and consider the alphabetic table as a convoluted and multilayered web of references where languages and numerals meet and call for disentanglement.

Tory's addition of "o," equally in its written form "nulla," as the final *letter* of the Arabic alphabet does, in fact, signify *something*. Zero was a cipher unknown to either the Greeks or the Romans and became, in the course of the Middle Ages, "the site of a systematic ambiguity between the absence of 'things' and the presence of signs."[86] It became first known in medieval Europe in two Latin translations of Al-Khwārazmī's ninth-century *Treatise of the Cipher*: one by the French mathematician Gerbert of Aurillac (945–1003), the future Pope Sylvester II who "introduced a new form of the abacus that employed tokens or *apices* bearing Arabic characters, an innovation in graphic forms that transcended words by providing symbols for numerical values";[87] and the other by Adelard of Bath. The treatise gained broader popularity with Leonardo Fibonacci's *Liber Abaci* (1202). Yet where the Arabic language used one word, "sifr" (with the meaning "empty"[88]) to denote zero, in Latin translations two words were used, "sifra" and "sefirum," disclosing starkly divergent meanings and triggering ambiguous pathways to think about this cipher:

> *Sifra* meant "a number," "a digit," whereas *sefirum* meant "nothing," and these meanings were brought out in the later development of both words: *sifra* took on an extended, more general meaning to denote any of the ten numerals and became the French *chiffre* and Italian *cifra* around the 14th century, the German *Ziffer* in the 15th century, and the English *cipher*, meaning "digit" and later, a hidden number, i.e. a secret code; *sefirum*, the word that meant "nothing" in Latin, on the other hand, developed into Italian to *zero*, which was adopted by French and also by English.[89]

While the medieval Christian Church rejected zero ("nothing") as the devil's work,[90] for in medieval theological philosophy God had filled the entire world and left no space for emptiness, the Arabic mathematical tradition appreciated the cipher "o" as an immensely effective and intrinsically spatial tool to produce numbers. Hindu-Arabic numerals, hinging on ten characters only, define the value of a number by a number's positionality. Unlike Roman numerals, "where each character denotes a fixed number"[91] (X is always ten, L is always fifty) and the number's ultimate value results from the addition of the symbols, in Arab numerals the value of a number depends on place: o (represented in Arab mathematics not by a circle, but by a dot) changes its meaning with its position. The Persian polymath Al-Khwārazmī, the inventor of algebra and also the author of the "earliest extant medieval Islamic geographical manuscript,"[92] the *Kitāb sūrat al-ard* (كتاب صورة الأرض, *Book of the Picture of the Earth*), thus writes about the productivity of zero:

> The zero ... perform[s] a certain kind of increase, but only by multiplying things tenfold: thus, by the grace of the Word, put a zero in front of one, and you get ten, put it in front of ten and you get a hundred, put it in front of a hundred and you get a thousand. And in this, you must

know, hides a great sacrament. For by that which is without a beginning and an end, is figured he who truly is Alpha and Omega, that is without beginning and end; and just as the zero neither increases nor diminishes, so he receives neither addition nor diminution; and just as it multiplies all numbers tenfold, so he multiplies them not only tenfold, but a thousandfold—or rather, to be more precise, from nothing he creates all, conserves all, and governs all.[93]

The opposite of "nothing," zero is here celebrated as a productive nucleus that stands for infinite increase rather than lack and emptiness. In 1510, the great mathematician and translator of Euclid Charles de Bovelles, whose work substantially impacted Tory's *Champ fleury*, published a small treatise titled "Liber de Nichilo," where he elaborates, also with an eye to cartography, on the possibility of conceptualizing "nothing" at once negatively and positively by using the linguistic strategy of a double negation: "nihil nusquam est neque in mente est neque in rerum natura, neque in intelligibili neque in sensibile mundo, neque in deo neque extra deum in ullis creaturis" (*nothing* is nowhere—neither in the mind nor in the nature of things, neither in the intelligible nor in the sensitive world, neither in god nor outside god in any other creatures).[94] Steeped in the tradition of negative theology, the treatise equally thematizes the potential of "nihil" (o) to unfold at once as *nothing* and *something*. For Jan Miernowski, Bovelles's *nihil* deploys its meaning in a fourfold manner—as the affirmation of a negation, the negation of an affirmation, the affirmation of an affirmation, and the negation of a negation[95]—which can be imagined as a logical square in which the proposition that nothing exists transitions into the proposition that everything is something.[96]

To visualize nothingness and God's creation "ex nihilo," Bovelles included a woodcut in his treatise (fig. 18) that illustrates the aporetic moment between nothing and something. The woodcut shows God, in the tradition of the *Liber Chronicarum*, presiding over the world from above and holding it firmly between his hands. The world is created out of elements leaving God's mouth through a conduit that is shaped like a trumpet—and itself evocative of the letter "I." The world is divided into heaven and earth and framed by a thick black circle reminiscent not only of the ocean on medieval *mappae mundi*, but also of the cipher "o." Most strikingly, "o" is that which inserts itself between God and the creation of the world: before the world comes into being, the "o" is already there. Itself empty and unmarked, it functions as God's material support, a table, upon which the

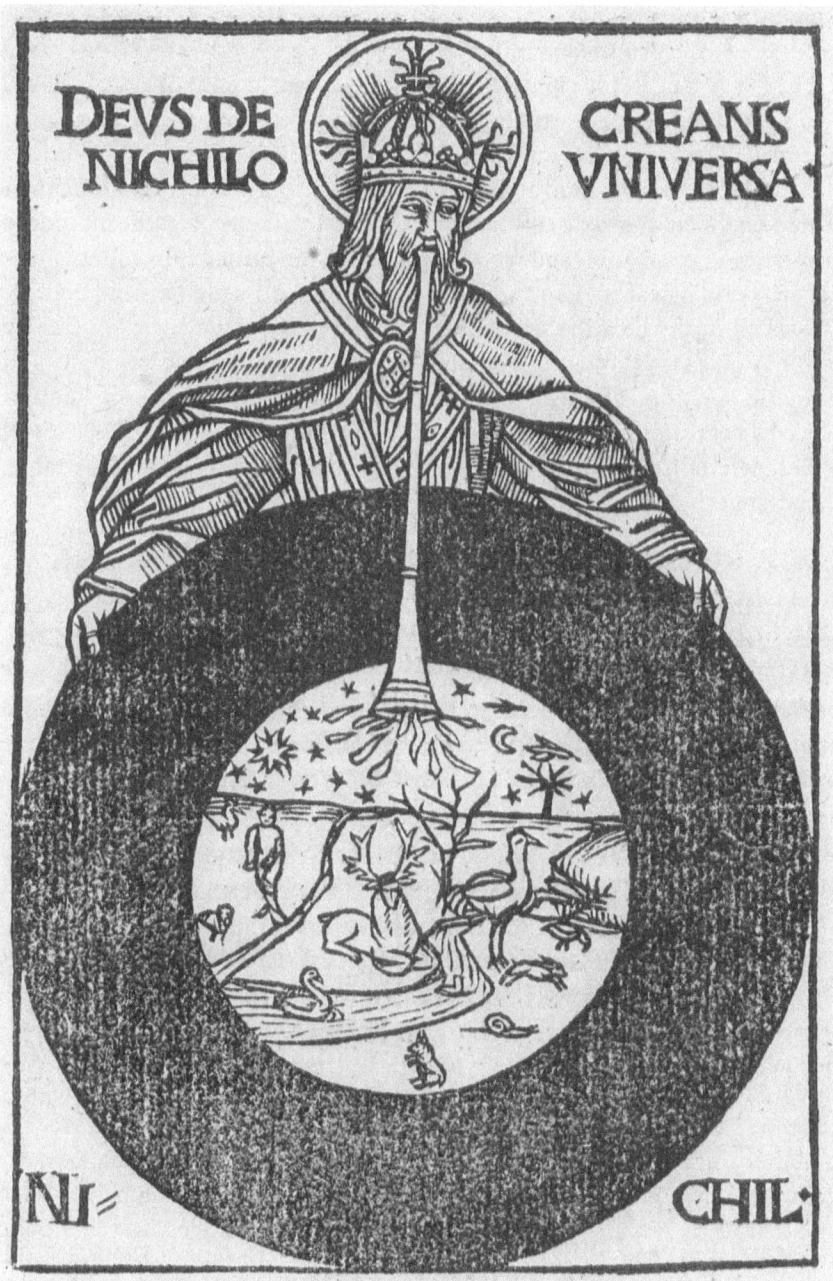

Figure 18. "Deus ex nichilo creans universa," in Charles de Bovelles, *Que hoc volumine continentur . . . Liber de nichilo* (Paris: Henri Estienne, 1510), n.p. [FC5 B6624 510q]. Photograph: Houghton Library, Harvard University.

elements pouring from his mouth are arranged into a coherent whole which takes the shape of a *mappa mundi*.

In this woodcut, the cipher "o" takes on a commanding position: it serves as a salient image of a cartographic table combining verbal and geographic features alike. Zero here is morphed into a cartographic device which brings the numbers 1 and 0 (and the letters I and O) to bear on the moment in which God's creation and cartographic production merge. Significant for the theorization of creation and representation, "o" is turned into a fundamental epistemological tool and the original framework which commandingly shapes all future creation. Here, Bovelles's illustration unearths o's faculty to function as a rigorous cartographic tool, a *tabula* enabling linguistic and numeric creation. Through its overwhelming visual presence, "o" anticipates by centuries Farinelli's claim that "it is not the table (that is, the map) which derives from numbers but it is number which derives from the table. We might take the case of the most elusive, mysterious and dangerous number: zero."[97] The "empty circle which today stands for zero," Farinelli contends by referring to the work of Robert Kaplan, "comes from the mark left by the circular pebbles on the surface of a calculating table covered in sand, so as to leave a trace, that is, a memory, of the calculation itself."[98] Just like the thick black circle on Bovelles's woodcut, "the zero stands for the part of the table that is not occupied by pebbles—thus, very simply, the zero stands for the table itself."[99] Mindful of the potent presence of the circle in Bovelles's "Liber de Nichilo," Tory's addition of the zero to the Arabic alphabet suggests not only a mathematical, but indeed a cartographic inflection of letters: they appear in the service of a larger cartographic preoccupation, which shifts the attention from philology and language to the very processes of mapmaking.

The added zero in the alphabetic table functions as a critical wedge transforming a treatise on the art of writing into a theory of (carto)graphic creation and representation. While the Arabic alphabet concludes with the zero (o), its first letter, the *Alif* ("ا"), oscillates in a versatile manner between letter and number. When looked at closely, Tory's design of the *Alif* makes the letter and the number 1 exchangeable. Tory thus furtively superimposes a numeric grid upon the Arabic letters, framing them not only mathematically with the Pythagorean perfect number 10 (as the Arabic alphabet's first and last sign), but also cartographically in that the graphemes 1 and 0 (or I and O) point to the shape of the globe and the boundaries within it. With the Latin letter A, the Arabic "ا" shares the phonetics and its privileged initial position as the generator of other letters. Angelo Piemontese has shown that within the Arabic alphabet the *Alif* undeniably occupies the most prominent position: "Just like the Greek Alpha and the Latin A, the *Alif* is the fountainhead of the Arabic alphabetic system." But in its graphic form, the *Alif* is more akin to the Latin letter I as it takes on, in Piemontese's words, the form of "a staff: |."[100] The tenth-century chancellor at the Abbasid court Ibn Muqla used the *Alif* as a measuring line when designing the letters of the alphabet according to rules of geometry and proportion.[101] Some Islamic authors imagined the *Alif* in its geometric symmetry as "the equator that divides the face and is represented by the nose."[102] The symmetrical conceptualization of the Arabic alphabet became the core of Islamic civilization from its inception in the eighth century CE. It was used in chancery, book calligraphy, and epigraphy and made a profound impact on the European art of writing. An expression of the unity between the human, the divine, and the world, the twenty-eight letters of the Arabic alphabet were endowed, especially in the esoteric tradition of Sufi mysticism, with a physical quality that linked them to body parts.

When Tory inserts a woodcut in which the staff-shaped "I" serves as a frontier-letter to separate the human face into two halves in the guise of the equator dividing the globe into two hemispheres (fig. 19),[103] one wonders if he relied solely on Vitruvius and Dürer as models for the art of spatializing letters and inscribing the human body within geometric figures[104]—or whether his sources might equally have included Arab manuals on calligraphy. The letter I discloses in the *Champ fleury* the functional proximity between the Arabic *Alif* and the Latin I. What Tory demonstrates is that the Latin letter I can take on the shape of the Arabic letter *Alif* "\"—and vice versa—when the linguistic and geographic perspective changes. The oscillation between the Arabic and Latin alphabets, with the centrality of the frontier-letter "\" (and "I"), comes here into sharp focus as a formulation of the question about the extension and limits of languages and territories. The transformability of the letter "I," its capacity to turn into an (*Alif*), powerfully inquires—here we come full circle to the myth of Hercules and the oscillation between the Parrhasians and the Parrhisians—about the limits of continents. When Tory at once provocatively juxtaposes and superimposes what we now call European and non-European languages and alphabets, he does nothing less than probing the continental boundary between them.

Tory unearths the translatability of letters across continents through the transformability of their graphic representation.[105] The *gramma* "I" is simultaneously the *grammen*, the dividing line, which separates languages and

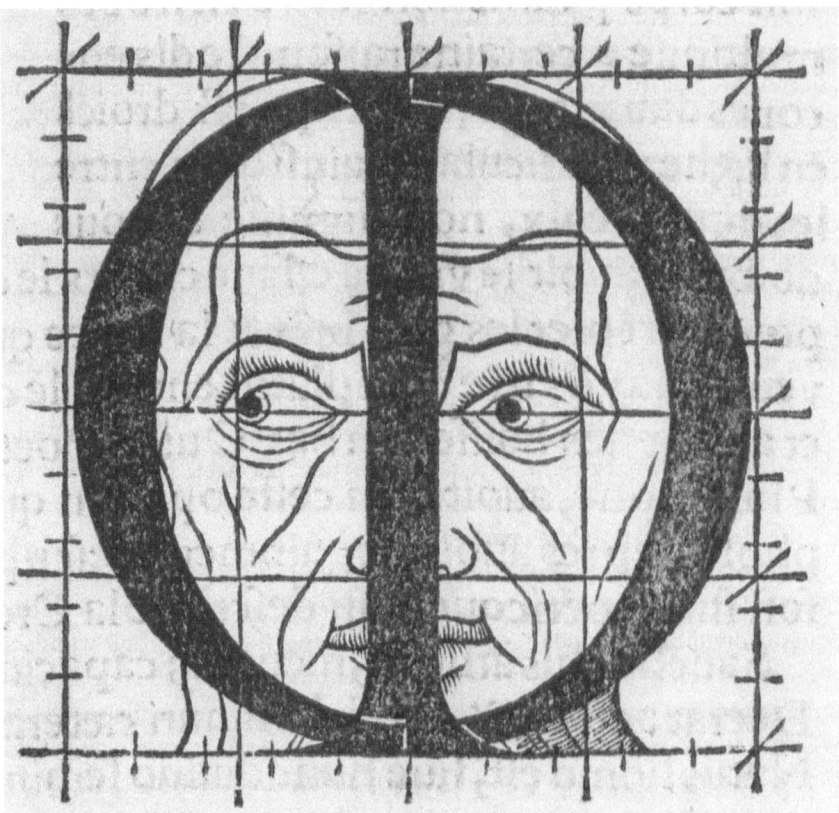

Figure 19. "Ordonnance de le I & de le O ensemble, au visage humain," in Geoffroy Tory, *Champ fleury* (Paris: Gilles de Gourmont, 1529), fol. 21v [Typ. 515.29.844]. Photograph: Houghton Library, Harvard University.

continents alike, reminding the reader that a single letter can readily contain multiple languages and translations. As Henri Meschonnic aptly put it: "The alphabet metaphorizes" (*L'alphabet métaphorise*).[106] Each letter already contains translations into different languages, idioms, and codes. Each letter harbors and bridges continental divides. To put it with Jacques Derrida: once embarked on the journey that unfolds with and as a letter, "we are within the multiplicity of languages and the impurity of the limit."[107] The dominant position of the frontier-letter "I"/"ı" in Tory's *Champ fleury* has the power to impose linguistic and alphabetical directionality and to set the reader on a path to meditate upon not only geographic boundaries but also the distinction between real places and places that exist as cartographic and poetic sites.

UTOPIA AS *TOPOGRAPHIA* AND *TOPOTHESIA*

The Arabic alphabet in the *Champ fleury* is not the only one with a cartographic patina, to be revealed by the reader's own arduous philological work. Tory also includes the Utopian alphabet, invented a few years earlier by Thomas More. In a letter addressed to Peter Giles on the topic of his new work *Utopia*, first published in 1516, More demonstrates, as Farinelli puts it, that "the island's proper names affirm their historic reality precisely to the extent that on the level of the *signifié* they 'did not correspond to anything', *nihil significantia.*"[108] Taking up the cartographic ambivalence of "nihil," Farinelli sees More advance the claim, not that toponyms are generally meaningless, but that "nihil" has a self-reflexive cartographic function: toponyms first come into being as cartographic signs on maps, before they emerge as names referring to real places. As Farinelli points out, More's emphasis on "nihil" epitomizes the very centrality of cartography in *Utopia* and the question of representation, expressed in the tension between physical and cartographic places, that the cipher "o" powerfully conveys. Initially a peninsula, Utopia became an island by artificial severance from the continent. Utopia's coming into being as an island is a paradox succinctly and effectively encapsulated by the cipher "o": at the moment of its "insulation" when it becomes representable as a circle (O)—or as a zero (o)—Utopia turns into a "nihil" and ceases to exist. Utopia's existence on a map coincides with More's repeated reminders within his narrative that the island is, indeed, fictional. Utopia's toponyms all bear the mark of negation: Ademo ("without people"), Amauroto ("barely visible"), and Acorii ("without region") are just a few examples of Utopia's (non-)places.[109] These toponyms vividly illustrate the immense power that cartography wields over those who look at a map and too easily forget that before them is the representation of a place, not the place itself. "The only country where there are rivers whose water one cannot drink and where there are trees that don't produce shadows," Farinelli suggests in his discussion of *Utopia*, "is the one represented on the map."[110] *Utopia* is a meditation on the emergence of places on maps set against the backdrop of the rising power of cartography as a tool that both establishes and critiques the tension between reality and representation. Considered by Jeanne Fahnestock as an "instanc[e] of topothesia,"[111] More's *Utopia* gestures toward the fragile boundary between *topographia* (the description of real places) and *topothesia* (the description of imagined places), while pushing the connection between insularity (suggested by Tory's parallel discussion of the letter O and the cipher o) and cartographic representation to its very limits.[112]

Erasmus, a contemporary to Tory and a correspondent to More, used the terms "topographia" and "topothesia" in *De copia* to define rhetorical figures of spatial amplification that express the tension between real and fictional places.[113] The less familiar word topothesia—literally "the setting of a place"—was first used in Servius's comments on the first book of Virgil's *Aeneid*, where it designated the island-shaped harbor on the African shore toward which the shipwrecked Aeneas swims.[114] Servius, however, confuses here the African city of Carthage with the Spanish city of Cartagena,[115] and in so doing, blurs the continental divide between Europe and Africa in a manner evocative of Hercules's mythological straddling of the two neighboring rocks Calpe and Abyla, which once united the western tips of Europe and Africa. The tension between real and imagined places that the terms topographia and topothesia ably convey was put into stark relief with the rise of cartography and the question of representing localities on maps. For Erasmus, the rhetorical power of topographia and topothesia to realize spatial amplification is intensified through mediating surfaces of visualization—such as the *tabula*: the best strategy to "bring before the eyes" real and imagined places, as we read in *De copia*, is to make sure that they are "viewed as though portrayed in color on a tablet, so that it may seem that we have painted, not narrated, and that the reader has seen, not read."[116] And it is precisely through the mediating interface of a "tablet," a *tabula*, that Tory arranges the alphabetic tables in those same years, be they tables of Arabic or Utopian letters.

If the third book of the *Champ fleury* follows an alphabetical order in its description and depiction of letters, the second book, Tory contends, arranges the letters not "en leur ordre Abecedaire" (in their alphabetical order), but "selon ma petite Philosophie" (according to my witty Philosophy). The letters are here "mingled together in such wise that they make a very beautiful and perfect work, which is called in Latin *Opus vermiculatum, Opus tessellatum et assarotum*, whereof Pliny, in his Natural History and Vitruvius in his book of Architecture speak amply."[117] The "opus vermiculatum" or "tesselatum," that Tory recalls, a method of laying mosaic tesserae, is akin to Cicero's often anthologized image of a mosaic into which words are fitted like small tiles.[118] For Tory, letters are fitted into the mosaic just as toponyms and territorial lines are inscribed in the gridded square, a (Ptolemaic) *tabula*.

Tory's incorporation of the fictional Utopian alphabet in the *Champ fleury* signals that imagination and *poiesis* cannot be entirely disengaged from cartographic and linguistic production. Topographia, in the *Champ fleury*, is nearly always accompanied by topothesia as a necessary reflection about the means of cartographic production itself. If for Farinelli More's *Utopia* constitutes a unique and paradigmatic example of the emergence of "place" as "the dream of conciliation between logic of space and logic of place,"[119] while teasing out the boundaries that separate real (physical) from fictitious (poetic) places, Tory offers such a dream of conciliation in his description of the insular city of Paris, which merges space and place in an insightful illustration of the tension between topographia and topothesia:

> Paris abunde en toutes especes de nobles metaulx, & est une droicte Grece en multitude de livres, ung vray pais Dinde en bonnes sciences & estude, une segonde Romme en poetes, unes Athenes en savans hommes Paris est la rose du monde, & le beaulme de luniversel firmament. Paris est une segonde cite de Sidon en tout aornement, abundant en toute maniere de victuailles & bons brevages. Riche en champs laborables. Fecunde en pur vin. Et doulce en ses habitans. . . . Entrelacee de sa belle riviere Seyne. Necte en son manoir. Forte en son seign[eu]r. Reverente & amyable a ses Roys. Gratieuse en son bel & doulx air. Delectable en son assiette. Bref, en Paris est toute venerable honestete, & tresor de tout bien, si fortune y visoit tousiours bien.
>
> (Paris abounds in all sorts of precious metals, and is a very Greece in the multitude of books; a true India in useful knowledge and study; a second Rome in poets; an Athens in learned men. Paris is the rose of the world, the balsam of the firmament. Paris is a second Sidon in outward splendour, abounding in all manner of food and pleasant beverages, rich in tilled fields, fecund in pure wine, & refined in her people. . . . Girt about by her beautiful river Seine. Spotless in her houses, strong in her Lord, venerable and lovable in her kings, delightful in her clear, soft air, delectable in her situation. In short, in Paris is to be found every respectable virtue and the treasure-house of all good, if Fortune choose to be always kind to her.)[120]

Paris is couched between Tory's sharp focus on one locale and the concomitant poetic dilution of geographic boundaries through his strategic incorporation of other cities as metaphorical comparands. At once contained and uncontained, the city of Paris follows the logic of islands "at [the] limits of the imagination or representation," as Simone Pinet put it in a different context, a logic that "makes them especially prone to links with the marvelous."[121] For Tory, Paris is indeed a "marvelous royal abode,"[122] which unfolds as both place (in its referential uniqueness) and space (in its metaphoricity that makes it translatable). Pinet notes that islands have long "come up in the context of margins, of limits, of cartography and juridical discourse, as a space produced in reality and in fiction."[123] The insularity of the city of Paris Tory describes takes on the shape of a miniature *oikoumene*, a place-space that seems to span the continents: Rome and India, Athens and Sidon, all places intimately tied to the origin of letters or numbers.

Tory's insular Paris is at once the endpoint of linguistic and cultural *translatio* and the origin of a new spatial and geographic order: "Paris is the rose of the world." The Paris of the *Champ fleury* recalls both the radiant letter O of the second book, in which Tory encloses a lyre-playing Apollo radiating the world's knowledge, and a compass rose on nautical charts. Paris is the point from which cardinal directions and rhumb lines—those that structure the world while exploring its limits—unfold. Paris is also the place of "tilled fields" the furrows of which are plowed like lines in boustrophedon, a poetic pattern allowing for the synchronous reading of verses from two different directions, not dissimilar to the European and non-European alphabetic tables included in the *Champ fleury*. Through this dynamic cartographic humanism, Tory undoes the city's sharp insular contours. As a copy or translation of Sidon ("a second Sidon"), Paris resets Europe's relationship with Asia and opens up new avenues for investigating continental liminality.

CHÔRA: THE CARTOGRAPHIC TABLE AS THIRD SPACE

Tory sets in motion a model to imagine not only the origin of letters and territories and their future unfolding but also their cartographic representation. His description of Paris is a gesture similar to the artificial intervention, in More's *Utopia*, of King Utopus, who created the island of Utopia by separating it from its maternal mainland. What brings Utopia and Paris together is the idea that both islands perform the birth of the matrix, thematizing the origin of the very cartographic support upon which the variegated linguistic, numeric, and geographic signs are cast as cartographemes.

One recalls Raphael Hythloday's medical terminology as he establishes a comparison between Utopia and a uterus:

> Utopiensium insula in media sui parte (nam hac latissima est) millia passuum ducenta porrigitur, magnumque per insulae spacium non multo angustior, fines versus paulatim utrinque tenuatur. hi velut circumducti

circino quingentorum ambitu millium, insulam totam in lunae speciem renascentis effigiant. Cuius cornua fretum interfluens, millibus passuum plus minus undecim dirimit, ac per ingens inane diffusum, circumiectu undique terrae prohibitis ventis, vasti in morem lacus stagnans magis quam saeuiens, omnem prope eius terrae alvum pro portu facit. magnoque hominum usu naves quaqua versus transmittit.

(The island of the Utopians extends in the middle (where it is broadest) for two hundred miles, and holds almost at the same breadth over a great part of it, but it grows narrower towards both ends. These ends form a circle five hundred miles in circumference, so that its figure is not unlike a new moon. Between its horns the sea comes in eleven miles broad, and spreads itself into a great bay, which is environed with land to the compass of about five hundred miles, and is well secured from winds. In this bay there is no great current; the whole coast is, as it were, one continued harbour, which gives all that live in the island great convenience for mutual commerce).[124]

The island of Utopia resembles a womb, a waxing moon (*luna renascens*), ready to grow and expand its contours. Shaped like a uterus, the island has two horns (*cornua*), a term that binds female anatomy and geography to one another. In fact, Hythloday describes the crescent-shaped island of Utopia as a place where medicine is held in great honor.[125] On two facing pages, the woodcut of the 1516 *editio princeps* exposes the island as a matrix, a maternal imprint. On the left page, a map of Utopia shows a uterus-shaped island, with a ribbon floating between the island and a ship passing by: an umbilical cord, the last element connecting the island to the mother-mainland. The facing page showcases a writing sample in the Utopian letters—the alphabet which Tory imports in his *Champ fleury*.

Imagined as a gendered matrix akin to a cartographic table, Utopia thematizes the very moment of territorial emergence, much in the same way Io marks the territory with her hoof. When in his preliminary epistle addressed to Thomas Lupset preceding *Utopia*, the French Hellenist Guillaume Budé termed Utopia a "seminarium"—a term akin to Tory's image of a flowery field denoting a generative ground comparable to a "nursery of young trees,"[126] which Jean Le Blond would translate into French as "pépinière" (breeding ground)[127]—he made the connection between the matrix and cartographic table even more explicit. But the association of a "breeding ground" with a map dates back, one might argue, to Plato, the first to put a name to the emergence of a cartographic model, referring to the breeding ground discussed in his *Timaeus* as chôra, a "third kind" between idea and copy.

Plato deems the dominant binary model between ideal form and copy insufficient: "We must... in beginning our fresh account of the Universe make more distinctions than we did before; for whereas then we distinguished two Forms, we must now declare another third kind. For our former exposition those two were sufficient, one of them being assumed as a Model Form [παραδείγματος εἶδος], intelligible and ever uniformly existent, and the second as the model's Copy, subject to becoming and visible."[128] Chôra—Plato calls it a "τρίτον γένος" ("triton genos" or "third paradigm")[129]—critically intervenes to soften the otherwise sharp binary opposition between model and copy. Plato describes chôra as both "baffling and obscure"; it is "the receptacle [ὑποδοχὴν], and, as it were, the nurse of all Becoming."[130] But most importantly, Plato's description of chôra uncannily dovetails with the very foundations of cartographic production:

> [Chôra] must be called always by the same name; for from its own proper quality it never departs at all; for while it is always receiving all things, nowhere and in no wise does it assume any shape similar to any of the things that enter into it. For it is laid down by nature as a moulding-stuff for everything, being moved and marked by the entering figures, and because of them it appears different at different times. And the figures that enter and depart are copies of those that are always existent, being stamped [τυπωθέντα] from them in a fashion marvellous and hard to describe, which we shall investigate hereafter.[131]

Hinting at the strong affinity—but also complex relation—between chôra and *tabula*, Stuart Elden in *The Birth of Territory* advances the claim that "the word *khora* is one of the most difficult words in the entire Platonic lexicon."[132] Just like a cartographic matrix, chôra is at once container and contained. It receives "entering figures"—the cartographemes that Plato calls "impressions," with a visual metaphor—then moulds these into a pattern, a map. For Rivaud, Plato's French translator, chôra becomes a "porte-empreinte," literally, an "imprint-carrier."[133] For Farinelli, chôra is akin to a map in that it "contains in itself not only all the elements and visible creatures, but also all possible figures."[134]

In his commentary on Plato's *Timaeus*, Ficino muses upon the possible different modes in which chôra ought to be translated into Latin: "Adduxit [Plato] in medium multa materiae nomina, scilicet speciem invisibilem,

Sinum informem omniumque capacem formarum. Naturam, Potentiam, Matrem mundi, Nutricem formarum. Subiectum, Receptaculum, Locum" ("[Plato] brought forth many names for matter, such as invisible species, formless bowl capable of all forms; nature, potentiality, mother of the world, nurse of forms; subject, receptacle, place").[135] But immediately after providing the catalog of possible translations of chôra (quite tellingly ending in "locus," "place"), Ficino shifts to cartography (a topic looming over Plato's *Timaeus*): praising the rendering of Ptolemy's *Geography* in Italian verse (the Dantesque *terza rima*) by his contemporary Francesco Berlinghieri,[136] Ficino rhetorically asks: "Quid dicam de Francisco Berlingherio nostro Nicolai filio? Nonne hic mathematicorum beneficio fretus de Cosmographia versibus scripsit egregie?" (What shall I say about Francesco Berlinghieri, the son of [our] Nicholas? Did he not write excellently, in verse, on Cosmography for the benefit of mathematicians?).[137] Ficino astoundingly bridges chôra and cartography and insists, furthermore, on the joint impact of cartography and poetics on mathematics. Neither idea nor copy, chôra is productive of a new "third kind" that gestures toward a new, cartographic, imagery. In the *Champ fleury*, chôra emerges as that "seminarium" which Tory envisions as "a flowery field," a fertile ground prepared for the cultivation and display of cartographemes in the form of linguistic, numeric, and geographic signs.

Plato describes chôra by recurring to the female figure of the nurse, a "conceptual support, a central characteristic of an 'idea' which is no longer sayable in terms of logic."[138] Just like the newly emerging entity of the continent, chôra is, in Alice Pechriggl's words, "not marked: despite all the passages of sensible forms it receives, this 'space' remains blank [*vierge*]."[139] "Is it not an old dream," Pechriggl asks, "to mark the virgin?"[140] While later titles would designate anthropomorphic maps of Europe as "Europa regina," they were originally titled "Europa virgo," thus presenting the allegorized territorial entity as newly crystallizing, still unmarked. The contours of the continent rose in tandem with the violence deployed in the myth of the Rape of Europa and articulated themselves as a (male) desire, Hercules-like, to mark and stake out virgin territory. The myth of Europa was part and parcel of the early modern editions of geographic texts from Ptolemy to Strabo, often used to adorn the margins; it became a popular subject for Renaissance artists from Albrecht Dürer to Titian and turned into a widespread literary topos, revived through the often illustrated editions of Ovid's *Metamorphoses*. In all these articulations, the myth of Europa corresponds to the cartographer's drive to demarcate the contours of a territory in the shape of a female body.[141]

Abducted from the same Phoenician coast, Io and Europa were often mentioned in tandem, among others by Galileo Galilei, who, in 1610, proposed naming the so-called Medicean stars, the four largest moons of Jupiter, Io, Europa, Ganymede, and Callisto.[142] Tory's account of Io is a companion myth to Europa, narrating the creation of the fragile boundary between Asia and Europe. What brings these two myths even closer together is the intimate link between territorial delineation, language, and naming. Io's foot provides Tory an opportunity to revisit the continental boundaries between Europe and Asia at the Ionian coast of Asia. Io—as I and O—can be read in light of what Emanuela Bianchi has described in Platonic terms as the "motile generativity of the receptacle/chôra, as a zone of creativity where dwelling, living, being as becoming, is always already taking place, ongoingly."[143] For Plato, in fact, chôra serves as a critical wedge that is driven between Being and Becoming:

> Let this, then, be, according to my verdict, a reasoned account of the matter summarily stated,—that Being [*on*] and Place [*chôra*] and Becoming [*genesis*] were existing, three distinct things, even before the Heaven came into existence; and that the Nurse of Becoming . . . exhibits every variety of appearance; but owing to being filled with potencies that are neither similar nor balanced, in no part of herself is she equally balanced, but sways unevenly in every part, and is herself shaken by these forms and shakes them in turn as she is moved.[144]

While "one Kind is the self-identical Form, ungenerated and indestructible," the "second Kind [is] generated, ever carried about, becoming in a place and out of it again perishing, apprehensible by Opinion with the aid of Sensation." Akin to a map, the third kind, chôra, then, "is ever-existing Place, which admits not of destruction, and provides room for all things that have birth, itself being apprehensible by a kind of bastard reasoning by the aid of non-sensation."[145]

The Platonic chôra dramatically overlaps with Farinelli's description of the *tabula* as an "undefeatable structure" (*inespugnabile struttura*), the "matrix of every transcendence" (*matrice di ogni trascendenza*).[146] Chôra (as cartographic representation) blurs the lines of demarcation and filiation—genealogical, philological, and territorial—and instead introduces a type of liminal thinking that Plato relegates to a bastard realm. Chôra, Bianchi contends, is constituted by its "excessive, bastard, shifting role as the unruly, disruptive, errant cause."[147] As a bastard, chôra challenges representation and dramatically exposes the tension between topographic and topothetic description by positioning itself on the cusp of philological and territorial indeterminacy. It stands for an aporetic moment: the continental ambivalence, territorial belonging, and philological desire that Tory deploys in his works, chief among them the *Champ fleury*. Here, chôra, rendered as the en(t)abling space, the flowery field, stands for the growing tension between Europe's acknowledged dependence on and desired detachment from Asia—an aporetic desire which, in the *Champ fleury*, manifests itself in the form of malleable cartographemes, strategically deployed to draw the emerging contours of the continent of Europe.

CHAPTER FOUR

Syphilitic Borders and Continents in Flux: Girolamo Fracastoro's *Syphilis sive Morbus Gallicus* (1530)

In 1507, the German cartographer Martin Waldseemüller invents the toponym "America" and uses it, for the first time, on a world map with the lengthy title *Universalis cosmographia secundum Ptholomaei traditionem et Americi Vespucii aliorumque lustrationes* (*Universal Cosmography according to the Ptolemaic Tradition with Discoveries by Amerigo Vespucci and Others*).[1] The map not only announces a complex co-presence of ancient geography and modern navigation but greatly honors Amerigo Vespucci, the Florentine navigator involved in decisive Portuguese and Castilian expeditions, the first to declare that the newly found land is a "mundus novus" (a new world). "Because it is well known that Europe and Asia were named after women," Waldseemüller explains, "I do not see why anyone would object to calling this fourth part Amerige, the land of Amerigo, or America, after the man of great ability who discovered it."[2] In the *Universal Cosmography*, the toponym "America" designates what is now the South American continent, then imagined as an island disconnected from the *oikoumene*, "found to be completely surrounded on all of its sides by sea."[3] America is depicted as a narrow sliver of land, couched between the Atlantic and Pacific Ocean—the latter still unknown and unnamed, but already projected, speculatively, on Waldseemüller's map. What Vespucci called in his famous letter to the Florentine Lorenzo Pietro de' Medici a "continens" (continent) and a "mundus novus" (new world)[4] was mapped, in 1507, as an elongated and narrow island located in the Southern Hemisphere.

In 1516, Waldseemüller publishes another map, the *Carta marina*, counting among the first printed nautical maps depicting the entire world.[5] Often praised as "superior in both the abundance and the quality of its artistry to [the] more famous . . . 1507 world map,"[6] it curiously undoes crucial information contained on the earlier map. Not only has Waldseemüller's projection method shifted but, more importantly, the toponym "America" has disappeared.[7] The "fourth part" of the world, previously named after Vespucci, is now called "Parias," the toponym used by Columbus. The American land mass is imagined no longer as an independent island but as attached to the Eurasian continent. What is now North America is defined here as "Terra de Cuba, Asie Partis" (the Land of Cuba, of the Asian Continent). Waldseemüller's *Carta marina* distances itself from the 1507 map—and, thus, from Amerigo Vespucci's vision of a "mundus novus"—and follows Columbus's theory that the newly encountered lands and the islands of the Caribbean belonged instead to the Asian continent. Aligning the *Carta marina* with Columbus's geographic vision, Waldseemüller identifies Hispaniola as the biblical island of Ophir ("Spagnolla que et Offira dicitur") and eliminates the Pacific Ocean, a body of water that had been, in the meantime, publicized following Vasco Núñez de Balboa's discovery in 1513.[8]

Taken together, Waldseemüller's two maps illustrate the nonlinear and convoluted process of continent making by drawing upon two divergent epistemological frameworks for imagining and mapping this "fourth part" of the world—all this, within the work of a single cartographer. While Waldseemüller's 1507 map imagines America as independent of Asia,[9] his *Carta marina* privileges the idea of a territorial continuity between the newly discovered lands and Eurasia. Waldseemüller's cartographic production, considered as a whole, shows that both geographical hypotheses not only sounded plausible to early modern cartographers but could indeed be pursued, side by side, despite their conflicting agendas.

Girolamo Fracastoro's neo-Latin poem *Syphilis sive Morbus Gallicus* (1530)[10] is the poetic equivalent of Waldseemüller's cartographic oeuvre. Here, the Veronese geographer, physician, and poet Fracastoro (1478–1553)[11] develops a cartographically ambitious program in what is the first poem fictionalizing the discoveries of America. Fracastoro not only coins the eponymous neologism for a new epidemic, "syphilis," but radically shifts the discussion of its origin from the widespread practice of national scapegoating to the question of continental mapping, which allows him to broach the question in a more nuanced manner, carefully peeling off its convoluted medical and geographic layers and leaving the matter of the epidemic's origin ultimately undecided. Syphilis originated, according to the two major theories, either in Europe in the last years of the fifteenth century, when the troops of the French king Charles VIII invaded the Italian peninsula, or upon Columbus's return to Europe from his first voyage to America in 1493.[12] With the question of continental connectivity sharply on the rise with the joint developments of the outbreak of the epidemic and the discovery of the New

World, Fracastoro juxtaposes two possible cures—the traditional mercury and the new guaiac tree cure—by first mapping them, respectively, onto the *oikoumene* and the New World before collapsing them in a complex, cartographically inflected poetry that defies clear-cut dichotomies. Pathbreaking for the history of immunology, *Syphilis* suddenly takes the shape of an unprecedented "cartography of disease,"[13] to borrow Tom Koch's term. By vigorously challenging national stereotyping and undoing any binary approach through the coinage of a new *terminus technicus* as a potent tool capable of resetting mindsets, Fracastoro steers away from the scale of the national and, instead, couches the question of the rapidly circulating infectious seeds ("semina"), which do not halt before territorial boundaries, within the larger cartographic question of a newly emerging scale: the continent.

SYPHILITIC LIMOLOGY

A poetic cartography of the *oikoumene* and America alike, Fracastoro's *Syphilis*—a succinct poem of 1,346 lines, written in dactylic Latin hexameter and divided into three books—offers an unparalleled insight into the complex geographic relationship between Eurasia and the New World. *Syphilis* was published contemporaneously with the first "prose cartographies"[14] of America, written by Spanish and Italian geographers and historians such as Peter Martyr d'Anghiera, Gonzalo Fernández de Oviedo y Valdés, Francisco López de Gómara, Bartolomé de Las Casas, and José de Acosta.[15] While these prose cartographers have in common a description of the newly discovered "Indies," as the New World was commonly referred to in those writings, their idea of how America related to the *oikoumene* starkly differed. Some of them insisted on an existing connection between America and the Eurasian continent. Oviedo imagined a geographic unity of the Eastern and Western Hemispheres "by way of a land bridge in the extreme North Atlantic," Las Casas had in mind a united West and East Indies, and Acosta hypothesized a land bridge in the North Pacific, the Strait of Anian (known today as the Bering Sea). Others, such as Gómara, imagined America, in the wake of Vespucci, as an autonomous island: "The land which we call the Indies is also an island like our own."[16] Against this backdrop of a wide range of geographic models available to him, Fracastoro's poem articulates a deeper and even more complex framework that blends poetics, medicine, and geography into an epistemological framework in its own right. It does so through the lens of an urgent medical matter that simultaneously upset and fostered continental thinking: the uncontrolled and unexpected spread of syphilis, long termed "the first disease of globalization."[17]

Prior to Fracastoro's poetic invention, the most frequently used term for the epidemic was "morbus gallicus," first documented before the first emergence of the disease in a manuscript written in Italian hand in the first half of the fifteenth century. It is probable, however, that the use of "morbus gallicus" here refers to another disease, not to syphilis, and that the name was transferred to syphilis after its first outbreak in 1493.[18] As early as 1496 or 1497, the German humanist and physician Joseph Grünpeck published the earliest extant treatises on syphilis, immediately followed by Niccolò Leoniceno's treatise with the telling title *De epidemia quam Itali morbum gallicum, galli vero Neapolitanum vocant* (*On the Epidemic that the Italians call the French Disease, and the French call the Neapolitan [Disease]*).[19] Translating the Latin-sounding expression "mala de franzos" (French disease) as "Böse Franzos," (literally "the nasty French"), Grünpeck turned the virulence of the disease into a vilifying characterization of a neighboring people. Recent scholarship has reiterated the "pejorative, even xenophobic" terminology,[20] turned into a lexical defense mechanism by which the French called syphilis the "mal néapolitain" (Neapolitan disease); the Turks might have labeled it the "male dei cristiani" (Christian disease);[21] the Spanish Christians, in turn, blamed the Arabs and Jews (*Marranos*) for spreading the epidemic;[22] and after the discovery of America, the "fourth part" of the world was included in the economy of circulating infections which did not halt before territorial borders, be they national or continental. Fracastoro's poetic cartography allowed him to unhinge the investigation into the origins and circulation of the new epidemic from the national scale and to raise it, for the first time, as a transcontinental problem.

A potent geographic intervention, Fracastoro's lexical invention is a critical wedge cast into what Étienne Balibar has theorized as the polysemy, heterogeneity, and multiplicity of the (continental) border.[23] Rather than being a straightforward division between two states or two continents, the border, Thomas Nail has recently suggested, "is not only its sides that touch the two states; it is also a third thing: the thing in between the two sides that touch the states. This is the fuzzy zone-like phenomenon of inclusive disjunction that many theorists have identified as neither/nor, or both/and. . . . States infinitely approach the limit in between them in the sense best described by the mathematical concept of "limit" in calculus. . . . Border theory is the study of this limit."[24]

A border theorist *avant la lettre*, Fracastoro inquires into this very asymptotic limit that separates the two continents and, conversely, those instances where continents overlap and their contours merge. He does so by tapping into and tarrying on those crucial moments of geopoetic "bifurcation"[25] in which the poet-geographer has the power to decide—through the careful choice of toponyms, stylistic features, and syntactic structures—between (continental) continuity and discontinuity. Already the poetic use of names such as Columbus and Vespucci can easily turn into a geopolitically and cartographically charged litmus test unearthing the specific epistemological model of geographic and cartographic thinking, which the poet folds into his lines. It is therefore quite telling (but not at all surprising) that Fracastoro refrains from ever naming the Spanish captain undertaking the first transatlantic voyage. He thus ingeniously leaves the question of continental connectivity open and privileges, instead, the view of the Caribbean islands as a fulcrum of continental indeterminacy. I would like to call Fracastoro's geopoetic exploration of continental borders (guided by the question on which continent syphilis first emerged), with a nod to Nail's concept of "critical limology"[26] which he develops in critical border studies to investigate the ever-changing techniques of bordering processes, syphilitic limology.

The emergence of a new global epidemic, the rapidly spreading syphilis, dramatically spurred the humanists' reflection upon the relation between America and Europe. Fracastoro's friend Andrea Navagero, a celebrated poet-diplomat who served as Venice's official historian until 1529,[27] had a firm grasp of the information flow between the New World and the Venetian intellectual elite (which included Fracastoro, Pietro Bembo, and Giovanni Battista Ramusio)[28] and was crucial in providing Fracastoro the latest news from the New World while residing in Spain. When Fracastoro was working on *Syphilis*—a project he might have begun as early as 1510–12[29]—Navagero, a translator of the first histories about the New World that reached Europe, was at work preparing Peter Martyr's *Libro Primo della Historia de l'Indie occidentali* (*First Book of the History of the West Indies*) and parts of Oviedo's *Sumario*, first published in 1526, for an Italian readership.

The uncertain origin and rapid dissemination of syphilis were accompanied by geopolitical reflections, anxieties of mapping, and desires to name and define the disease. Its emergence was, as Roberto Esposito observed, of a "topological order," becoming an unprecedented threat that, for the first time, suggested that a disease could not be "endogenous, generated from within the body politic"[30] of the *oikoumene*, but instead—so it was feared—from without. The spread of syphilis was coterminous with attempts at isolation and boundary making. It coincided with the European discovery of America and the Treaty of Tordesillas of 1494, the first to draw an arbitrary line ("raya") across the Atlantic Ocean, which, on maps such as the Cantino planisphere (1502), resembled a continental divide between Europe and the New World, even though it separated intra-European possessions (Spanish from Portuguese). If read retrospectively, against the backdrop of the history of immunology, the poetic lines of *Syphilis* can easily be construed as setting the stage for what Esposito has identified as Europe's rising "obsession with self-protection" and "the need, increasingly emphasized, for immunitary barriers, protection and apparatuses aimed at reducing, if not eliminating, the porosity of external borders to contaminating toxic germs."[31] But once *Syphilis* is understood as the unprecedented result of a joint geographical, poetic, and philological venture, in which the history of immunology plays but one part, albeit an important one, it emerges as an astonishingly ambiguous poem focused at investigating geographic indeterminacy with novel poetic means and new lexical coinages.

When Fracastoro published *Syphilis* in 1530, his elegant neo-Latin work quickly turned into "perhaps the most famous Renaissance Latin poem,"[32] and by the twentieth century *Syphilis* had gone through over one hundred editions in several languages.[33] The three books of the "bella Operetta" (nice little work),[34] as Pietro Bembo called the poem, engage with the epidemic's origins and symptoms (book one) before turning to the two alternative cures (books two and three), mercury and the guaiac tree. In the course of the sixteenth century, both cures were popular. The benefits of the guaiac-tree cure had been described by one of its earliest beneficiaries, the German Protestant humanist Ulrich von Hutten, in his short treatise, *De guaiaci medicina et morbo gallico* (*On the Medicine of Guaiacum and the French Disease*).[35] In contrast to Hutten, Paracelsus, outspoken in his defense of the mercury cure, harshly criticized the mass importation of the "Indian wood," as the guaiac tree was known, into Europe, decrying it as a great business scheme promoted by the wealthy German Fugger family deeply involved in the guaiac trade.[36]

In the second book the fictitious Syrian hunter Ilceus, the keeper of Diana's sacred groves, descends to the underworld beneath the Mediterranean Sea to retrieve mercury, a metal needed for his rapidly advancing venereal disease. There, he discovers a subterranean world of mining, a lightless underworld exposing the fluvial connections among the three parts of the *oikoumene*, Europe, Asia, and Africa. In the third book, the shepherd Syphilus (Fracastoro's invention), an inhabitant of the Caribbean island of Hispaniola (identified here, as in Waldseemüller, as the biblical island of Ophir) and the island's first victim of syphilis, mobilizes the myth of the sunken island of Atlantis to lay out his complex genealogy and arcane geographic belonging, suggesting that America is both connected to and disconnected from the *oikoumene*. Both books unfold as a clever play with (poetic and geographic) sameness and difference and the process of continental division.

FRACASTORO'S SYPHILITIC CARTOGRAPHY

While located in two different parts of the world and sporting two unrelated protagonists, the stories of Ilceus and Syphilus unfold in an uncannily similar manner. In many ways, the second story is intended as a geographically dislodged variation of the first. In both cases, human hubris causes the epidemic affliction, redeemed by a performed sacrifice. In the case of Ilceus, the sacrifice takes the shape of a *catabasis*, the descent to the dark depths of the underworld where he witnesses the artificial production of precious metals (mercury, gold), while Syphilus's sacrifice, performed in a hollow valley in Hispaniola (Ophir), is not only productive of the cure, but also of the epidemic's new technical medical term (syphilis) as the result of the sacrifice. Both stories plot unsettling cartographies that resist representation as they expose the pressing question of continental delineation in a time of Europe's expansion and new epidemics.

Fracastoro defined his poem as a "lusus," a playful and funny joke—a misleading designation that has led scholars to mistakenly conclude that, unlike his later medical writings such as *De contagione*, *Syphilis* did not contain any substantial information about the new epidemic. Fracastoro himself seems to have taken his poem very seriously, going as far as coining here a new *terminus technicus*, "syphilis," which he used and promoted in his subsequent medical writings. Fracastoro argues that his later treatise on syphilis, *De contagione*, elaborates "medically" (*medice*) on the epidemic, while his earlier poem *Syphilis* discusses the same topic "poetically" (*poetice*).[37] Fracastoro dedicated *Syphilis* to one of the most outstanding Italian humanists, Pietro Bembo, whose influential book on the emergence and definition of the Italian vernacular language, *Prose della volgar lingua*, was printed roughly at the same time, in 1525. Bembo was not only an exquisite philologist, but also, as Denis Cosgrove reminds us, "a collector of globes and maps and an intimate correspondent of cartographers and geographers,"[38] such as Oviedo. Bembo was a crucial philological and geographical compass for Fracastoro, so much so that his poem is framed by explicit references to Bembo: *Syphilis* both opens with an apostrophe to Bembo ("Bembe"), nestled in one of the poem's opening lines, and ends with "Bembus" as the poem's concluding word.[39]

Bembo praised the Lucretian verve with which Fracastoro opens the first book: "Il primo libro ha molte cose belle anzi tutte, e parmi che scriviate in verso cose tolte di mezzo la Filosofia molto poeticamente e molto più graziosamente, che non fa Lucrezio molte delle sue" (The first book has many beautiful things, even all of them, and it seems to me that you are writing in verse about things taken out of Philosophy, very poetically and more graciously than Lucretius does in many of his [lines]).[40] In stiff competition with the stylistic luminaries of ancient poetics such as Virgil and Lucretius, Fracastoro's poem is striking, Bembo muses, because of its "vaghezza Virgiliana" (Virgilian beauty), recognizable in its "copious and abundant style" (*copia & abondanza*),[41] which could not create a starker contrast to the nauseating content of the first book: a thorough description of the "conditions and symptoms of this sad contagion," reminiscent of Lucretius's description of the pestilence of Athens in the closing book of *De rerum natura* (*On the Nature of Things*). Fracastoro describes the unfolding of the epidemic in a human body with these words:

> Quippe, ubi per cunctas ierant contagia venas,
> Humoresque ipsos, et nutrimenta futura
> Polluerant, Natura malum secernere sueta
> Infectam partem pellebat corpore ab omni
> Exterius: verum crasso quia corpore tarda
> Haec erat, et lentore tenax, multa inter eundum
> Haerebat membris exanguibus, atque lacertis.
> Inde graves dabat articulis extenta dolores.
> Parte tamen leviore, magisque erumpere nata,
> Summa cutis pulsa, et membrorum extrema petebat.
> Protinus informes totum per corpus achores
> Rumpebant, faciemque horrendam, et pectora foede
> Turpabant.

(When the contagion had passed through all the veins and had polluted even the humours and what was meant to feed the body, Nature whose wont is to reject what is harmful, attempted to expel the infected part from the whole body towards the surface. But because this matter had a lighter element, more naturally inclined to erupt, which, as it was expelled, made for the surface of the skin and the limbs' extremities.

Immediately unsightly sores broke out over all the body and made the face horrifyingly ugly, and disfigured the breast by their foul presence.)[42]

The contagion "pastur[es] deep within our limbs"[43] and invisibly flows through the veins, through bodily canals that lead the corrupted material to the body's surface, where it finally erupts. The process of cutaneous eruption that renders previously invisible material visible, the skin that functions as a liminal layer separating the body's hidden realm from its exterior strata, and the surface of the body, upon which the discharged liquids come to rest, all imagine the human body in geological terms, as a telluric stratification. The figure of hypotyposis used here, Fracastoro's not only vivid, but truly visceral description of the putrid material's discharge on the skin's surface, its flow from the inside to the outside generative of pustules "gaped wide open," likens the surface of the human body to a geological upheaval, a true volcanic eruption. Bembo, whose first poem was a description of the Sicilian volcano Etna,[44] was clearly taken with Fracastoro's framing of medical symptoms as geological phenomena when he described his evocative images as "marvelous." Recurring twice to the semantics of marvel in one single sentence, Bembo notes that Fracastoro's poem is "mirabile" (wondrous) and "fa maravigliare, chiunque il legge" (makes marvel whoever reads it).[45]

The second book (on the hunter Ilceus)—to which, in contrast, Bembo outspokenly objected—was added after the completion of the third book (on the shepherd Syphilus).[46] Here, Fracastoro charts an alchemistically inflected itinerary across and beneath the Mediterranean to explore the production of mercury and quicksilver, "a Divine gift" culled through "man's great skill"[47] (subterranean mining) as a "more successful" cure than traditional Hippocratic and Galenic diets. In an elaborate "digression," the poet's voice tells the fabulous story of the origins of the mercury cure: "Syriae nam forte sub altis / Vallibus, . . . Callirhoe qua fonte sonans decurrit amoeno" (In a deep Syrian valley . . . where Callirhoe runs down murmuring from her pleasant fountain),[48] the hunter Ilceus offended the goddess Diana by killing her sacred stag. Diana punished the hunter by inflicting upon him the venereal disease, but specified that "quare tellure sub ima, / Si qua salus superest, caeca sub nocte petenda est" (if any salvation exists it must be sought in the depths of the earth below, beneath night's darkness).[49]

The transition from the second to the third book of *Syphilis* is a temporal as it is a spatial leap: from a timeless forge to a specific moment in time, 1492, and from the vertical axis of a chthonic mining site to the horizontal expanses of the Transatlantic and the vegetal riches of the Caribbean.

The third book traces the recent transatlantic voyages of the Spanish fleet to the Caribbean, where the New World's herbal cure is found: the guaiac tree, bestowed upon mankind, just like the mercury cure, as a "gift from heaven."[50] This is the plot that Bembo most approved of: "quella [favola] del legno mi sodisfa & empie l'animo maravigliosamente. Senza che per essere il legno cosa nuova, ella vi sta più propriamente; che non fa quella dello argento, che è cosa trita & ad ogni uno famigliare, come sapete" (the [fable] of the wood [guaiac tree] gives me great satisfaction and fills my mind with wonder. Besides, being a new thing, it fits more appropriately, unlike the [fable] about mercury which is a trite thing and familiar to everyone, as you know).[51] Here, the Caribbean islands emerge as a pivotal site, what I call a suture line of continental differentiation, marked by the emergence of a new lexeme: "syphilis," a poetically and geographically ambitious neologism which spans the Mediterranean world of Virgilian *Georgics* and the "hollow valley"[52] of the Antilles where it is first forged.[53]

Scholars of the Italian Renaissance tend to forget that during his lifetime Fracastoro was celebrated as one of the greatest experts in geography. It is to Fracastoro that Giovanni Battista Ramusio, who corresponded with Fracastoro on various occasions on geographic and cartographic topics ranging from projection methods to the rising levels of the Nile,[54] dedicated his monumental three-volume anthology *Navigazioni e viaggi*, whose first volume appeared in 1550 in Venice. Ramusio confidently characterizes Fracastoro as one of the greatest geographic authorities of his time:

> Non truovo uomo a chi la debba piú convenientemente raccomandare . . . salvo che l'Eccel. Vostra, percioché nessuno penso che la possa meglio di lei giudicare, o che con maggiore affezione la desideri di leggere, o che col chiaro nome suo gli acquisti e piú credito e piú lunga memoria . . . ch'è tanto instrutta della geografia.
>
> (I cannot find a man to whom I could recommend [the *Navigazioni e viaggi*] more appropriately . . . than to your Excellence [Fracastoro]. I think that no one could judge it better, or desire to read it with greater affection, or give it more credit and a longer memory through his illustrious name . . . than you, who are most learned in geography.)[55]

The uniqueness of *Syphilis* resides not only in the surprisingly complex and multifaceted picture it offers at once of the subterranean space beneath the Mediterranean and the hollow valleys of the Caribbean islands, but in particular in their complicated geographic relationship to one another. The story of the hunter Ilceus injects a geological depth into the question of continental connectivity, which is calibrated in the images of the flat surfaces that dominate the world of the Antilles. Together, the two books combine both the *oikoumene* and the New World into a site of productivity, creativity, and porosity which cannot be, but at the same time needs to be, disentangled against the backdrop of a rising pressure and increased practice of drawing cartographic lines. Before delving deeper still into the separate, and at the same time intertwined worlds of books two and three, a few words about Fracastoro's guiding concept, the Lucretian seed (or atom) as the smallest generative unit, are in order.

ITINERARIES OF ERRANT SEEDS

Syphilis opens with an explicit nod to Lucretius's *De rerum natura* and proceeds by strategically tying together the intricate workings among geography, poetry, and medicine:

> Qui casus rerum varii, quae semina morbum
> Insuetum, nec longa ulli per saecula visum
> Attulerint: nostra qui tempestate per omnem
> Europam, partimque Asiae, Libyaeque per urbes
> Saeviit: in Latium vero per tristia bella
> Gallorum irrupit, nomenque a gente recepit:
> Necnon et quae cura, et opis quid comperit usus,
> Magnaque in angustis hominum solertia rebus,
> Et monstrata Deum auxilia, et data munera coeli,
> Hinc canere, et longe secretas quaerere causas
> Aera per liquidum, et vasti per sydera Olympi
> Incipiam: dulci quando novitatis amore
> Correptum, placidi Naturae suavibus horti
> Floribus invitant, et amantes mira Camoenae.

(What were the varied accidents of matter, what the seeds which brought on an unaccustomed disease through long centuries seen by no one: which in our time raged through all Europe, parts of Asia and through the cities of Africa: it burst into Italy with the unhappy French wars and took its name from that people: further what was the cure and what the resource experience and man's great skill in straitened circumstances discovered, and the help shown by the Gods, and the gifts bestowed by heaven, these I shall now begin to sing and through the liquid air and through the stars in vast Olympus I shall search for causes far hidden: since Nature's quiet gardens with fragrant flowers and the Muses who love wonders invite me, gripped by a sweet love of this strange event.)[56]

While words such as "semina" and "rerum" loudly echo the atomism of Lucretius's *De rerum natura* and his focus on the seeds of things, the set of questions with which *Syphilis* unfolds allows Fracastoro to shift, from the poem's first lines, from medical matter proper to the realm of geography. Foregrounding the creative, transformative, and hidden nature of the "semina morbum" (seeds of the disease), Fracastoro focuses on the contingency of their circulation across the *oikoumene*: Europe, Asia, and Libya (a common synonym for Africa). The opening questions do not confine the disease to a specific locale, but rather create the condition of possibility for unpredictable epidemic cartographies to form in a more comprehensive manner.

Already the opening lines of *Syphilis* lead the reader *in medias res* into the poet's blacksmith shop in which liminal spaces and ambiguous worlds are forged. Geoffrey Eatough has long insisted that "*Syphilis* has too often been viewed from an almost exclusively medical angle which has obscured even the medical realities."[57] Pushing against the Europeans' frenzy to unambiguously locate the origins of syphilis and pinpoint the moment of its emergence, Fracastoro insists on the disease's ubiquity and, surprisingly, its ancient origins in a complex poetic operation that goes beyond facile medical explanatory models. The chronotope the poem plots is dual: the epidemic's alleged French origins (*per tristia bella gallorum*) is accompanied by Fracastoro's observation that the disease is, instead, universal. Distancing himself from the widespread doxa that syphilis had a French origin, he claims that the "unaccustomed disease" (*morbus insuetus*), ravaging at the same time through the entire *oikoumene*, has polygenetic foci. The twinned title itself—*Syphilis sive Morbus Gallicus*—offers two alternative models to apprehend the epidemic: through the lens of either the national or the transcontinental scale (forged by Fracastoro). Here, the initial position of "syphilis," which stands for the unlocalizable and universal, pushes back against the national. The idea of one sole geographic origin is questioned, diffused, and morphed into a global preoccupation, instead.

At the same time, Fracastoro dismantles the specific temporal anchoring of the epidemic's origin: "in our time" (*nostra tempestate*) is undone by the claim that the disease was known "through long centuries" (*longa per secula*). While syphilis allegedly erupted only during his own lifetime, its roots might well be, Fracastoro muses, ancient and have just remained invisible and hidden for the past centuries (*nec longa ulli per saecula visum*).

In conceiving of the disease as an ever-existing, but periodically dormant, matter, in which the seeds are inactive and invisible in certain periods, while active and visible in others, Fracastoro centers on the tension between the concealed and the unconcealed, significantly shifting the poetics of the poem toward a complex, cyclical chronotope, in which matter—the Lucretian seeds—appears under different guises across time and space. If the disease existed in antiquity and has now reemerged after laying dormant for centuries, the same could be true, Fracastoro's audacious hypothesis (which we shall revisit later on) suggests, for newly discovered continents: while they now appear new and disconnected, they might have previously been connected through the workings of changing sea levels. Plato's myth of Atlantis provides Fracastoro the prototypical model for such theory.

The Lucretian concept of the seed (*semen*), compared in *De rerum natura* to both atoms and letters, is the generative principle of bodies, be they living, inanimate, or poetic. As Pablo Maurette has pointed out, for Lucretius the seeds and "texturas rerum" (textures of things)[58] constitute a sweeping explanatory model "from the generation and corruption of bodies to the origins of human civilization, the movement of celestial bodies, and the causes of meteorological phenomena."[59] The Lucretian seeds are in constant and regular motion. At times, however, they take an inexplicable and unpredictable swerve (*declinatio*) from their course, allowing for new contacts to unfold in the form of "blows" (*plaga*):

> corpora cum deorsum rectum per inane feruntur
> ponderibus propriis, incerto tempore ferme
> incertisque locis spatio depellere paulum,
> tantum quod momen mutatum dicere possis.
> quod nisi declinare solerent, omnia deorsum,
> imbris uti guttae, caderent per inane profundum,
> nec foret offensus natus nec plaga creata
> principiis: ita nil umquam natura creasset.

> (While the first bodies are being carried downwards by their own weight in a straight line through the void, at times quite uncertain and uncertain places, they swerve a little from their course, just so much as you might call a change of motion. For if they were not apt to incline, all would fall downwards like raindrops through the profound void, no collision would take place and no blow would be caused amongst the first-beginnings: thus nature would never have produced anything.)[60]

Lucretius's unexpectedly swerving seeds produce unforeseen contagions and new contacts just as they generate unprecedented itineraries and novel cartographies, coterminous with what Fracastoro considered to be the often erratic navigations across the (inclined) surface of the Atlantic. The textures of things, imagined in Lucretius as an expanse of seminal contact, are crystallized in Fracastoro's poem as a cartographic texture, akin to a map upon which sailors travel with the randomness of errant seeds charting unexpected, swerving itineraries. For Fracastoro, the transatlantic crossings of early navigations were indeed utterly unpredictable, ever subject to contingency. In the poem, the anonymous Spanish captain's role as a leader—he is referred to as "dux"[61] (captain, leader) or "rex" (king),[62] on equal footing with the "rex" of the New World, the king of Hispaniola—is seriously cast in doubt, in the moment we learn that he serves as the "dux errantis . . . classis,"[63] the captain of an errant fleet who appears to cross the earth's surface with the swerving contingency of a traveling seed. "Often driven far astray" (*longis erroribus actae*), Fracastoro's captain steers his ship to the Western Hemisphere "without knowledge of his course" (*ignaraeque viae*).[64] The "plaga" (blow) that Lucretius refers to in several instances in *De rerum natura*—such as in "plagas / gignere quae possint genitalis reddere motus" (blows which can produce generative motions)[65]—suddenly acquires a new geographic valence, one that Renaissance poets would not have missed: that of a "hemisphere" (*plaga*), a common usage of the Latin noun readily available to sixteenth-century humanists.

The intimate tie Lucretius establishes between the semantically capacious noun "plaga" and the concept of creation ("plaga creata" and "plagas / gignere") transpires in Fracastoro's poem as a driving force aimed at comprehending the newly emerging transatlantic space of hitherto untrodden itineraries. Read in a cartographic key, then, the Lucretian generative motions are not produced by "blows," but, perhaps, by the productive and mobile forces of "hemispheres," which Fracastoro's poem unhinges from a millennial immobility. The unfolding poetic maps of *Syphilis*, a work that conjures up the generation and circulation of seeds as early as its opening lines, is akin to the generative power of the new land, pregnant not only with lifesaving flora, but also with new lexemes. A compound of seed (*semen*), the word "seminarium," would likely encapsulate the joint workings of medicine and the soil: before it acquired the meaning of "seedbed of contaminations," "seminarium" appeared in Roman treatises on agriculture, where it referred to the seeds sown on fertile ground.[66] In *Syphilis*, the Caribbean islands are transformed into that fertile ground productive not only of heretofore unknown healing plants such as the guaiac tree, but also of unprecedented words such as "syphilis," which rises with the force of a toponym—albeit one which does not stand for a place, but rather, theorizes the very liminality of places that is the subject of *Syphilis*. Here, the new noun syphilis "hide[s] an unplaceable place,"[67] to put it with J. Hillis Miller, namely the complex suture line between two continents.

CHANGING SEA LEVELS, TELLURIC SHIFTS

From the smallest level of a single word to the grandest geographical maneuvers, Fracastoro bases much of his interventions on the concepts of contact, connection, and flux. For Fracastoro, all earthly phenomena exist through a causal relationship between sympathy (*consensus*), antipathy (*dissensus*), and contagion (*contagio*)—the latter understood as a constant necessary contact (*con-tagio*) among beings and things: "The first type of sympathy, common to all things and worthy of our admiration is that of the universe itself. All bodies strive to be connected and almost stuck to one another, so much so that there is no force capable of separating and dividing the external parts with which they touch each other; so much so that there is no void space between them."[68] Rejecting the idea of vacuum, Fracastoro imagines that bodies are always in direct contact with one another—an idea with consequential ramifications for the conceptualization of continental boundaries. The cosmos is, in Fracastoro's neo-Platonic conception of the world, "like a living animal, whose members are related thanks to ties of *sympathia*."[69] Sympathy is a "power" (*vis*) of nature whose task it is to hold the parts of the world together.[70] It is through the lens of the concept of sympathy that Fracastoro formulates his theory of the necessary connectivity of the different parts of the world.

Fracastoro's concept of universal sympathy precedes current ecocritical preoccupations by several centuries. His engagement with and understanding of the ecosphere were driven by the "fundamental premise," formulated by Cheryll Glotfelty with an eye to modern times, "that human culture is connected to the physical world, affecting it and affected by it."[71] As Fracastoro reformulated his theory of contact in a geographic key, he came to imagine all land masses as connected to one another, albeit not always in a visible way: some connections occur above ground, some under water. The connectivity of the different parts of the world can only be comprehensively understood when looked at in a deep and diachronic way, Fracastoro surmises, when the geological, telluric time of continental transformation is included in the evaluation of the ever-shifting contours of land masses.

What might appear to the naked eye as two detached land masses, Fracastoro hypothesizes, may likely be connected underwater—and, conversely, may become visible aboveground when sea levels change. Fracastoro insists on the telluric power that holds land masses together, making clear-cut continental divisions difficult. It is no coincidence, then, that the word "tellus" ("earth") appears twenty-one times in the first two books alone.[72]

Fracastoro had a pronounced interest in the changing nature of sea levels and the fluctuating elevation of the globe's surface, constantly moved and transformed by tides and volcanic eruptions. According to Charles and Dorothea Singer, Fracastoro either renewed the doctrine of the secular changes of land and sea—"already . . . set forth in the thirteenth century by the Arabian writer Qazwini, whose views must have been widespread since numerous MSS. of his *Wonders of Nature* have survived in both Arabic and Persian"—or alternatively, "at least, introduced [this view] to the West." Fracastoro is indeed credited for being "the first to hold that Western European land and water were subject to secular changes of elevation, so that an area now dry and even raised to mountainous height may once have been submerged."[73] In his cosmographic *Homocentrica*, Fracastoro writes, "if a man consider[s] how islands and mountains come into being, he will recognize that time was when they were built out from the sea and that time will be when land now covered by the waves will be habited and tilled, and yet again in future time will be again hidden by Ocean."[74]

The sixteenth-century Flemish scholar Johannes Goropius Becanus contended that for Fracastoro "all the mountains were made of the sea, by amassing and mixing together a lot of sand with its waves, and where now are mountains, there once was the sea. The mountains were left on dry land once the water retreated."[75] Fracastoro's geologic thought has enjoyed a long legacy over the centuries: one finds his theory of the changing connectivity of land masses, produced and altered by the movements of the sea, reformulated in Deleuze's description of the reversibility of mountains and islands in time:

> The ark [of Noah] sets down on the one place on earth that remains uncovered by water, a circular and sacred place, from which the world begins anew. It is an island or a mountain, or both at once: the island is a mountain under water, and the mountain, an island that is still dry. Here we see original creation caught in a re-creation, which is concentrated in a holy land in the middle of the ocean. This second origin of the world is more important than the first: it is a sacred island.[76]

Fracastoro's imagined map plots the continuous change of geology, geography, and cartography over time. His theory of the secular changes of land and sea, which he explored in his cosmographic and poetic works alike, studies the shifting surface of coastal lines which rigorously registers and continuously modifies the telluric movements subject to eternal change.

The undulating movement of the sea which constantly transforms the surface of the globe is folded into the very texture of the poem: the noun "unda" (wave) along with the adverb "unde" (where, whence), which relates to the waves as a playful pun (its homophone, "undae," is the plural of "unda"), are carefully inscribed at the beginning, middle, and end of several verses closely grouped together. A visualization of the movement of the waves, Fracastoro's poetics is also a map thematizing its own fleeting nature:

> Dextera sed sacri fluvii te sistet ad *undam*,
> Argento fluitantem undam, vivoque metallo,
> *Unde* salus speranda. Et jam aurea tecta subibant,
> Rorantesque domos spodiis, fulgineque atra
> Speluncas varie obductas, et sulphure glauco.
> Jamque lacus late *unda*ntes, liquidoque fluentes
> Argento juxta astabant, ripasque tenebant.
> "Hic tibi tantorum requies inventa laborum,"
> Subsequitur Lipare, "postquam ter flumine vivo
> Perfusus, sacra vitium omne reliqueris *unda*."

(But the right-hand path will set you by the waters of the sacred river, waters flowing with silver and living metal from which you can expect salvation. And now they approached the golden halls and the mansions dripping with lead slag and caverns covered with streaks of black soot and grey sulphur. And now they were standing by wide swelling lakes flowing with quicksilver where they kept to the bank. "Here is the relief found for your great toils," continued Lipare. "After you have been steeped three times in the living stream you will leave all this corruption behind in the sacred waters.")[77]

As seen in the attention paid to images of undulation created through careful positioning of words, Fracastoro's theory of changing sea levels unlocks new avenues for rethinking the plasticity of a word's literal and metaphorical meaning—and metaphoricity itself.

Fracastoro's technical mimicking of the movement of the waves affords readers an opportunity to see in his meditation on waves a prime example of the fluidity of his cartographic thought. When read synchronically, in a particular moment in time, the expression "plowing the ocean" (as in "totum potuit sulcare carinis / Id pelagi, immensum quod circuit Amphitrite" [to plough with its ships all the ocean, that immensity circled by Amphitrite]),[78] early on in book two, serves as a metaphor for crossing a body of water; however, when read diachronically, across the extended period of the secular changes of land and sea, when the ocean's low elevation exposes dry and arable land, the expression becomes literal. Akin to Fracastoro's concept of changing sea levels, Deleuze recalls that for the nineteenth-century artist Georges Seurat painting was "the art of ploughing a surface" in that painting "raises, accumulates, piles up, goes through, stirs up, folds. It is a promotion of the ground.'"[79] If for Deleuze the "new powers of texture" arise with modern art, when the painter "no longer paints 'on' but 'under',"[80] a similar process of what I perceive as "mapping under" takes shape in the work of Fracastoro, who creates poetic cartographies from below, from the ocean floor, as a means to map the groves above. His poetic cartography, in constant motion and transformation, incorporates both geographic and temporal changes as it plots new poetic textures and tarries on the hermeneutics of reading, the intricacies of style, and the limits of representation.

Fracastoro's complex engagement with the fleeting surface of the seas, which impacts, establishes, and shifts lands and borders alike, articulates itself in a twofold manner, explored in greater detail below, in the second and third books of *Syphilis*. In the former, the hunter Ilceus descends into a spectacular hollow cave under a Syrian grove, a subterranean "globus"[81] (globe), where he witnesses the alchemistic production of metals necessary to cure syphilis, while in the third book the myth of Atlantis, imagined as a prosthetic land bridge between America and the *oikoumene*, is revived. While Waldseemüller's Ptolemaic world map of 1507 detached the New World from the *oikoumene* and his 1516 *Carta marina* suggested a continuity between them, Fracastoro's theory of the secular changes of land and sea conjoins the two models in a convoluted display of Europe and America which focuses on their intricate relation of simultaneous connection and detachment.

SUBTERRANEAN CARTOGRAPHIES

The outbreak of syphilis in the late fifteenth century saw a decisively cartographic turn taken by the medical inquiry into different methods of producing and applying traditional metals—mercury, quicksilver, gold—for therapeutic purposes and newly concocted artificial metals to match the demand. In his *Ars et theoria transmutationis metallicae* (*Art and Theory of the Transmutation of Metals*), the Venetian priest, philologist, and alchemist Joannes Augustinus Pantheus offers perhaps the most incisive visual synthesis of the affinity between the alchemistic study of the production and transformation of metals, the malleability and transformativity of words, and the cartographic writer's exploration of continental contours. A great source of inspiration for Paracelsus's work on alchemy and John Dee's *Monas Hieroglyphica* (1564), Pantheus might well have served as a touchstone for Fracastoro, whose affinity with Venetian alchemists of his time such as Giovanni Aurelio Augurelli has long been recognized.[82] *Ars et theoria transmutationis metallicae*, first published in Venice in 1518, was superseded in philological and alchemistic complexity in 1530 with the publication of *Voarchadumia contra Alchimiam, ars distincta ab archimia et Sophia*, where Pantheus proposed what he believed to be a clear-cut distinction—encapsulated in a single consonant—between the practice of (fraudulent) *Alchimia* and (rightful) *Archimia*. Pantheus's explanation of the origin of his neologism "voarchadumia" follows a complex transcontinental logic. While claiming it to be "Chaldaic" in origin, the author explains that the word "combines an 'Indian' word *voarh*, which supposedly means 'gold', and a Hebrew locution *me-ha-adumot*, taken to mean 'two red ones', referring to two full processes of cementation."[83]

When in 1550 both treatises were (posthumously) published in one volume in Paris,[84] a map in the style of a *mappa mundi* adorned the center of the frontispiece (fig. 20). Alchemy had taken a decidedly cartographic turn. The frontispiece offers a potent statement of the scope of alchemy as an art and as a theory of transmutation and presents the work as an inquiry not solely into the metamorphosis of metals but also into the transformation and transformability of continents. The center of the frontispiece is reserved for a woodcut illustration featuring a circular map reminiscent of the familiar T-O map, but divided into four parts.[85] Unlike Africa and Asia, the continent of Europe here is unnamed and noticeably reduced in size, almost falling off the map's rim. Here, the "fourth part" of the world, studded with a few small islands, is left largely unmapped. Instead of offering an image of the newly discovered territories, Pantheus's visualization of the fourth part of the world invites onlookers to fill the indeterminacy of the page with their own spatial imagination—to connect it to the *oikoumene* or not.

By the early modern period, alchemy had become the epitome of interdisciplinary thinking. Sparking enthusiasm since Apuleius's *Metamorphosis*, it embraced the ideas of transformation, translation, and the origins of things in their broadest possible sense, catalyzing fields of knowledge as

ARS ET THEO
RIA TRANSMVTATIONIS ME-
tallicæ cum Voarchadúmia, proportionibus, numeris, & iconibus rei
accommodis illu-
strata.
IOANNE AVGVSTINO PAN-
THEO VENETO AVTHORE.

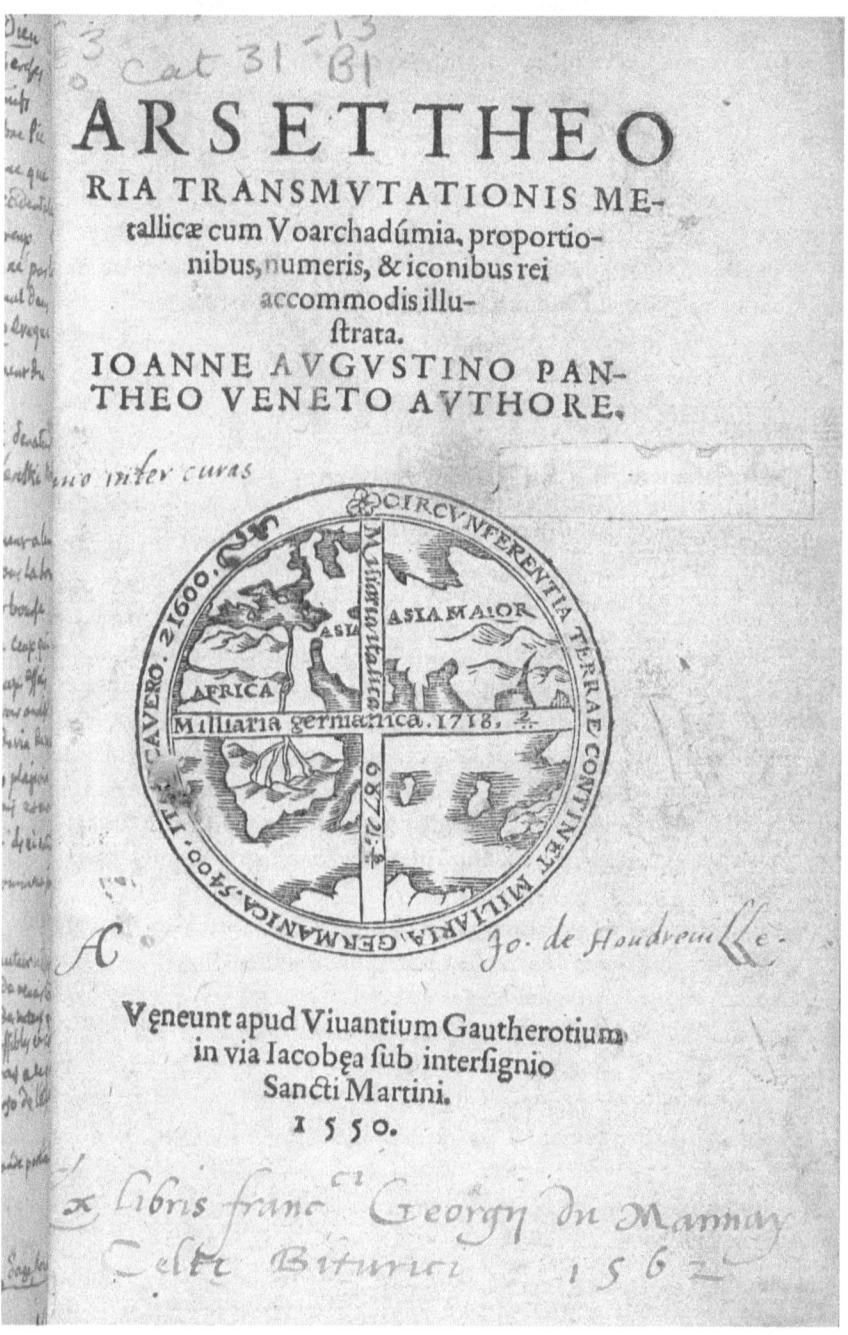

Veneunt apud Viuantium Gautherotium
in via Iacobęa sub intersignio
Sancti Martini.
1550.

Figure 20. Johannes Augustinus Pantheus, *Ars et theoria transmutationis metallicae cum Voarchadumia* (Paris: Vivant Gualtherot, 1550), title page [*IC5 P1959 B550a]. Photograph: Houghton Library, Harvard University.

diverse as geography, geology, mining, hieroglyphics, etymology, and poetry, with a specific focus on the coining of new words. Given the changing contours of territories and the unstable nature of toponyms—always subject to changes by translation and dislocation—early modern transatlantic voyages and the topography of newly encountered territories were often imagined through the lens of alchemy's transformative power. Alchemy offered a tool to reconcile epistemologically divergent—or incompatible—frameworks and to bring seemingly distinct disciplinary domains into a strategic and effective alignment.

When Bembo, the dedicatee of Fracastoro's *Syphilis*, was looking for a specialist to proofread the first volume of his *Prose della volgar lingua* (1525), he asked his friend Augurelli, the author of an alchemistic treatise on the secret technique of the "aurifera ars" (gold-mining art) titled *Chrysopoeia* (1515), to undertake the revision.[86] Having previously published Petrarchan love poems, Augurelli equally professed his deep interest in the joint workings of geography and alchemy. Once approached by the Venetian editor of the famous 1511 edition of Ptolemy's *Geography* with an invitation to contribute to the volume, Augurelli crafted the opening poem, in which he praises, with a nod to alchemy, the "figura . . . mutata,"[87] the ever-changing nature of the earth (and of the map).

When in the second book of *Syphilis* Fracastoro powerfully collapses alchemy and geography, he does so with Augurelli in mind. At first, Fracastoro details the different forms of the commonly applied cure for syphilis, what was believed to be a beneficial blend of mercury and quicksilver:

> Sunt igitur styracem in primis qui, cinnabarimque,
> Et minium, et stymmi agglomerant, et thura minuta,
> Quorum suffitu pertingunt corpus acerbo,
> Absumuntque luem miseram, et contagia dira.
> . . .
> Argento melius persolvunt omnia vivo
> Pars major: miranda etenim vis insita in illo est.
> . . .
> Cujus et inventum medicamen munere Divum
> Digressus referam.

(There are some who first of all heap together styrax, red mercuric sulphide and lead oxide, antimony and grains of incense, with whose bitter fumes they envelop the body entirely and destroy the deplorable disease, the dreadful contagion. . . . The majority prove more successful by using quicksilver to loosen everything completely; there is a miraculous power implanted in it. . . . I shall digress, and recount the discovery [of mercury treatment], through a Divine gift.[88]

What immediately follows is an elaborate, geographically informed "digressio" (digression), already alluded to, which mythologizes the discovery of the mercury cure. It is worth halting before the myth of the hunter Ilceus, punished with syphilis for killing Diana's stag. Diana, the "Goddess of the Crossroads"[89] (Fracastoro calls her "Trivia," literally, "Goddess of the Three Roads"), punished the hunter's hubris but mitigated the curse by specifying that "if any salvation exists it must be sought in the depths of the earth below, beneath night's darkness" (Si qua salus superset, caeca sub nocte petenda est).[90] In a *catabasis* modeled on Augurelli's *Chrysopoeia* (whose protagonist Lynceus, a quasi-anagram of Ilceus, equally descends to the underworld) as well as on the sixth book of Virgil's *Aeneid*, Ilceus descends to the underworld, where he is guided through the subterranean caverns by the nymph Lipare, who discloses to the subterranean voyager the vastness of the underworld as a "lightless globe," a dark replica of the world above:

> Hoc quodcunque patet, quem maxima terra est
> Hunc totum sine luce globum, loca subdita nocti,
> Dii habitant: imas retinet Proserpina sedes,
> Flumina supremas, quae sacris concita ab antris
> In mare per latas abeunt resonantia terras.
> In medio dites Nymphae, genera unde metalli,
> Aerisque, argentisque, aurique nitentis origo . . .

(All this which opens up before you is Earth in all its vastness: the whole of this lightless globe, realms in thrall to night, Gods inhabit: the lowest habitations are kept by Proserpine, the highest by rivers which roused from their sacred caves pass echoing through the wide earth to the sea. In the middle are the rich Nymphs from whom are born the families of metals, the course of copper and silver and gleaming gold . . .)[91]

Fracastoro produces an underground itinerary of the *oikoumene* reminiscent of Plato's *Phaedo*, in which Socrates describes the subterranean world and its rivers below Sicily, "connected underground with each other by channels in many directions, both narrower and wider," and "where a great deal of water flows from one to another as if into mixing bowls."[92]

Fracastoro's subterranean "globe" is a world in constant flux, where new materials are created and improved, where artifice emulates nature, and where the transformative power of the underworld exceeds and at the same time inverts the activities of the world above. The "lightless globe" Ilceus discovers in the depths of the earth hides from the world above the origins of the world's creative forces, which manifest themselves as a panoply of precious metals, here produced with great effort and artifice.

Lipare "discloses" (*pateat*) the underworld to Ilceus as if it were a newly discovered territory, a *mundus novus* silenced by the underworld's darkness yet eloquent in its production. In a Platonic tension between light and darkness, Lipare reveals the mute underworld as a paradoxical repository of protological elements and powerful cures. While the hunter Ilceus experiences the symptoms of an unknown disease under the blazing Syrian sun, his human ailment fades as he enters the hollow subterranean spaces untouched by daylight, where remedies are created in silence (*opaca silentia terrae*). While Fracastoro's use of the word "terra" (earth) binds the world above to the world below, lightlessness and silence characterize the realm under ground. Lipare's revelatory power gives her access to the production site of medical cures but also of names. Though Ilceus does not know the name of his disease, its name as well as his own name are known to the underworld nymphs, the bearers of a *prognosticon*, a foreknowledge that is not only therapeutic but also philological in nature.

Propelled by the reference to Diana as Trivia, Fracastoro's poetics foster the image of the underworld rivers as three liquid pathways that simultaneously converge in and unfold from the center of the *oikoumene*. Akin to a subterranean navel, Diana-Trivia subtends the *trivium*, the three domains of language-based knowledge, as well as the three parts of the known world.[93] If medieval *mappae mundi* were centered on Jerusalem, near the Syrian region from which Ilceus reaches the underworld, the protagonist's descent to the lightless globe brings together the hydrography of the *oikoumene* in a subterranean knot. Half a century later, Torquato Tasso would remember Fracastoro's underworld geography in his *Gerusalemme liberata* (*Jerusalem Delivered*),[94] when the two Christian envoys Charles and Ubaldo are granted an insight into the subterranean caves below the Mediterranean where "the veins from which our brooks and rivers teem" (ogni vena / la qual rampilli in fonte, o in fiume vago)[95] originate:

> E veder ponno onde il Po nasca ed onde
> Idaspe, Gange, Eufrate, Istro derivi,

ond'esca pria la Tana, e non asconde
gli occulti suoi principi il Nilo quivi.
Trovano un rio piú sotto, il qual diffonde
vivaci zolfi e vaghi argenti e vivi;
questi il sol poi raffina, e'l licor molle
stringe in candide masse e in auree zolle.

[And [Charles and Ubaldo] could see the birth springs of the Po.
The Danube, the Euphrates, the Ganges,
The secret sources of the Nile below,
And of the Don. But further down than these
They saw a stream of active sulfur flow
And living silver, lovely mercuries,
Which the sun purified and shaped and rolled
Into silvery nuggets or lumps of gold.][96]

What Tasso offers here is a nod to the alchemistic workshop below sea level so eloquently forged in the second book of Fracastoro's *Syphilis*, which designs an alternative cartography of the *oikoumene* that reveals new hidden spaces and transcontinental connections to those who venture beyond the surface of the earth. The subterranean itinerary Fracastoro designs provides him with a framework to imagine the origins of matter, trace the roots of the process of mapping, and disclose the aquatic navel of transcontinental connections.

ATLANTIS

For the geographer Strabo, writing in the first century BCE, Plato's story of Atlantis demonstrated in an exemplary manner that the visibility and existence of land masses depended upon changing sea levels:

[Poseidonius] correctly sets down in his work the fact that the earth sometimes rises and undergoes settling processes, and undergoes changes that result from earthquakes and the other similar agencies.... And on this point he does well to cite the statement of Plato that it is possible that the story about the island of Atlantis is not a fiction. Concerning Atlantis Plato relates that Solon, after having made inquiry of the Egyptian priests, reported that Atlantis did once exist, but disappeared—an island no smaller in size than a continent [ηπειρον, *epeiron*].[97]

Strabo's comparison of Atlantis to an *epeiron*, a (limitless) mainland-continent, provided early modern cartographic humanists an ideal model to think continentally. When he introduces the myth of Atlantis, home to the ancestors of the cursed shepherd Syphilus, Fracastoro is not only tapping into but also reviving a tradition of narrative and visual renderings related to the mythological god Atlas in its manifold articulations and cognate forms (such as Atlantis) that would span the sixteenth and seventeenth centuries—a tradition that will give rise, among others, to the atlas as a new cartographic genre (in 1570) and to Francis Bacon's *New Atlantis* (1626).[98] In the wake of New World discoveries, Plato's myth of Atlantis, first formulated in his *Timaeus*, was central to the imagining of a (lost) land bridge between Europe and America. Revived by Marsilio Ficino's Latin translation and commentary in 1484,[99] the *Timaeus* had many subsequent editions, among others in 1532, when the Swiss humanist Simon Grynaeus published two works: Plato's *Opera omnia* and a comprehensive travel anthology, *Novus orbis regionum ac insularum veteribus incognitarum*, which in turn inspired Ramusio's *Navigazioni e viaggi*.

With the launch of transatlantic voyages the urgency to locate Atlantis as an explanatory model to apprehend the newness of the New World became more pressing. How exactly Atlantis and America were related was debated by Renaissance humanists in manifold ways. For those who took Plato's story literally, Atlantis was an island between Europe and America that had disappeared over time. In 1527, Bartolomé de Las Casas declared that "Christopher Columbus could be reasonably sure that at least part of the continent described by Plato had been unaffected by the disaster."[100] In his *Historia del descubrimiento y conquista de la provincia del Perú* (1555), Augustín de Zárate describes Atlantis as a long-lost bridge between Europe and the New World, used by the descendants of Noah to cross the Atlantic Ocean from the *oikoumene* to Peru. For others, America was Atlantis, ancient Greece's defeated antagonist. López de Gómara describes America as such in his *Historia general de las Indias* (1552–53) in order to defend Hernán Cortés's territorial claims in Mexico, newly conquered for the Spanish Crown.[101]

The myth of Atlantis was so persuasive and pervasive that in 1665 the German Jesuit Athanasius Kircher included Atlantis as a woodcut map to illustrate his *Mundus Subterraneus*,[102] a book on "the divine work of the subterranean world" under the auspices of Proteus's "all-encompassing forms" (πανταμοφφον, *pantamorphon*).[103] Oriented to the south, Kircher's "Insula Atlantis" figures as an oversized island, competing in size with Africa and America, and serves as an allusion to a possible land bridge between the *oikoumene* and America.[104] But Kircher's core idea about a possible connectivity between the New World and the Old had already been rendered much earlier in verse: Fracastoro's *Syphilis* mobilizes the myth of Atlantis precisely at the moment of encounter between the Spanish sailors and the indigenous population of Hispaniola, as they set out to tell their story:

> Forsitan Atlantis vestras pervenit ad aures
> Nomen, et ex illo generis longo ordine ducti.
> Hac et nos, longa serie, de stirpe profecti
> Dicimur, heu quondam foelix et cara Deum gens,
> Dum coelum colere, et superis accepta referre
> Majores suevere boni: sed, numina postquam
> Contemni coeptum est luxu fastuque nepotum,
> Ex illo quae sint miseros, quantaeque secutae
> Aerumnae, vix fando unquam comprendere possem.
> Insula tum prisci regis de nomine dicta
> Ingenti terrae concussa Atlantia motu
> Corruit, absorpta oceano: quem mille carinis
> Sulcavit toties, terrae regina marisque.
> Ex illo et pecudes, et grandia quadrupedantum
> Corpora, non ullis unquam reparata diebus
> Aeternum periere: externaque victima sacris
> Caeditur, externus nostras cruor imbuit aras.

(Perhaps the name of Atlas has reached your ears and those who are descended from him in a long family column. We too are said to have issued from this stock in a long chain, alas once a happy race dear to the Gods, as long as our good ancestors continued to worship heaven and acknowledge what they had received from the Gods; but after the divine powers began to be held in contempt through the luxury and pride of their prodigal descendants, I could scarcely ever encompass by my speech the nature and magnitude of the troubles which subsequently attended the wretched people. Then the island, called Atlantia from that venerable first king, collapsed, shattered by a huge earthquake, swallowed up by the ocean which she, queen of earth and sea, had so often ploughed with a thousand keels. From that date cattle and four-footed creatures of great bulk perished forever, never to be restored at any time: it is a foreign victim which is slaughtered at our ceremonies, foreign blood soaks our altars.)[105]

Given the reference to Atlas, associated with Africa's northwestern shores, Atlantis takes on the shape of a geographic prosthesis and epistemological tool, still intimately related to Fracastoro's theory of changing sea levels, to simultaneously connect and disconnect the local population from the *oikoumene*. As the historical inhabitants of Ophir—the biblical gold-bearing island of King Solomon—the population is, from a diachronic perspective, part of the *oikoumene* (its genealogical tree reaches back to Atlas). But as the current dwellers of the island of Hispaniola, they are disconnected from it. The island of Atlantis thus epitomizes the critical dividing line between continental connection and separation and also serves, through the divine curse, as a mnemotechnic image for human hubris. Atlantis is crystallized here as a replica of the Tower of Babel: just as the *oikoumene* was colonized by Noah's three sons after the fall of the tower, the scattered island of Atlantis triggered a second dispersion, at once disconnecting the inhabitants of the Caribbean from and connecting them to the *oikoumene*, their former place of dwelling.

LIMINAL GROVES

Gardening has long been, Denis Cosgrove contends, "a trope for Europeans' imaginative domestication of a new, global spatiality that was characterized above all by the disruption of previously established spatial, ethnographic and conceptual boundaries. And we know that fixing a boundary between the wild and the cultivated is the primary act of gardening."[106] Fracastoro's description of two groves, associated each with precious metals (mercury and gold), binds together the two extremities of the Mediterranean while framing the second book of *Syphilis*. Here, the far-flung groves and gardens of the eastern Mediterranean, cultivated by the hunter Ilceus and described with the alchemistically sounding adjective as "glaucus groves"[107] (*nemora glauca*) and "sacred gardens" (*sacri horti*), are juxtaposed with the Gardens of the Hesperides, the daughters of the evening, who keep the mythical Golden Apples (which Fracastoro refers to, with a nod to Giovanni Pontano, as "citrus" or "citron" trees) at the western limits of the Mediterranean, in the vicinity of the Pillars of Hercules. When Fracastoro describes the groves, he launches an enthusiastic encomium of the citron:

> Sed neque carminibus neglecta silebere nostris
> Hesperidum decus, et Medarum gloria citre
> Sylvarum: si forte sacris cantata poetis
> Parte quoque hac medicam non dedignabere Musam.

Sic tibi sit semper viridis coma, semper opaca,
Semper flore novo redolens: sis semper onusta
Per viridem pomis sylvam pendentibus aureis.
Ergo, ubi nitendum est caecis te opponere morbi
Seminibus, vi mira arbor cithereia prastat.
Quippe illam Citherea, suum dum plorat Adonim
Munere donavit multo, et virtutibus auxit.

(But you too, citron, grace of the Hesperides and glory of Persian forests, will not be passed over in silence, neglected by our poem—that is, if after being sung by sacred poets you do not disdain the Muse of medicine in this role too. So may your tresses be ever green, ever shady, ever fragrant with new flower; through your green wood may you be ever laden with pendent golden fruit. Therefore, when you must strive hard to oppose the disease's hidden seeds, the Cytherean tree is there to help with its wonderful powers. For the goddess of Cythera, while she wept for her Adonis, endowed it with great properties and increased its virtues.)[108]

Fracastoro's apostrophe to the citron is a nod both to Virgil, who described, in the *Georgics*, the bitter-tasting "citrus" as a tree from Media (Persia),[109] and to Giovanni Pontano, whose neo-Latin poem *De hortis Hesperidum sive de cultu citriorum* (*The Gardens of the Hesperides or On the Cultivation of the Citron*) traces the miraculous and alchemistic rise of the "sacred" citron from Adonis's dead body.[110] In Pontano's version of the myth, the citron was brought to Italy when the goddess Venus, mourning the death of her beloved Adonis,[111] noticed, as the subtitle of Pontano's first book indicates ("De conversione Adonidis in Citrium" [The Conversion of Adonis into a Citron]), that his dead body was transformed into a citron tree.[112] The creation of a golden fruit and new living matter out of a dead, decaying body is a profoundly alchemistic process.[113] From Virgil's *Georgics*, where Aristaeus's new beehives are created out of bugonia (a cow's carcass), to early modern alchemistic treatises, putrefaction has firmly been associated with the "artis origo," the origin of art. No work demonstrates this more forcefully than Fracastoro's *Syphilis*, where the neologism syphilis is forged, in the third book, simultaneously with the "sure treatment" (*medicamina certa*), the guaiac cure, out of a white heifer's dead body,[114] in the liminal space of the Caribbean grove straddling the continental fault line between the *oikoumene* and the *mundus novus*.

In the early years of the sixteenth century, Elio Antonio de Nebrija, Spain's foremost lexicographer, grammarian, and humanist, whose famous claim, launched in 1492, that language is always the companion of empire,[115] was at work on his *Dictionarium medicum*, in which he celebrates the citron with the same consciousness to continental liminality as Fracastoro did, in those very years, in *Syphilis*. Decried by contemporary physicians as a work full of "barbarisms and arabisms,"[116] Nebrija's *Dictionarium* located the citron tree in the vicinity of the Atlas Mountains, as an arboreal marker of the critical continental juncture between Africa and Europe: "Citrus, arbor nobilissima, iuxta Athlantem sola nascens" (Citrus, a most noble tree, grows by itself next to [the] Atlas [Mountains]).[117] Nebrija conceives of the citron tree as a plant that, having previously traveled from Asia to Africa, is now couched between Africa and Europe almost with the force of a toponym indicating a continental boundary that is at once natural and artificial.

The Hesperides Islands of the Atlantic close to the African shore, intimately linked to Hesperus, the Evening Star and in early modern times often associated with the Cape Verde Islands,[118] have long been speculatively projected by ancient geographers. Pliny the Elder refers to navigational reports and voyages by the Carthaginian Hanno when alluding to the existence of the Hesperides (the "Ladies of the West") off the African shore:

> Traditur et alia insula contra montem Altantem, et ipsa Atlantis appellata; ab ea II dierum praenavigatione solitudines ad Aethiopas Hesperios et promunturium quod vocavimus Hesperu Ceras, inde primum circumagente se terrarum fronte in occasum ac mare Atlanticum. contra hoc quoque promunturium Gorgades insulae narrantur. . . . Penetravit in eas Hanno Poenorum imperator. . . . Ultra has etiamnum duae Hesperidum insulae narrantur.

> (There is also reported to be another island off Mount Atlas, itself also called Atlantis, from which a two days' voyage along the coast reaches the desert district in the neighbourhood of the Hesperian Ethiopians and the cape mentioned above named the Horn of the West, the point at which the coastline begins to curve westward in the direction of the Atlantic. Opposite this cape also there are reported to be some islands, the Gorgades. . . . These islands were reached by the Carthaginian general Hanno. . . . Outside the Gorgades there are also said to be two Islands of the Ladies of the West.)[119]

What in the ancient geographic imagery had been tied to the African continent was dislodged in the course of transatlantic navigations, when

Gonzalo Fernández de Oviedo's *Historia* (1535) transformed the Hesperides into a powerful colonizing tool spearheading an unprecedented *translatio* of the Hesperides from the *oikoumene* to the Caribbean islands. Heavily relying on the myth of king Hesperus brought to life with Annius of Viterbo's forged *Antiquities*,[120] published in Rome in 1498 by the Pope's own theologian, Oviedo was the first to (dis)locate the Hesperides, identifying them with the Antilles. Pierre Vidal-Naquet contends that

> in it [the *Historia*] he declared that the Antilles, that is to say the land of the Hesperides, had already long been possessions of the Spanish Crown. Charles V, informed in advance of these investigations and discoveries, on 25 October 1533 expressed his keen satisfaction at learning that "for three thousand and ninety-one years these lands have been among the royal possessions of Spain, so that it is not surprising that, after so many years, God has restored them to their owner."[121]

In analogy with Oviedo's push to the west, the third book of *Syphilis* performs that very leap from the Gardens of the Hesperides to the Antilles. Here, the citron tree's liminal location at the Pillars of Hercules is substituted by America's "happy groves of another world . . . beyond the Pillars of Hercules." Here, the "sacred" guaiac tree is celebrated as a novel citron of sorts, a new cure and gift from heaven:

Sed jam me nemora alterius foelicia mundi,
Externique vocant saltus: longe assonat aequor
Herculeas ultra metas, et littora longe
Applaudunt semota. Mihi nunc magna Deorum
Munera, et ignoto devecta ex orbe canenda,
Sancta arbos, quae sola modum, requiemque dolori,
Et finem dedit aerumnis.

(But now the happy groves of another world and foreign woodlands summon me; the sea resounds to me from afar, beyond the Pillars of Hercules, and from afar distant shores echo their applause. I must now sing of the Gods' great gifts and of the sacred tree brought from an unknown world, which alone has moderated, relieved and ended suffering.)[122]

In his encomium to the guaiac tree (and the book's second encomium to a plant), the poet deplores that the "sure treatment," endemic to the Caribbean islands, cannot be successfully transplanted to Europe:

Salve magna Deum manibus sata semine sacro,
Pulchra comis, spectata novis virtutibus arbos:
Spes hominum, externi decus, et nova gloria mundi:
Fortunata nimis, natam si numina tantum
Orbe sub hoc, homines inter gentemque Deorum
Perpetua sacram voluissent crescere sylva.
Ipsa tamen, si qua nostro te carmine Musae
Ferre per ora virum poterunt, hac tu quoque parte
Nosceris, coeloque etiam cantabere nostro.

(Hail great tree sown from a sacred seed by the hand of the Gods, with beautiful tresses, esteemed for your new virtues: hope of mankind, pride and new glory from a foreign world; most happy tree, if only the holy powers had wished you to have been born under our heaven and to grow amid this race of men belonging to the Gods, sacred with everlasting wood. Yet you yourself shall be known even in these parts and you will also be sung under our heavens, wherever through our song the Muses can make you travel by the lips of men.)[123]

The guaiac tree epitomizes the complex relationship not only between the *oikoumene* and the New World, between sameness and difference, but also between original and copy: the tree's uniqueness (its belonging to the vegetal world of the Caribbean) is paired with the Europeans' fantasy to reproduce it back home. While the Antilles were understood, with Oviedo, as a translation of the Hesperides, the guaiac tree, in contrast, is genuinely Caribbean in origin and, what is more, resists any movement of transplantation. The colonizers' desire to copy it in Europe is poetically rendered through the repeated use of the rhetorical figure of gemination, the doubling (literally, the "twinning") of words within a verse:

Arbore, *voce vocant* patrii sermonis Hyacum.
Ipsa teres, *ingens*que *ingentem* vertice ab alto
Diffundit *semper* viridem, *semper* que comantem.

(they call this in the sounds of their native speech Guaiacum. The tree itself is smooth and being huge spreads from its high top a huge canopy ever green.)[124]

The triple gemination, lost in the English translation, epitomizes the process of translation and transplantation in a gesture that firmly holds sameness and difference, the original and the copy, together. It inscribes in the very texture of the poetic line the oscillation between the tree's singularity and a desire to multiply it. Fracastoro's ingenious gemination discloses the iterative mode, a tension created through repetition and, in two instances, a slight alteration. The geminated cluster enveloping the theme of the guaiac tree unearths the poet's obsession to mark the very moment of bifurcation, when a poetic (and geographic) distinction between continuity and discontinuity, between sameness and difference, emerges. Altered through repetition, Fracastoro's gemination binds the indeterminacy of poetic iteration to the ambiguity generated by an excess of geographic liminalities. The Antilles and Hispaniola, in particular—simultaneously identified as the Hesperides and coterminous with Ophir, respectively—are generated through an overload of twinned toponyms which point to continental ambiguity.

Fracastoro's poetic gemination loudly echoes the process of colonization as double toponymic naming, whereby previously existing indigenous toponyms are violently overridden and European toponyms (motivated by the Bible, classical sources, and contemporary politics) are superimposed upon a newly colonized place in their stead. Suddenly, Fracastoro's poetic gemination gestures toward new "geminus orbes" (twin worlds), a term used by Pseudo-Seneca in the second century CE to describe the relationship between Imperial Rome and its overseas colony, Britain. In his epigram titled *Laus Caesaris*, these twin worlds are imagined as two dislocated and yet conjoined parts of a single political body separated by the "oceanus" which "medium venit in imperium" (was placed in the middle of the Empire).[125] While Pseudo-Seneca's reference to what is now the English Channel echoes the bounded space of the Mediterranean surrounded by the Roman Empire, Fracastoro's Atlantic is a more oblique reference to the expanding world of the Spanish Empire. Resisting a clear-cut movement of *translatio imperii*, it serves as a transformative pivot which does not reflect a given reality as much as it challenges, distorts, and inverts translated geographies.

TRANSLATION AND THE PHILOLOGY OF FAULT LINES

The transition from the second to the third book of *Syphilis* is a passage from the Gardens of the Hesperides cultivating the Golden Apples to the Caribbean groves where the shepherd Syphilus, together with the local population of Hispaniola, cultivates the guaiac trees while attending to his flock of sheep. What takes shape as the vast expanse of the Atlantic Ocean vertiginously implodes when looked at through the lens of philology: suddenly, the apples and sheep separated by the ocean collapse into a single Greek homonym: *mēlos*. In his *Rerum rusticarum libri tres*, the Roman writer Marcus Terrentius Varro argued that "ut in Libyam ad Hesperidas unde aurea mala id est secundum antiquam consuetudinem capras et oves [quas] Hercules ex Africa in Graeciam exportavit. Eas enim [a] sua voce Graeci appellunt mêla" (the Golden Apples of the Hesperides were, in reality, sheep and goats brought from northern Africa [to Greece]. The misunderstanding arose . . . through a confusion of the Greek μῆλον, or sheep, with the Latin malum or apple).[126] In *Syphilis*, the two extremities of the Atlantic are eclipsed through a philological ploy that, however, discloses its full strength exclusively in the original Greek language, which, to the readers of Fracastoro's neo-Latin poem, could have remained hidden. But what Fracastoro is aiming at, throughout *Syphilis*, is to get at the bottom of philological as well as geographic connectivity and to pursue both pathways as a joint and inseparable endeavor.

Navigating across the Atlantic, the Spanish crew sails past islands mythical, fictional, and real,[127] before reaching the safe harbor of Hispaniola, which Columbus, in his *Book of Prophecies* and in a letter to Pope Alexander VI from 1501, had already termed Ophir.[128] As she accompanies the sailors in their transatlantic voyage, the goddess Urania notices that Hispaniola is "fertile in gold, but made far richer by one tree—they call this in the sounds of their native speech Guaiacum."[129] Before the Spanish sailors set foot on the island they observe a flock of lovely birds flying in the air. Pointing their cannons at the birds, they start to shoot randomly and see the birds drop dead to the ground,[130] a metaphor, as Timothy Reiss shows, "for the places, victims, and perpetrators of colonial and imperialist aggression."[131] In turn, the sailors are cursed by a speaking bird, who prophesies that the Spanish dominion over the islands of the New World will not go unpunished: the new epidemic, syphilis, shall be a divine punishment for the invaders' colonial project.

Following the aggressors' belligerent landing and the subsequent curse by the prophetic bird, the captain of the fleet encounters the leader of the indigenous population, with whom he "joined hands upon the shore and secured their friendship" ("inter sese reges . . . / jungunt dextras, et foedera firmant").[132] While the word "foedus" (meaning "treaty" or "pact") is included twice in this sentence, emphatically underscoring the two kings' creation of a legal *confoederatio* based on equality, the guaiac tree instead appears as a "gift from heaven," disrupting the potentially smooth dynamics of exchange and the economy of gift giving between the two leaders. In the company of the local king, the sailors penetrate the island's dense and opaque vegetation (*opacis sylvis*) before arriving in a "convalle cava" (hollow valley), in the midst of the island's lush vegetation, in which the encounter between the newly arrived colonizers and the local population takes place.

Here, the king of Hispaniola introduces the Spanish sailors to the local festivities, where the European newcomers observe that the entire local population is "tristes" (melancholy)[133] due to the terrible epidemic afflicting them. Ophir's population, Hispaniola's king recounts to the amazed sailors, was punished, a traditional local legend has it, with syphilis because of the blasphemy committed by their ancestor Syphilus, a descendant from the sunken island of Atlantis, who "longa serie, de stirpe profecti" (issued from this stock in a long chain).[134] Syphilus was punished because he cursed God for inflicting the island with unbearable heat and decided to worship a king instead. He was sentenced to death but was ultimately saved by an act of sacrificial substitution: instead of executing Syphilus, the population of Hispaniola had to "offer a white heifer to mighty Juno" and "slaughter a black cow to the mighty Earth" ("niveam magnae mactate juvencam / Junoni, magnae nigrantem occidite vaccam / Telluri").[135] The sacrifice led to an annually performed rite, in which a member of Ophir's population reenacts the sacrifice of Syphilus. It is in this very moment of the performed sacrifice that Fracastoro introduces, for the first time, the neologism "syphilis." Hispaniola's hollow valley suddenly turns into a site where the origin of the disease and its effective cure, the sacred guaiac tree, converge. The cure for the venereal disease originates, in a manner akin to alchemy, in and through the heifer's putrid, corrupt, and decaying flesh, which Fracastoro claims morphs into the new, life-giving seeds of the salvific guaiac tree. Here, the site of the poet's act of name giving—the lexical matter of the death-bringing epidemic, what Lucretius would call the *textura rerum*—and the site of the life-giving plant collapse into the hollow valley which takes the shape of a continental interstice between Europe and America. The neologism syphilis emerges against the backdrop of liminal cartography, continental indeterminacy, and linguistic productivity.[136]

In one of the first medical treatises on the French disease, *De epidemia quam Itali morbum gallicum, galli vero Neapolitanum vocant* (1497), the physician Leoniceno claimed that "que quidem ambiguitas nominum & de re ipsa quoque dissensio multos suspicari fecit novam hanc esse luem nunquam a veteribus visam atque ideo a nullo medico vel Graeco vel Arabe inter alia morborum genera tactam" (the ambiguity of and the disagreement about the name and the thing [syphilis] itself led many to suspect that this epidemic was new and that the ancients had never seen it before, and hence no Greek or Arab physician had ever touched upon it [when discussing] other diseases).[137] In contrast to those voices, Leoniceno himself believed that a disease that on the surface appeared as new (such as syphilis) might well have been known by the ancients—but under different names. Less a medical text than a manual on philology, his treatise proposes that philological care and increased attention to lexical inquiry and the workings of translation could help unearth the hidden links between the ancient and modern medical nomenclature upon which, ultimately, both diagnosis and cure depend. For Leoniceno, medical mistakes predominantly originate in the "nominis ambiguitate" (ambiguity of the word) and the "corruptum vocabulum" (corrupt name),[138] contorted syllables and letters which make an even familiar terminology unrecognizable.

Fracastoro understood the word "syphilis" itself as a philological "luxation" of a previously existing word—a lexical limb dislocated by twists of letters across the transatlantic space—with a strong bearing on spatial transformation. In his poem, syphilis not only mobilizes philology, but paves new avenues for thinking about the specific locale of the origin of both the contagion and the cure at the intercontinental tension between Europe and the New World. At the end of the poem, the Spanish sailors return to Europe only to discover that "late (proh fata occulta Deorum) / Contagem Europae coelo crebrescere eandem, / Attonitasque urbes nullis agitare medelis" (widely—oh the hidden fates of the Gods!—this same contagion was spreading in the sky of Europe, and harassing cities dismayed by lack of remedies).[139] Fracastoro's coinage of the word "syphilis" at the very suture line between the *oikoumene* and the New World underscores the word's power to serve at once as a potent barrier and bridge intervening in and shaping a space at the tension between connection and detachment. The neologism emerges here as a *mise en abyme* of philological and geographic liminality. It marks a movement across (*trans-*), an "Übersetzung" (a word which in German denotes both translation and carrying across), not only across languages, but also across continents.

THE FANTASY OF MULTILINGUALISM: SYPHILIS AS *STILUS TRANSUMPTIVUS*

Fracastoro's coinage of the pseudo-Greek or pseudo-Latin word "syphilis," created in the tropical setting of the Caribbean islands, gestures toward the condition of possibility (and fantasy) of a dynamic multilingual space—a space in which Latin and Greek partake in a process of translation and contamination with novel local idioms (a route, however, not explored in *Syphilis*)—and the emergence of generic miscegenation, ranging from new forms of epics to travel accounts to alchemistic treatises. But the neologism also crystallizes as a token of linguistic imperialism and the desire to implant a classicizing nomenclature in the emerging New World.[140] The word syphilis takes on the shape of a unique linguistic marker that rouses the reader's curiosity about the geographic indeterminacy that unfolds throughout the poem and the role of language played in it. As a disruptive intervention—it was produced in a sacrificial moment in which death and birth coincide— the lexeme redirects the reader's attention to a different register, to a metalinguistic reflection on the status of poetic creation, translation, and textual and geographic transformation. "Syphilis" explicitly gestures toward poetic creation and geographic location; it is a figure reflecting upon itself and its new place. In ancient and medieval rhetoric, this self-reflective mode was known as *transumptio* or *metalepsis*, a figure intimately associated with multilingualism and the reflection upon language.

Fracastoro made *transumptio* one of the most prominent rhetorical figures in one of his lesser-known works on poetics, *Naugerius sive de poetica*, dedicated to his friend Ramusio.[141] The dialogue, between the personas of Fracastoro's friends Andrea Navagero and Gian Giacomo Bardulone, centers on the tension between the simultaneous subjugation to and mastery of language and poetics as experienced by the orator (*orator*) and the rhetorician (*rhetor*). Fracastoro introduces *transumptio* in a moment of linguistic and rhetorical tension, when Navagero and Bardulone negotiate the indeterminate space between linguistic mastery and subjugation to (a dominant) language as a reflection upon rhetorical, poetical, and linguistic mastery.

The master-orator, who has a complete command of words, knows which words are "propriae et impropriae, quae novae, quae antiquae, nostrates, peregrinae, . . . compositae, transsumptae" ("proper and improper, which are new, antiquated, domestic, foreign, . . . composite, transumptive").[142] The perfect rhetorician-orator is a philologist with a well-mapped vision of poetic delineations and limits; it is someone who penetrates and discovers the depths of language as well as the depths of the world and knows how to examine (*perpendere*), change (*mutare*), and transpose (*transponere*) the word's textures (*texturas vocum*). A master is someone, one is tempted to continue Fracastoro's thought, who knows when "plowing the ocean" ought to be taken literally.

According to Quintilian's *The Orator's Education*, *transumptio* (which he refers to as metalepsis) "involv[es a] change of meaning" (*aliter significant*). It serves as "a path from one thing to another. . . . It is the nature of Metalepsis to be a sort of intermediate step between the term transferred <and the thing to which it is transferred>; it does not signify anything in itself, but provides the transition."[143] *Transumptio* marks transformation, transition, and translation. Scholars have identified *transumptio* as a synonym for "plurilingualism."[144] For Leonard Barkan, *transumptio* is "the figure that renders into explicit consciousness the cultural activity of figuration and thus the existence of a diachronic cultural tradition as well as the gap across which figures may be constructed."[145] Denoting a "movement across," it "may mean either *traversing* or *skipping*."[146] *Transumptio* is a metatrope that serves as a pivotal figure between disparate terms, bridging dislodged times and places. Fracastoro's coinage of a neologism serves a similar purpose: mindful of the word's liminal location, the reader halts before the indeterminacy and translational, transformative power of a lexeme that can simultaneously straddle and divide continents. As an instance indicating bifurcation, *transumptio* is the ancients'—and Fracastoro's—equivalent of what I have termed syphilitic limology.

A text like *Syphilis*, written in a *stilus transumptivus* and showcasing a neologism with ramifications that far transcend its medical meaning, thus discloses the nature of the poet-rhetorician who creates a tension between the poem's visible surface and the invisible depths of the text's subterranean workings. It lays bare the boundaries, be they poetic, medical, or geographic, which the poet mobilizes in the creation of a text that, for the first time, sets out to apprehend the contours of what will become the New World. It insists on the very locus where differentiation emerges. The coinage of the word "syphilis" as an unprecedented example of *transumptio* captures the very moment of differentiation which, as Quintilian put it, "provides the transition." The word syphilis operates within a potent cartographic and poetic framework: it mobilizes its own invention as the philological equivalent of the sunken island of Atlantis, a metatrope that functions at once as boundary and bridge between dislodged times and places, between the Old World and the hesitantly emerging contours of America that, in the years Fracastoro was composing his poem, were believed to be at once a part of the *oikoumene* and a *mundus novus*.

CHAPTER FIVE

Cartographic Curses: Europe and the Ptolemaic Poetics of *Os Lusíadas* (1572)

In 1552, the Portuguese historian João de Barros published the first volume of his history of Asia, Africa, and South America, titled *Década primeira da Ásia* (*Of Asia*),[1] where he offers what is perhaps one of the most intriguing accounts of the power of maps. A map—what Barros refers to as a "painting" (*pintura*)—has the capacity to produce "shocking imagery" (*espantosa imaginação*), even trigger physical reactions. To look at the immensity of a coastline delineated on a navigational chart is nauseating to the mind's eye, Barros contends, especially if the onlooker happens to be less familiar with maps in the first place.[2] Barros was deeply disturbed by the impact a map can have on its users:

> E ainda a muytos [Mouros], vendo somente na carta de marear huma tão grande costa de terra pintada . . . fazia nelles esta pintura huma tão espantosa imaginação, que lhe asombrava o juizo. E se esta pintura fazia nojo á vista, ao modo que faz ver sobre os hombros de Hercules o mundo que lhe os poetas posseram, que quasy a nossa natureza se move com affectos a se condoer dos hombros daquella imagem pintada: como se não condoeria hum prudente homem em sua consideração, ver este reyno (de que elle era membro) tomar sobre os hombros de sua obrigação hum mundo, não pintado, mas verdadeiro, que as vezes o podia fazer acurvar com o gran peso da terra, do mar, do vento & ardor do sol que em sy continha.

> (Even for many [Muslims] this painting [map] produced such a shocking imagery that their judgment was obfuscated by simply beholding such a large, painted coastal line on the navigational chart. . . . And if the

sight of this painting [map] was nauseating—just as is the vision of the world which poets imagine to sit on Hercules's shoulders, so that we ought to be moved and sympathize with the painted image of Hercules's shoulders—how much more should one commiserate a prudent man in all his care, when he takes this kingdom [of Portugal] (of which he is a member) upon the shoulders of his duty, a world not painted, but real, which could easily weigh him down with the great weight of the earth, sea, wind, and heat of the sun contained in it.)[3]

The effect of the map on the human body is so powerful, Barros contends, that it provokes a physical reaction ranging from dizziness to compassion—just as if one felt, like Hercules, the weight of the world on one's own shoulders. What creates the map's uncanniness is its power to condense the immensity of the world into a single graphic image and to epitomize the evident but all-too-often-forgotten slippage between reality and representation, a world real and pained. Although Barros subsequently moves on to describe the responsibility of sailors and navigators as the true bearers of the kingdom's weight, the unsettling image of the map lingers in his readers' minds. The map "produces its own horizon of imagination" (*Vorstellungshorizont*), as Jörg Dünne characterized its power, and competes in significance and presence with the actual geographic experience.[4]

The uncanny slippage between direct and mediated geographic experience substantially determined the early modern European encounter with the new, the not yet familiar. A hallmark of early modernity, the map turned into a surface mediating between the onlooker and the world, a critical intervention determining the experience of the world. Scholars have pointed to the map's power to shape experience as a distinctive marker setting the Middle Ages apart from early modernity. While medieval travelers, scholars of geography contend, embraced the unfamiliar and the marvels of a journey directly as it unfolded, early modern voyagers experienced newness in a radically different perspective: the marvels of the unknown world were first and foremost those already featured on the map. Columbus's travels to the New World, for instance, were determined by contemporary maps (such as those created by Paolo Dal Pozzo Toscanelli) and the increasing authority of an abstract system of measurement and reckoning in which "geographic representation has already taken over the world, space has already encompassed and absorbed all places, and the map has already substituted that which it represents to the point of anticipating its nature and features, prefiguring its very existence."[5] Acquiring the skill, then, to read and navigate a map became the very precondition of apprehending, colonizing, and dominating the world in a time of territorial expansion.

No literary work demonstrates more forcefully the anticipatory function of the map and its power to prefigure the forms, shapes, and features of territories yet to be discovered than Luís de Camões's ten-canto poem, *Os Lusíadas* (Lisbon, 1572),[6] which tells the story of Vasco da Gama's successful circumnavigation of the Cape of Good Hope in 1497–98 and his subsequent crossing of the Indian Ocean ending in his arrival in India. The poet, soldier, and knight Luís Vaz de Camões (ca. 1524–80), who spent a part of his life in India, has been known, in Portugal and abroad, as the "Prince of the Poets"—and the "Prince of Spanish Poets"[7] after 1580, when Portugal lost its nationhood to Spain for a sixty-year period. In his "Essai sur la poésie épique" ("Essay on Epic Poetry") (1733), Voltaire portrayed Camões's poem as having "ouv[ert] une carrière toute nouvelle" (open[ed] up an entirely new course),[8] hinting, through the double meaning of "carrière" as "course" and "quarry," at the weight of Camões's groundbreaking work. In his historical narration of Gama's voyage, removed by over half a century, Camões relied on Barros's *Of Asia* and Fernão Lopes de Castanheda's *História do descobrimento & conquista da India pelos portugueses* (1559)—so much so that more unforgiving nineteenth-century critics have described Camões's poem as "mera prosa" (mere prose) contending that numerous passages amount to little more than simple translations from the historian Barros.[9] Since the work's publication, the two historical moments and narrative strands encapsulated in *Os Lusíadas*—the historical discourse of Gama's venture to India at the height of Portugal's territorial expansion with the establishment of the "Estado da Índia" (the Portuguese Overseas Empire), in 1505, and the nation's decline during Camões's own lifetime, exacerbated by the death of the young king Dom Sebastião in the Battle of al-Qasr al-Kbīr in 1580—have unfurled an incessant string of attempts at arresting the poem's meaning. Terry Cochran has quite suggestively termed the lingering question of the work's unity and the built-in discrepancy between the poem's means of representation and the content it represents, epitomized in the central canto by the appearance of the disfigured figure (*figura . . . disforme*)[10] of Adamastor, as "figural disparity."[11] But for Cochran, not only the work, but the author himself has become "the figure of modern figuration."[12] As I will argue in this chapter, the groundbreaking figuration Camões deploys in his poem cannot be dissociated from the increasingly imposing presence of the map—for which humanists used the Latin word "figura." Cartography occupies a privileged place within Camonian poetics as it dislodges the question of figuration from history and rhetoric to that of mapping.

CAMÕES'S "PAINTED WORLD"

In his monumental *Cosmos: A Sketch or a Physical Description of the Universe* (1845–62), the German scientist and explorer Alexander von Humboldt famously described Camões as "a great sea painter."[13] Camões's depiction of the circumnavigation of the Cape of Good Hope by Gama in 1498, the crossing of the Indian Ocean (with the often unacknowledged aid of a Gujarati Muslim pilot), Gama's subsequent arrival in the Indian city of Calicut, and his felicitous return to Portugal, Humboldt argues, heighten "the animated impression of the greatness and truth of the delineations."[14] Humboldt's assessment of Camões's detailed observation of the nature of the world's oceans, stemming from the latter's firsthand experience as "an active seafarer, not an armchair geographer,"[15] strongly affected the reading of *Os Lusíadas* (1572) in the centuries that followed. Camões is generally credited with having been a watershed poet, who replaced the traditional descriptions of the *oikoumene* known to the epic poets of ancient times with the newness of an increasingly globalized world shifting the emphasis from the Mediterranean world to that of the Indian Ocean. His preference for the naturalistic depiction of the sea is considered central to this shift: "Few other sixteenth-century instances of epic poetry," Bernhard Klein contends, "are so heavily invested in the culture of seafaring."[16]

Yet already Humboldt senses Camões's predilection for the more familiar world of the Mediterranean in the poet's imagining of the Indian Ocean, so deeply anchored in the vegetal world of the Mediterranean: "The whole poem bears no trace of graphical description of tropical vegetation, and its peculiar physiognomy. Spices and other aromatic substances, together with useful products of commerce, are alone noticed. The episode of the magic island certainly presents the most charming pictures of natural scenery, but the vegetation, as befits an *Ilha de Venus* (Isle of Love), is composed of 'myrtles, citrons, fragrant lemon-trees, and pomegranates', all belonging to the climate of Southern Europe."[17] But Humboldt identifies a divide between land and sea descriptions: while Camões's seascapes seem to voraciously incorporate the novelty of an ever-expanding world, his landscapes (here exemplified by the Isle of Love) appear trapped in the vegetal framework of familiar European epics centered on the Mediterranean. Scholars thus contend that Camões's emphasis on the aquatic element, his orientation of the poem toward the "open, uncertain spaces of the world's oceans" is an innovative break with the "enclosed Ptolemaic *orbis terrarum*"[18] of earlier epics, grounded in the Mediterranean space of Virgil and Homer.

However, a careful reading of the poem reveals a surprising paradox, which I shall investigate in greater detail: by rehearsing a rich inventory of rhetorical figures (ekphrasis most prominent among them), Camões designs the textuality of Os Lusíadas (in particular the poem's second half) not as an open space, as the unknown world of the Indian Ocean, but as a (Ptolemaic) map, a powerful idiom performing a twofold function: the transformation of the immediate experience of the Indian Ocean and India into a strategically mediated European gaze upon newly "disclosed" lands and the concomitant appropriation of cartographic mapping as a distinctive (and superior) tool in the hands of the European navigator-colonizer. Camões shows that the vigorous intervention of the (Ptolemaic) map, uncannily looming over Os Lusíadas, dictates the Portuguese navigators' perception of the Southern Hemisphere and the Indian Ocean. With ingenious poetic means turned into a consistently deployed cartographic imagery, Camões insists that the power to map the Southern Hemisphere has turned into the foremost key to its apprehension and domination.

Quite disturbingly, the capacity to master and comprehend the impact of cartography crystallizes here—as it does in Barros—as a distinctive feature of the Europeans, the Portuguese in particular. As Michael Wintle aptly puts it, "Eurocentrism was not only mirrored in, but was actually encouraged and facilitated by the cartography of the time."[19] If Barros—and after his death in 1570, Diogo do Couto—had aimed at articulating what Vincent Barletta has termed "a philosophy of Portuguese nationalism that revolve[d] in large part around Portugal's long and thorny relation to Islam,"[20] Camões's poetic project flips the underlying religious tension into a question of "carto-literacy": the century-old accomplishments of classical Arabic cartographers (intimately familiar with the Indian Ocean), which to early sixteenth-century Portuguese sailors must have served as a repeated reminder "of their comparative intellectual backwardness in the face of Islam's cosmopolitan world civilization,"[21] were eclipsed, and cartographic models popular in Europe advanced in their stead. While traditional definitions of carto-literacy emphasize the synchronous development of knowledge and cartography (maps and atlases),[22] Camões's poem is much more daring: it is perhaps the first to historicize cartography by aligning Gama's late fifteenth-century voyage to the cartographic models which circulated during his lifetime: Os Lusíadas, then, mobilizes not only the navigational chart (as scholars have heretofore contended) but also the zonal and, most importantly, the Ptolemaic map.

What emerges in Os Lusíadas is the idea that the Southern Hemisphere was colonized by the map before it was "discovered" by navigators such as Gama, the Portuguese captain major (*capitão-mor*). If Josiah Blackmore has pointed to the "metaphorics of Africa" that Camões develops in *Os Lusíadas*, whereby Africa becomes "one of the founding discourses of European imperialism,"[23] I mobilize here the idea of the "metaphorics of Europe": despite the continent's then uncertain and shifting contours, Camonian Europe is instrumentalized as a metaphor becoming a model to be exported and imposed upon other parts of the world. Detaching itself from the Northern Hemisphere, Europe is here transformed into a cartographic commodity serving, in the form of a map, as a transportable tool of colonization. What Camões repeatedly calls "proud Europe" (*Europa soberba*)[24] takes shape as cartography's power to superimpose images of itself upon the world. Concomitantly, "proud Europe," described in great geographic detail by Gama for the ears of his East African interlocutors, evokes the anthropomorphic image of Europe as a queen, advanced by Sebastian Münster in editions of his *Cosmography*, which circulated in the very years Camões was writing his poem. Here, "o Reino Lusitano" (the Lusitanian kingdom) serves not only as Europe's crowned head but also turns into an undisputed marker of a continental divide:

> Eis aqui, quase cume da cabeça
> De Europa toda, o Reino Lusitano,
> Onde a terra se acaba e o mar começa
> E onde Febo repousa no Oceano.
> Este quis o Céu justo que floreça
> Nas armas contra o torpe Mauritano,
> Deitando-o de si fora; e lá na ardente
> África estar quieto o não consente.

> (And here, as if crowning Europe's
> Head, is the Lusitanian kingdom,
> Where the continent ends and the sea begins,
> And where Phoebus reclines in the ocean.
> By Heaven's will she prospered
> Against the unworthy Mauritanians,
> Driving them out; and in their hot garrison
> In Africa has not ceased to harass them.)[25]

Os Lusíadas, written in a liminal moment in Portuguese history, when the kingdom's unquestioned dominance and colonial power was coming to a close, heralds, with a nod to Walter Mignolo,[26] the "dark side" of cartography and its enabling of the all-too-easy conflation, still operating today, of globalization and Europeanization.

When, in the poem's fifth and central canto, Gama makes landfall at St. Helena Bay (on the northwestern shore of the Cape of Good Hope), he sets foot in the Southern Hemisphere for the very first time. Gama's first concern is not to investigate the land and its inhabitants, but to look at the map—the only "'real' map [that] appears in the poem"[27]—hoping to acquire correct insight into the place:

> Desembarcamos logo na espaçosa
> Parte, por onde a gente se espalhou,
> De ver coisas estranhas desejosas,
> Da terra que outro povo não pisou.
> Porém eu, co'os pilotos, na arenosa
> Praia, por vermos em que parte estou,
> Me detenho em tomar do Sol a altura
> E compassar a universal pintura.
>
> (We went ashore at an open stretch,
> Where our men quickly scattered
> To reconnoitre this welcome land
> Where no one seemed to have ventured;
> But I, eager to know where I was,
> Stayed on the sandy beach with the pilots
> To measure the sun's height, and use our art
> To fix our bearing on the universal picture.)[28]

In order to locate himself, Gama does not turn to "experience" but, with a hefty dose of ostentation, to the authority of a "universal picture": to cartography, a familiar tool in his epistemological toolbox. From the midpoint of his journey, from the moment he enters the Southern Hemisphere—what Barros called "[o] caminho d'outro novo [mundo]" (the pathway to the new world)[29]—the captain major turns into a cartographer who reads the unfamiliar space through the authoritative interface of the map. In doing so, Camões transforms the description of the Southern Hemisphere (the focus of the second part of Os Lusíadas) into a set of cartographic representations. Gama's tarrying on the shore in St. Helena's Bay and his explicit refusal to explore the new land with his own eyes mark an instance of great import: from the second half of the poem on, the Southern Hemisphere is perceived and registered as a map, as a "universal painting" (*universal pintura*). Yet the "universal pintura," the epistemological shift toward cartography as a substitution for experience, occurs contemporaneously with the imposition of dominant European cartographic models as a "universal" idiom determining the representation of regions lying outside of Europe and displacing, as has been said, non-European conventions of mapping and apprehending the world. In the poem's second part, the scale of European and global imagery collapses.

Gama understands the Southern Hemisphere as an empty surface—a *tabula rasa*—that "appears uninhabited to him" (*inabitada a terra lhe parece*).[30] Sailing across the Southern Hemisphere with a European cartographic framework in mind, Gama's encounter with regions populated and societies in full bloom is akin to a gaze upon a map: instead of the physical ground he registers empty land, a "spacious part" (*espaçosa parte*), as he calls it, which stands ready to be operationalized according to the mapmaker's fancy—and thus colonized. In the course of his voyage across the Indian Ocean, therefore, nothing new awaits the captain major. His travel unfolds as what Farinelli has termed a preexisting "diagram within which the existent has already colonized all the forms of the future."[31] The omnipresent cartographic framework dictates Gama's navigation across the Indian Ocean and conditions the encounters between Europeans and unfamiliar peoples whom Camões recurrently refers to as "estranha gente" (strange people)—peoples, one is inclined to say, extraneous to the European map.

The predominant cartographic model behind Gama's voyage is not Barros's route-enhancing[32] (and thus potentially open-ended) "carta de marear," heretofore considered to be the central cartographic reference in *Os Lusíadas*, but—at first glance, paradoxically—the equipollent Ptolemaic map, among the most popular and widespread cartographic models in the years when Gama was rounding the Cape of Good Hope and sailing across the Indian Ocean. Jacques Le Goff described the Indian Ocean as a *hortus conclusus*, a bounded "repository of dreams, myths, and legends for the medieval mentality . . . an Eden in which raptures and nightmares were mixed."[33] Quite tellingly, from the time Ptolemy's *Geography* was brought to Italy in the last years of the fourteenth century to the end of the sixteenth century, cartographers across Europe reiterated Ptolemy's design of the Indian Ocean as a landlocked sea, as the popular 1482 Ulm edition of Ptolemy's *Geography*, almost contemporary with Gama's voyages, illustrates. In sixteenth-century editions of the *Geography*, two parallel (and mutually exclusive) versions of the Indian Ocean were published side by side: an "ancient" (*antiqua*) world map (enclosed Indian Ocean) and a "modern" (*moderna*) map (open Indian Ocean). Ayesha Ramachandran noted that in the course of the sixteenth century, Ptolemy's *Geography* manifested itself as "a peculiar mix of ancient and modern," with maps grounded in Ptolemaic projections "coexist[ing] uncomfortably with maps based on . . . the latest information gleaned from explorers and voyagers to distant places."[34] Such is the case of Sebastian Münster's popular *Cosmography*, which, from the *editio princeps* in 1544 until its last edition in 1628, knew forty editions in six languages,[35] turning into a standard geography manual typical of a transitional period in which Ptolemy was contemporaneously upheld and dismissed. Every edition published the two diverging depictions of the Indian Ocean (fig. 21 and fig. 22) next to one another. In directing his journey across the Indian Ocean with a Ptolemaic imagery in mind, then, Gama does not chart unknown maritime itineraries but rather (ful)fills the outlines and contours already predetermined by the map. With Ptolemy as a principal guide, the navigational support Gama receives from experienced African sailors *en route* is readily dismissed as unsubstantial or, in some instances, even treacherous. Ptolemaic cartography turns in the poem's second part into a commanding epistemological tool, a Deleuzian "table of information,"[36] which overrides local knowledge as it guides the reader's attention to what has already been projected and drawn on European maps.

Contrary to what *Os Lusíadas* makes its readers believe, the Indian Ocean, known to and traveled by Muslim (Mamluk and East African) and Chinese merchants throughout the Middle Ages, had been "disclosed" centuries before Gama undertook his voyages. Marco Polo described the Indian Ocean in his travel narrative, *Il Milione*,[37] and the Venetian monk Fra Mauro registered the busy international networks crisscrossing the Indian Ocean on his famous *mappamundi* (completed before 1459). With regards to the islands lying in the Indian Ocean, he writes that "there is a lot of new information that I do not give here. The islands lie between the Scirocco and the north-west wind—that is, across almost the whole of the Sea of India. The ships that sail here avoid going too close [to these islands] because of the obvious danger."[38] But these European accounts were preceded by a robust corpus of classical Arabic geographies, some illuminated, which had already widely described and mapped the Indian Ocean. Before the Portuguese (and the Ottomans) explored it, the Indian Ocean was dominated by the Egyptian Mamluks, who established a strong maritime presence with the purpose of expanding their spice trade. In contrast to their elaborate maps, which allotted ample space to the (mostly open) Indian Ocean, the

Figure 2.1. "Ptolemaisch general tafel begreiffend die halbe kugel der weldt" (Ptolemaic hemispheric map) in Sebastian Münster, *Cosmographey oder beschreibung aller länder . . .* (Basel: Heinrich Petri, 1564), n.p. [Typ. 565 64.584]. Photograph: Houghton Library, Harvard University.

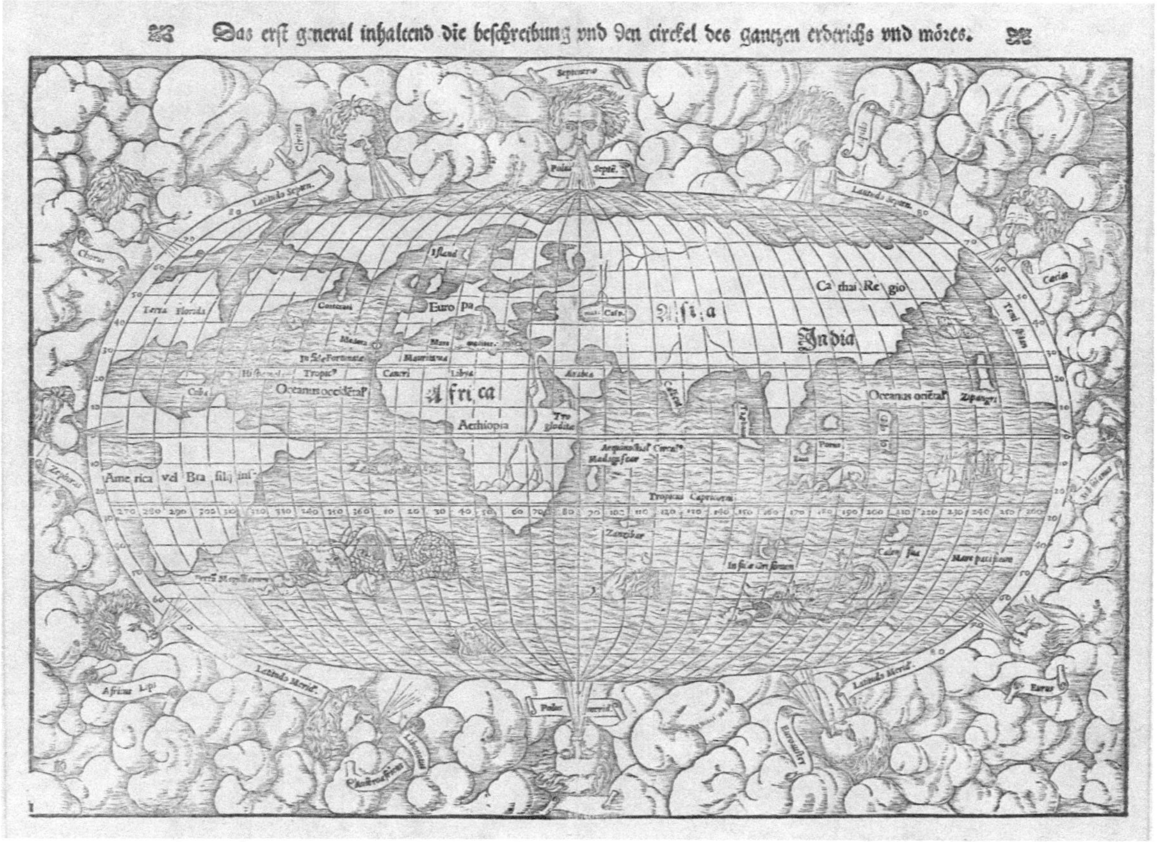

Figure 22. "Das erst general inhaltend die beschreibung und den circkel des gantzen erdtrichs und moeres" (modern world map), in Sebastian Münster, *Cosmographey oder beschreibung aller länder . . .* (Basel: Heinrich Petri, 1564), n.p. [Typ. 565 64.584]. Photograph: Houghton Library, Harvard University.

medieval European T-O maps, centered on the Mediterranean Sea, did not even foresee a space for the Indian Ocean: the *mare nostrum*, the Nile, and the Don feature here as the only bodies of water. As Giancarlo Casale put it, "the history of the discoveries is the story of Europe's emergence from the state of intellectual otherworldliness graphically represented by these *mappaemundi*."[39]

It might seem counterintuitive that Camões would mobilize the imagery of a landlocked Indian Ocean to tell the story of what he perceived as Gama's daring inaugural venture to India. But it is precisely through the image of an enclosed ocean advanced by the Ptolemaic framework that Gama acquires in *Os Lusíadas* an enhanced heroic status: it allows Camões to paint the Portuguese captain major as the first sailor disclosing what had hitherto been imagined as bounded. Gama's breaking through the currents at the southern tip of Africa, past the Cape of Good Hope, is a repeated physical effort to pass beyond the pillars (*padrões*) cast in 1488 by Bartolomeu Dias during his less-successful circumnavigation of the Cape. In his *História do descobrimento* Castanheda writes: "The fleet again passed along the same course that it had passed, taking advantage of a strong stern wind that lasted three or four days and with which it broke through the currents [*rompeo as correntes*] that they feared they would be unable to pass through. And thus they all proceeded, very pleased to have passed beyond where Bartolomeu Dias had reached, and Vasco da Gama pressed them on, saying that therefore God wanted them to find India."[40]

Dias and subsequent explorers such as Gama used a stone marker (*padrão*) to indicate the possession of new land. Deriving from the Portuguese word for stone (*pedra*), *padrão* was a cognate of *pater* (father), *patrão* (boss), and *patria* (fatherland). But it acquired a new meaning with the rise of cartography, when it came to signify a "model, pattern"[41] of a map. While Dias's voyage ended with the placing of a stone pillar east of the Cape of Good Hope, the gist of Gama's travel consists in his "breaking through the currents" (strategically placed in the center of the poem) and charging beyond the pillars. Camões paints Gama's breakthrough moment in clear analogy to the labors of Hercules, who, as Gama's closest comparand, separated the two rocks, Abyla and Calpe. Camões did not have to look back to antiquity (Pomponius Mela, in particular) to revive this Herculean image. Also his near-contemporary, the Italian poet Giovan Battista Giraldi Cinzio, mobilized the image of Hercules as the creator of a continental divide between Europe and Africa in great detail in *Dell'Hercole* (*On Hercules*) (1557). Giraldi Cinzio pauses before the very moment of Hercules's disclosure of the Mediterranean Sea, extended over several stanzas of his ottava rima:

> Or, giunto al fin del mare il forte Alcide,
> Per voler gir ne l'Oceano à Gade,
> Da un'altißimo monte chiuse vide
> A poter più oltre andar tutte le strade,
> Che il monte un mar da l'altro si divide
> . . .
> Così l'Europa, & l'Africa, che unite
> Eran, sin da principio, per natura,
> Furo per sempre allhora disunite,
> Tanta vi pose industria Hercole, & cura,
> Ne le parti del mondo dipartite,
> Di se lasciar memoria eterna cura,
> Et due colonne affige su quell'Alpe,
> L'una Abila chiamata, & l'altra Calpe.
>
> (Now strong Hercules arrived at the end of the sea,
> In Gades, and as he wanted to go toward the Ocean,
> He saw that all the pathways were blocked
> By a very high mountain which, dividing one sea from the other,
> Did not allow him to go beyond
> . . .
> Just as Europe and Africa were united,
> From the beginning by nature,
> They were forever disunited,
> By Hercules' industry and care,
> And are now divided parts of the world.
> He cared to leave an eternal memory of himself,
> And fixed two columns on that Alp,
> One called Abyla, the other Calpe.)[42]

Camões, cognizant of the powerful image of Hercules's creation of a continental divide, portrays Gama in analogous terms: as a Hercules of the Southern Hemisphere heroically opening the landlocked Indian Ocean. Just like Castanheda, Camões uses the verb "romper" (to break through) to mark Gama's physical effort in the act of disclosing "[o] liquid estanho" (the liquid pool) of the Indian Ocean:

> Assim, com firme peito e com tamanho
> Propósito vencemos a Fortuna,
> Até que nós no teu terreno estranho
> Viemos pôr a última coluna.
> Rompendo a força do líquido estanho,
> . . .
> A ti chegámos. . . .
>
> (And so, with steadfast hearts and great
> Ends in view, we conquered Fortune
> And reached your distant country
> To plant the last of our stone columns.
> Breaking the strength of the liquid pool
> . . .
> We have voyaged to your court.)[43]

Camões's choice of the word "estanho," derived from the Latin "stagnum," is a cartographically inflected pun: it signifies both tin (an alloy of silver and lead) and pool (an enclosed body of water).[44] By using the image of the breaking through of a "líquido estanho," Camões gestures toward Gama's strength in breaking not only the stillness of a formerly calm sea—whose surface resembles liquid metal—but also the bounded nature of the Indian Ocean. Once the sea is disclosed, Gama, a second Hercules, can proceed to cast his "last column" (última coluna).

The comparison Camões establishes between Gama and Hercules brings the Indian Ocean and the Mediterranean Sea into intimate proximity. *Os Lusíadas* indeed entertains a vision (reinforced by the analogy between Gama and Hercules) of the Indian Ocean as a reflection of the Mediterranean Sea. Here the Southern Hemisphere is transformed into a copy of its northern counterpart. In his *Cosmos*, Humboldt celebrated *Os Lusíadas* as a watershed moment in the transition from previous, *oikoumene*-centered epics to a truly global poem exploring the openness of the sea.[45] But, as this chapter goes on to show, the poem creates a much more complex, Janus-headed vision of the globe (and its continents) against the backdrop of the making of Europe: *Os Lusíadas* is probably the last poem in a long tradition of ancient and medieval epics that use Ptolemy's landlocked imagery of the Indian Ocean. At the same time, it is also among the first epic poems to use cartography as a powerful colonizing tool and to self-confidently export the idea of an autonomous and hegemonic Europe across newly "discovered" continents.

THE SUPERIOR AND THE INFERIOR SEA

The Mediterranean Sea and the Indian Ocean, the two great seas of Antiquity, have always been cartographically, culturally, and economically entangled—so much so that scholars have designated their vast shared space as the "Indo-Mediterranean."[46] Ancient and medieval geographers have termed the two seas in analogous terms, referring to them sometimes as the "superior" and "inferior"[47] and sometimes as the Western and Eastern seas.[48] Characterized by a specular relationship, the *mare nostrum* and the Indian Ocean have faced each other "in an ambiguous relationship of conflicted confrontation."[49] Most historical events related to the ancient world, Carlo Saccone estimates, can indeed be traced back to the "confrontation between the two seas and the various civilizations that emerged around them."[50] In *Of Asia*, Barros further tightens the geographic mirror image of the two seas by comparing "the islands of Ceylon and Sumatra to Italy's Sicily."[51]

Also in the Islamic cartographic tradition, both seas were often depicted as specular bodies of water, mapped almost geometrically around a pivotal axis. In the Qur'ān, the geographic hinge is referenced as the "confluence of the two seas" (*majma' al'bahrayn*), possibly alluding to the isthmus of Suez and to Moses, whose words are thus recorded in *sura* 18: "I shall not pause until I reach the place where two seas meet, even if I journey for years to come."[52] Medieval and early modern Islamic maps, such as Al-Idrisi's twelfth-century world map or the popular encyclopedia by the fifteenth-century Arab historian Ibn al-Wardī, titled *Kharīdat al-'ajā'ib wa farīdat al-gharā'ib* (*The Unbored Pearl of Wonders and the Precious Gem of Marvels*), illustrate the close connection and vivid exchange between the two seas by approximating their latitudes. Several manuscript maps, among others the fourteenth- to sixteenth-century south-oriented copy of the Al-Idrisi world map, stylize the Indian Ocean and the Mediterranean Sea as elongated and semi-bounded seas occupying almost the same latitude. Here, the designations Western Sea and Eastern Sea disclose their full geographic meaning, and the artistic manipulation with which the two seas are brought together hints at the political and cultural drive uniting those spheres under a real or desired Islamic rule.[53]

Some depictions of the Mediterranean Sea and the Indian Ocean in the medieval Islamic tradition of Istakhrī's tenth-century *Kitāb al-masālik wa-al mamālik* (*Book of Roads and Kingdoms*) acquired, over the course of the fifteenth and sixteenth centuries, an increasingly political and religious value, as becomes visible on the so-called Ottoman cluster maps. Karen Pinto observes, with regards to one of the earliest exemplars of these maps, that "the exaggeration of the distinctly menacing, hook-like end of the Indian Ocean—that is, the Red Sea—" appears "to have broken through the meager land barrier that separates it from the Mediterranean."[54] In the context of rising tensions between Muslims and Christians over dominion in the Indo-Mediterranean realm, this design may have alluded to the Ottoman desire to expand its imperial boundaries. "Such an adjustment to the Indian Ocean's outline," Pinto surmises, "would have been the perfect visual metaphor for Mehmed [II]'s territorial ambitions: the 'Muslim' Indian Ocean first menaces then overwhelms the 'Christian' Mediterranean."[55] But it is primarily with the Ottoman sultan Selim and his invasion of territories belonging to the Mamluk empire in 1516 that "the Ottoman Age of Exploration," to use Casale's terms, commenced and the Indian Ocean was "pulled . . . into the Ottoman orbit."[56]

This brief excursion into the realm of Islamic mapping and the subsequent rise of Ottoman explorations serves to provide a backdrop for *Os Lusíadas*, which was completed in September 1571, during one of the most tumultuous periods in early modern Mediterranean naval warfare, when the island of Cyprus was fiercely embattled by the antagonistic forces of the Christian Holy League and the Ottoman Empire. Refusing to cede the island to Selim II, the Venetians saw Cyprus invaded by the Ottomans in 1570—and subsequently liberated, from the Holy League's perspective, in the dramatic and highly symbolic Battle of Lepanto on October 7, 1571, only one month after Camões entrusted his manuscript to the press.[57] The completion of *Os Lusíadas* coincides with the time of a heightened naval war, in a moment of utmost suspense shortly prior to the victory of the Holy League. The victory that ensued became emblematic for Europe's sense of superiority and emboldened its poets to craft boisterous epics and songs about the victory of Christian troops over the Turkish fleet—before the latter was back to full strength not long after.

The Battle of Lepanto triggered a considerable poetic response internationally, among others by the Portuguese poet, soldier, and painter Jerónimo Corte-Real. In his epic poem *Felicissima victoria*, published in Spanish in 1578—with large portions completed as early as 1569, thus prior to the publication of *Os Lusíadas*[58]—the Battle of Lepanto symbolized a significant victory of Christianity over the Ottoman Turks. The site of the battle, Cyprus, was a historical *pomme de discorde* between the Portuguese and Ottoman armies and took on, in both Corte-Real's and Camões's poems, the allegorical shape of the Cyprian goddess Venus, who intervenes with specific missions. In *Felicissima victoria*, Venus shows a shield to the sleeping Don

Juan, the brother of Phillip II, as she foretells him his future victory. In *Os Lusíadas*, Camões transposes Venus from her native island in the Mediterranean to the Indian Ocean, where she becomes crucial for preparing, in canto nine, not a shield, but an entire island: the Isle of Love, where the love-hungry and homebound sailors will find a repose. The Cypriot goddess serves as a shuttle connecting not only Portugal's two most eminent epic poets, but also, in Camões, the Mediterranean Sea and the Indian Ocean.

The comparison between the Mediterranean Sea and the Indian Ocean *Os Lusíadas* establishes with the figure of Venus unfolds as part of a more elaborate symmetry between the Northern and the Southern Hemispheres. Since Herodotus, geographers, philosophers, and poets had been contemplating the possibility of the existence of a Southern Hemisphere beyond the "torrid zone" around the equator, named "antipodes" or "antichthones." It was up for debate if the Antipodes existed (Augustine rejected the idea) and, if so, whether the (potential) inhabitants of the other hemisphere were similar to the people inhabiting the *oikoumene*. Eratosthenes, whose fragments constitute the first extant geographic writings from ancient Greece, used the adjective "antipodes" to imagine "a strict symmetry across the equatorial plane."[59] In contrast, Strabo and the Stoics assumed that if the Southern Hemisphere is inhabited at all, its inhabitants must be "entirely dissimilar."[60] When Barros describes how Portuguese sailors set out "to enter the path to the other world,"[61] he mobilizes a hemispheric imagery which Camões pursues under the aegis of his complex cartographic framework. Camões was familiar with the ancient and medieval idea of the Antipodes—he explicitly refers to them in canto eight:

> Mas já a luz se mostrava duvidosa,
> Porque a alâmpada grande se escondia
> Debaixo do Horizonte e, luminosa,
> Levava aos Antípodas o dia,
> Quando o Gentio e a gente generosa
> Dos Naires da nau forte se partia,
> A buscar o repouso que descansa
> Os lassos animais, na noite mansa.

> (But now in the divided light of dusk
> as the sun's great beacon sank
> slowly beneath the horizon, to dawn
> in the antipodes with another day,

> The Catual and the noble company
> Of Nairs took leave of the warship,
> Seeking that rest which furnishes respite
> To weary creatures in the peace of night.)[62]

What here undergirds the idea of the existence of the Antipodes is the division of the world in zones. Cicero's *Republic* and subsequently Macrobius's *Commentary on the Dream of Scipio*—the "first and perhaps most important source on geography known directly and widely during the Middle Ages"[63]—took up the idea of a north-south division of the globe into climatic zones and a Northern and Southern Hemisphere. The idea persisted long into the early modern period and is firmly anchored in Camonian poetics. When Gama describes Europe to the population of Africa's eastern coast, he deploys an image of the world that dovetails with a Macrobian zonal map:

> Entre a Zona que o Cancro senhoreia,
> Meta Setentrional do Sol luzente,
> E aquela que por fria se arreceia
> Tanto, como a do meio por ardente,
> Jaz a soberba Europa, a quem rodeia,
> Pela parte do Arcturo e do Ocidente,
> Com suas salsas ondas o Oceano,
> E, pela Austral, o Mar Mediterrano.
>
> (Between the zone governed by Cancer,
> The bright sun's most northern track,
> And that region shunned for cold
> As much as the equator is for heat,
> Lies noble Europe, confronting both
> The Arctic and the Occident. To the north
> And west the Atlantic is its boundary,
> And to the south the Mediterranean Sea.)[64]

When in canto seven Gama meets the Samorin, the king of Calicut, the Indian city is described as the "head of the [Indian] Empire" (*cabeça de Império*) in analogy to Portugal, framed as the "head crowning all of Europe"[65] (*cume da cabeça / De Europa toda*):

> Aqui de outras cidades, sem debate,
> Calecu tem a ilustre dignidade

De cabeça de Império, rica e bela;
Samorim se intitula o senhor dela.

(Here, of all other towns, Calicut
Is undisputed head, beautiful
And prosperous, a city to glory in;
Its ruler is known as the Samorin.)[66]

Yet the apparent symmetry between the two hemispheres belies a built-in hierarchy: the constant use of ekphrastic devices, repeatedly deployed in the poem's second part, poetically encodes the Southern Hemisphere as a projection and replica of the Northern Hemisphere. Seen through the eyes of a cartographer and the medium of a map, the newly discovered maritime space of the Indian Ocean is not crafted as new and different but, on the contrary, as a copy of the familiar Mediterranean, and more specifically European, space. In *Of Asia*, Barros had advanced an extravagant geographic analogy between the numerous islands of the Indian Ocean that, taken together, would constitute a "body" so great that it would exceed the size of Europe: "as nossas navegações, a conquista daquella parte, a que propriamente chamamos Asia, não se contém sómente na terra firme, . . . mas ainda comprehendem aquellas tantas mil Ilhas a esta terra de Asia adjacentes, tão grandes em terra, e tantas em numero, que sendo juntas em hum corpo, podiam constituir outra parte do Mundo, maior do que he esta nossa Europa" (our navigations, the conquest of that part [of the world] which we appropriately call Asia, does not limit itself to the mainland . . . but also includes those thousands islands adjacent to the land of Asia which are so great in size and so numerous that, if joint into a single body, they could constitute another part of the World, greater than this our Europe).[67] Here, the islands of the Indian Ocean, still fragmented, will become comparable to the (mapped) continent of Europe once they are imagined as a whole, mappable, body.

Os Lusíadas, rather than being the first global epic oriented entirely toward the open Indian Ocean, looks back and presents the unknown hemisphere as a mirror image of the Northern Hemisphere. The body of Asia is constituted in analogy to the body of Europe. "Europa soberba," as Camões recurrently characterizes the European continent, becomes a model to conceive of the Southern Hemisphere—and of the Indian Ocean, in particular. *Os Lusíadas* proceeds by developing a poetics that aims at intertwining a hierarchical vision of the superior and inferior seas while carefully calibrating the Northern and Southern Hemispheres in a relationship of comparability. Camões does so through the use of artful and subtle inversions, variations, and chiastic structures. A powerful case in point is the succinct description of the Portuguese shoreline which separates land from sea as it marks Europe's territorial limits:

> Eis aqui, quase cume da cabeça
> De Europa toda, o Reino Lusitano,
> Onde a terra se acaba e o mar começa.
>
> (Here you see, as if crowning the head
> Of all of Europe, the Lusitanian Kingdom,
> Where the land ends and the sea begins.)[68]

The image of a territorial dividing line is mimetically rendered by Camões's use of a single poetic line: "Onde a terra se acaba e o mar começa." Camões replicates this line twice, in cantos five and eight, when he describes the Southern Hemisphere: in both cases he introduces an almost imperceptible variation. In canto eight, the Samorin describes the Portuguese as coming from a country "Donde a terra se acaba e o mar começa" (from / Where the land ends and the sea begins).[69] Cunningly, Camões marks repetition while alluding to hemispheric difference by using a different phoneme, "d," the smallest unit of distinction: the genitive "whence," "from where" (*donde*) is substituted for the local adverb "where" (*onde*). This poetic line is also inserted, in another slight variation, in the central canto five to indicate the transition from the Northern to "the new hemisphere" (*novo Hemisfério*):

> Vimos a parte menos rutilante
> E, por falta de estrelas, menos bela,
> Do Pólo fixo, onde inda se não sabe
> Que outra terra comece ou mar acabe.
>
> (We saw new heavens, less sparkling
> And, for lack of stars, less beautiful
> Nearing the pole, where no one apprehends
> If another land begins or the sea ends.)[70]

The moment of hemispheric transition, strategically positioned in the poem's central canto, takes the shape of a complex and condensed poetic structure. Dominated by a set of negations—"menos," "falta," "não"—the

Southern Hemisphere is presented as an inversion—or a "negative" version, an echo or copy—of its northern counterpart. The poetic line with which Camões had drawn the continental contours of Europe in canto three is here transformed in a twofold manner. What had been a geographic affirmation in canto three is here turned into a doubtful hypothesis about the geographic contours of the Southern Hemisphere. Far from having the geographic knowledge to assess territorial delineations, Camões's Gama blurs the very distinction between land and sea. What is more, Camões uses the rhetorical figure of antimetabole, a form of chiastic inversion, to transform the first occurrence of "Onde a terra se acaba e o mar começa" into a poetic *rochade*: "Que outra terra comece ou mar acabe." With regard to canto three, the two verbs "to begin" and "to end" are meaningfully exchanged, thus gesturing toward hemispheric "inversion," accentuated by the subjunctive mode, which ushers in a hypothetical key (as opposed to the indicative used for the description of the Northern Hemisphere). If the Greek word "μεταβολή" (*metabolē*) signifies "change" by conjuring up the vivid image of "throwing beyond" (*meta-boleō*), "antimetabole" complicates this image further by introducing a counterpart (anti-) that suggests geographic inversion through the maintenance of syntactic parallelism and linguistic rigor. Despite the change of hemispheres, Gama and the Portuguese sailors still hold on to the image of antipodic symmetry.

As *Os Lusíadas* unfolds, the poem generates a set of hierarchies that critically determine the relationship between the Northern and Southern Hemispheres. Following the Latinate etymology of "inferior" (the word "inferus" is related to the underworld), Camões rehearses the vast semantic field of "inferior" that ranges from the Antipodes (8.44) to the Acheron (1.51). Manuel de Faria e Sousa, probably Camões's most famous commentator, was the first to detect multiple "lugares de Dante"[71] (Dantesque places) and to compare specific episodes of *Os Lusíadas* to the *Inferno*'s "punitive and Dantesque universe."[72] Steeped in a paradigm that conceives of the Southern Hemisphere as an underworld, Camões conflates the horizontal and vertical axes of "inferus" when he describes Gama's voyage across the Indian Ocean in canto six: here, the lively image of Bacchus, the antagonist of the Portuguese, diving deep down undersea to visit Neptune in his underwater palace, described through an extended ekphrasis, is substituted for narration of the actual maritime voyage:

No mais interno fundo das profundas
Cavernas altas, onde o mar se esconde,
Lá donde as ondas saem furibundas

Quando às iras do vento o mar responde,
Neptuno mora e moram as jucundas
Nereidas e outros Deuses do mar, onde
As águas campo deixam às cidades
Que habitam estas húmidas Deidades.
. . .
As portas de ouro fino, e marchetadas
Do rico aljôfar que nas conchas nace,
De escultura formosa estão lavradas,
Na qual do irado Baco a vista pace.

(In the deep chambers of the innermost
Vaulted caverns where the sea retreats,
There, whence the waves leap in fury
When the sea responds to the winds' challenge,
Is Neptune's home, and the cheerful
Nereids', and other gods the ocean
Recognizes, granting its damp deities
Enough space for their underwater cities.
. . .
The doors were of gold, richly inlaid
With those pearls that are born in shells,
And were worked with gorgeous carvings
On which angry Bacchus feasted his gaze.)[73]

Bacchus's dive under the Indian Ocean, dramatically rendered by repeated movements of undulation phonetically inscribed in the poetic lines ("fundo," "profundas," "onde," "esconde," "donde," "ondas," "furibundas"), at once mirrors and inverts Venus's dive under the Mediterranean Sea in the eighth canto of Corte-Real's *Felicissima victoria*: while these two bodies of water are, once again, imagined as a mirror-image, the scope of the underwater descent is diametrically opposed: in *Felicissima victoria*, Venus visits Neptune's underwater abode with the goal to implore him to quell, not provoke, the sea storm: "Venus se quexa a Neptuno su hermano, el manda a Triton que vaya al cabo de las Columnas y aplaque el furor y violencia delas ondas" (Venus complains [about the storm] to her brother Neptune, who sends Triton to go to the end of the Pillars and quell the furor and violence of the waves);[74] a succinct summary announces the canto. In *Os Lusíadas*, the presence of Venus blurs the boundary between the Mediterranean Sea and the Indian Ocean. When Venus sets out, in canto nine, to create the Isle of Love, she exclaims: "Quero que haja no reino Neptunino, / Onde eu nasci, progénie forte e bela" (I wish to populate Neptune's realm / Where I was born, with strong and beautiful progeny).[75] Here, Camões brings the horizontal and vertical axes together as he condenses the Mediterranean Sea (Venus's birthplace) and the Indian Ocean, with an allusion to the underworld, in just two lines. But Neptune's underwater realm is not the sole image of the underworld that is correlated with the Southern Hemisphere. When Gama reads the carvings, statues, and images displayed in an Indian temple, rendered as a series of ekphrases, they are referred to as "infernal," as the work of "the Devil" (Demónio):

Ali estão das Deidades as figuras,
Esculpidas em pau e em pedra fria,
Vários de gestos, vários de pinturas,
A segundo o Demónio lhe fingia.
Vem-se as abomináveis esculturas,
Qual a Quimera em membros se varia
Os cristãos olhos, a ver Deus usados
Em forma humana, estão maravilhados.

(Inside were images of their gods
Fashioned in wood and cold stone,
The faces and colours as discordant
As if the Devil had devised them.
The carvings were repulsive, like
The Chimera with its different members;
The Christians, used to seeing God portrayed
In human form, were baffled and dismayed.)[76]

In order to make sense of the images, Gama undertakes the work of translating what are for him the unfamiliar deities of Indian mythology and Hinduism into the familiar framework of the Mediterranean, in particular Egyptian, mythology:

Um na cabeça cornos esculpidos,
Qual Júpiter Amon em Líbia estava;
Outro num corpo rostos tinha unidos,
Bem como o antigo Jano se pintava;
Outro, com muitos braços divididos,
A Briareu parece que imitava;

Outro fronte canina tem de fora,
Qual Anúbis Menfítico se adora.

(One [deity] had horns protruding from its head,
Like Jupiter Ammon of Libya;
Another had two faces on one trunk,
As Janus was shown in Roman times;
Another had so many different arms
It seemed modelled on Briareus;
Yet another had a dog's proboscis
Like the idol of Anubis at Memphis.)[77]

What takes shape in the latter half of *Os Lusíadas* is an uncanny co-presence—part superimposition, part juxtaposition—of the Mediterranean Sea and the Indian Ocean, not only through the translational process by which the unfamiliar Southern Hemisphere is uprooted and inserted within the familiar framework of the Northern Hemisphere, but also through the juxtaposition of different cartographic models. Written in a watershed moment, Camões's poem anticipates what Michel Foucault has described in reference to the modern period, as an "epoch of juxtaposition, the epoch of the near and far, of the side by side, of the dispersed."[78] The two bodies of water, central to the imagery of the Northern and Southern Hemispheres, do not overlap smoothly or seamlessly, but produce uncomfortable twin images of original and double; truth and falsehood; and reality and fiction, which organize and subtend the poem.

CARTOGRAPHIC CURSES

While Camões's rich cartographic imagery draws from different sources throughout the poem, the groundbreaking moment when Ptolemaic cartography prominently enters the scene and takes over his poetics is Gama's sighting of the monstrous Adamastor, the anthropomorphized representation of the Cape of Good Hope at the southern tip of the African continent. As David Quint has suggested, "the middle of Camões's poem enacts the twofold dynamics of the typical narrative middle, that indeterminate space where the repetition that constitutes narrative either may become purely, compulsively repetitive and hence collapse back upon itself or may move forward, repeating with difference, toward a predetermined goal."[79] Raising his monstrous body in front of the Portuguese sailors in that pivotal "indeterminate space" of the poem, in the middle of canto five, Adamastor embodies the critical juncture of mere repetition and repetition with difference in a powerful cartographic key. The first sentence he utters rehearses a catalog of names of ancient geographers unfamiliar with the Cape, while, in contrast, the following sentence reveals his own function as South Africa's Promontory:

> Eu sou aquele oculto e grande Cabo
> A quem chamais vós outros Tormentório,
> Que nunca a Ptolomeu, Pompónio, Estrabo,
> Plínio e quantos passaram fui notório.
> Aqui toda a Africana costa acabo
> Neste meu nunca visto Promontório,
> Que para o Pólo Antárctico se estende,
> A quem vossa ousadia tanto ofende!
>
> (I am that secret and vast promontory
> You Portuguese call the Cape of Torments,
> Which neither Ptolemy, Pomponius, Strabo,
> Pliny, nor any authors had knowledge of.
> Here I end Africa's entire coast
> In my never seen Promontory,
> Which extends toward the Antarctic Pole
> And which your daring so much offends!)[80]

Adamastor's body—now a Promontory—assumes the cartographic task to mark the tip of the African continent at the confluence of the Atlantic and Indian Oceans. And yet, there is a geographic paradox encapsulated in what might appear, at first glance, as Adamastor's faulty cartography: his body at once marks the "Stormy Cape," what would become the Cape of Good Hope, while simultaneously extending, in contradiction to his previous claim, toward the Antarctic Pole. Adamastor's body indicates continental completion while alluding toward continuation. It is monstrous because it conflates, I argue, different cartographic templates that, when considered together, produce conflicting (if not mutually exclusive) geographic images—just like the two mutually exclusive versions of the Indian Ocean in Münster's *Cosmography*, at once enclosed and open. When brought together, the foundational geographic figures from Greek and Roman Antiquity that Adamastor conjoins in a boisterous gesture of geographic knowledge—Ptolemy, Pomponius (Mela), Strabo, and Pliny—indeed produce a cartographic monstrosity. Adamastor's body is a container of disparate and contradictory geographic knowledge, announced already in his very first words, that resists the representation of intact and coherent geographic contours. If for Cochran Adamastor epitomizes Camões's "question of figuration,"[81] it does so by vigorously establishing cartographic figuration as the very basis of poetic figuration and by gesturing toward the etymology of "figura"—a noun used, in Latin antiquity, alongside "forma," "orbis pictus," or "orbis terrarum descriptio,"[82] to designate a map. Adamastor's reference to Ptolemy and his insistence on the potential continuity of the African coast toward the Southern Pole usher in a cartographic image that dramatically informs the latter half of the poem: the possible figuration of the Indian Ocean as a bounded body of water[83] and Gama as a Herculean hero disclosing a maritime space that Ptolemaic maps imagined as a landlocked sea. In doing so, Camões equally points to the complexity of cartographic representation always composed of different models stemming from a variety of epistemological frameworks.

The rhetorical lexicon and poetic toolbox deployed to describe the Indian Ocean, too, conjure up images of an enclosed body of water. When the Muslim sailors along Africa's eastern coast, intimidated by the uproar coming from the approaching Portuguese ships, jump into the water, Camões compares them to frogs jumping into a pond:

> Assim como em selvática alagoa
> As rãs, no tempo antigo Lícia gente,
> Se sentem porventura vir pessoa,
> Estando fora da água incautamente,
> Daqui e dali saltando (o charco soa),
> Por fugir do perigo que se sente,
> E, acolhendo-se ao couto que conhecem,
> Sós as cabeças na água lhe aparecem:
>
> Assim fogem os Mouros; e o piloto,
> Que ao perigo grande as naus guiara,
> Crendo que seu engano estava noto,
> Também foge, saltando na água amara.
>
> (As in a pond deep in the countryside,
> Frogs, those one-time people of Lycia,
> If they happen to be out of the water
> When they sense someone approaching,
> From here, there, and everywhere plop
> Back where they feel safest,

Dissolving in the element they know,
But above the surface their heads still show;

So the Muslims scampered; and the pilot
Too, who took the ships into danger,
Believing his plot was discovered,
Fled by jumping in the brackish water.)[84]

Stemming from Greek antiquity, the frog simile is evocative of Aristophanes's play *The Frogs* and Plato's *Phaedo*. In *The Frogs*, the god Dionysus—the Greek term for Bacchus, the discoverer of India, sung by the Greek epic poet Nonnus in the longest-surviving poem from Antiquity, *Dionysiaca*[85] (ca. 400 CE), and fierce antagonist of the Portuguese sailors in *Os Lusíadas*—descends to the underworld where he engages in a mock debate with a chorus of frogs while traversing Lake Acheron. With the underworld as their home, the frogs are not cognizant of the existence of an upper world. In *Phaedo*, Socrates describes the *oikoumene* as a metonymy of the world at large: the earth, Socrates contends, "is a thing of enormous size and we inhabit a small portion of it, from the Phasis to the Pillars of Heracles, living around the sea like ants or frogs around a pool."[86] While Aristophanes and Plato both find the frog simile helpful to allude to the Mediterranean dwellers' ignorance of another, larger world, Camões strategically transposes the frog simile from the Northern to the Southern Hemisphere with the purpose to strengthen the idea of the Indian Ocean as an "inferior" landlocked body of water and to dislodge the lack of knowledge about the greatness of the "real" world from the Mediterranean region to the Indian Ocean.

The image Adamastor advances of a coastal body extending toward the southern pole signals to the Portuguese captain that his task of "opening" the new sea is not exclusively a navigational metaphor, but truly a cartographic process of disclosing what was thought to be enclosed according to centuries-long Ptolemaic cartographic knowledge.[87] Gama's pioneering voyage is matched by Camões's corrective rewriting of cartography. When Gama encounters the Muslim populations inhabiting the Indian Ocean rim, he proudly exclaims: "Do mar temos corrido e navegado / Toda a parte do Antárctico e Calisto, / Toda a costa Africana rodeado; / Diversos céus e terras temos visto" (We have crossed and navigated / The entire part of Antarctica and the Great Bear / We have rounded the coast of Africa / Seeing different heavens and lands).[88] Gama's act of opening up a new body of water is in tune with the cartographic ruse used by Martin Waldseemüller, a few years after Gama's voyage, in his Ptolemaic world map of 1507 (fig. 23): here, Africa's

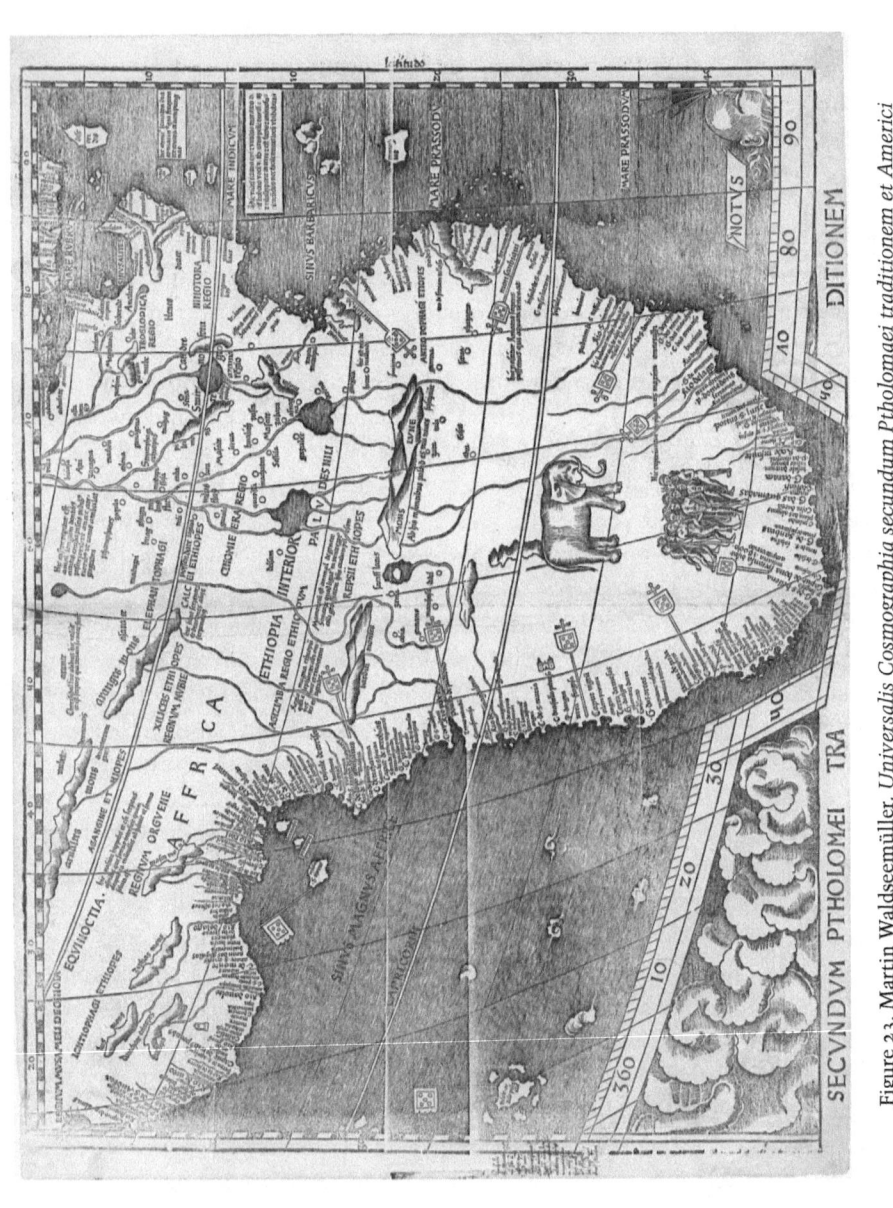

Figure 23. Martin Waldseemüller, Universalis Cosmographia secundum Pthtolomaei traditionem et Americi Vespucii aliorumque lustrationes (St. Dié, 1507), detail of South Africa [LC-G3200 1507.W3]. Photograph: Courtesy of the Library of Congress.

southern coast is disconnected from the South Pole only through the drastic intervention of the mapmaker, who creates the necessary pathway into the Indian Ocean only by "breaking through" the map's frame.

Adamastor's curse against Gama, then, is tied to the captain major's forceful manipulation of the Ptolemaic vision of the Indian Ocean. Gama's transgression is cartographic: he disturbs an authoritative model, valid since antiquity, which advanced the notion of the bounded nature of all seas:

> Pois os vedados términos quebrantas,
> E navegar meus longos mares ousas,
> Que eu tanto tempo há já que guardo e tenho,
> Nunca arados de estranho, ou próprio lenho:
> . . .
> E da primeira armada, que passagem
> Fizer por estas ondas insofridas,
> Eu farei de improviso tal castigo,
> Que seja mor o dano que o perigo!
>
> (Because you have breached the forbidden boundaries,
> Daring to navigate my lengthy seas,
> Which I have protected and kept for so long
> And no ship, foreign or one's own, has ever plowed,
> . . .
> The very first armada that attempts
> To plot a passage through these uncharted waves,
> I punish quickly and instantaneously
> So that damage will be greater than danger.)[89]

Adamastor admonishes Gama to keep the Ptolemaic image of the Indian Ocean intact and not to "breach the forbidden boundaries"—an unequivocal allusion to the Pillars of Hercules in the Strait of Gibraltar, described by Dante in canto 26 of the *Inferno*: "Ercule segnò li suoi riguardi / acciò che l'uom più oltre non si metta" (Hercules marked off the limits / warning all men to go no farther).[90] Hercules has supreme control over the western limits of Europe, be it through his act of foreclosing travel beyond the Pillars or of opening the previously landlocked sea through the separation of Abyla and Calpe. Repeating Hercules's daring gesture of manipulating continental divides, as Gama sets out to do, will have to be punished by a curse.

Adamastor's body is a capacious absorber of past cartographic forms as it is a generator of new ones. In its paradoxical blend of the two extremes of form and formlessness, it contains all imaginable shapes. His body, at once "disforme" (formless) and "uma figura . . . robusta e válida" (a robust and strong figure),⁹¹ raises itself as an emblem of disparate figuration. Adamastor's formless form, a monstrosity often taken for melancholy, stands for "the far-reaching transplantation of matrices of political, linguistic, and cultural patterns of order that lies at the root of all expansionist endeavors."⁹² But rather than "stand[ing] for the place where maps lose their potency,"⁹³ as Lawrence Lipking put it, Adamastor represents the vigorous confluence of accumulated cartographic knowledge, past and present. Adamastor's excessive association with soil, ground, and earth—"Medonha e má e a cor terrena e pálida; / Cheios de terra e crespos os cabelos, / A boca negra, os dentes amarelos" (Its complexion earthy and pale, / Its hair frizzy and matted with clay, / Its mouth coal black, teeth yellow with decay)⁹⁴—is an allusion to the very origins of Portuguese (and European) colonization, narrated by Gomes Eanes de Zurara in his *Crónica da tomada de Ceuta* (*Chronicle of the Conquest of Ceuta*, 1450), when the Portuguese embarked on a colonizing mission along the west African coast and were met with resistance from the local Muslim population. In one episode, Zurara describes how during the siege of Ceuta a Portuguese colonizer, Vasco Martins, was hit on the head by a stone or a piece of earth⁹⁵ launched by a "mouro grande e crespo todo nu" (tall and frizzy Moor, all naked) whose body was "todo negro assim como um corvo e os dentes mui grandes e alvos e os beiços mui grossos e revoltos" (all black as a crow and his teeth very big and white, and his lips very thick and protruding).⁹⁶ But Adamastor's telluric appearance also articulates itself as a liminal marker of the underworld:

> Tão grande era de membros que bem posso
> Certificar-te que este era o segundo
> De Rodes estranhíssimo Colosso,
> Que um dos Sete milagres foi do mundo.
> C'um tom de voz nos fala, horrendo e grosso,
> Que pareceu sair do mar profundo.
>
> (So towered its thick limbs, I swear
> You could believe it is a second
> Colossus of Rhodes, that giant
> Of the ancient world's seven wonders.
> It spoke with a coarse, gravelly voice
> Booming from the ocean's depths.)⁹⁷

Monstrously present, Adamastor joins the disjointed bodies of the seas with the marvelous strength of the Colossus of Rhodes—one of the seven wonders of antiquity—whose massive limbs brought together the distant edges of the port city of Rhodes. Towering over the port's entrance, the Colossus surveyed the incoming and outgoing ships with the same vigor as Adamastor presiding over the tip of the African continent and exerting supreme control over the approaching vessels.

Adamastor's coarse voice itself seems to come from the underworld in its announcement of the poem's second half as an inversion (or a copy) of the first. What unfolds in canto six recommences as a dis-joined replica, a negative image, of the first half of the poem. Canto six thus opens by inverting Adamastor's affirmative "eu sou" (I am) into the negation "não" [no(t)] as the canto's very first word:

> Não sabia em que modo festejasse
> O Rei Pagão os fortes navegantes,
> Para que as amizades alcançasse
> Do Rei Cristão, das gentes tão possantes;
> Pesa-lhe que tão longe o aposentasse
> Das Europeias terras abundantes
> A ventura, que não no fez vizinho
> Donde Hércules ao mar abriu caminho.
>
> (Not knowing in which manner to entertain
> The brave navigators, the Pagan king
> Set out to gain the friendship of the Christian king
> And of such strong people.
> He deplored that he was lodged so far
> From the abundant lands of Europe,
> Lamenting that fortune had not placed him closer
> To where Hercules had opened the path to the sea.)[98]

Whatever unfolds in the Southern Hemisphere occurs under the aegis of the negative adverb "não" which sharply separates the two hemispheres in a gesture that distinguishes between the model and its copy. The more Gama ventures into the Indian Ocean the more the maritime space closes in on him and the more the specter of "the abundant lands of Europe" becomes (op)pressing.

David Quint has pointed out that "Camões's monster, born of the initial encounter of Portuguese imperialism and its native subjects, is the first in a line of specters haunting Europe."[99] Adamastor's presence announces the power of cartography as a specter-to-be: the projected forms, figurations, and shapes of the lands to be discovered—specters that the Greeks termed "eidola"—precede the voyage itself as a series of competing cartographic models. Adamastor's curse—and perhaps the epic curse writ large—unfolds as an inherently cartographic malediction: uttered from a liminal shore whose delineation is uncertain, it emblematically announces a dramatic revision of Ptolemaic cartography. Adamastor's cartographic curse disturbs and disrupts chronology in that it prefigures the navigators' voyage and the contours of the land they are yet to disclose in a joint act of navigational and cartographic heroism. Preoccupied with form, reproducibility, and figuration, the map as a specter is, in Jacques Derrida's terms, "the hidden figure of all figures."[100] Adamastor's curse is inextricably bound up with the cartographic question of the reproduction and multiplication of Europe's image beyond Europe's boundaries—which comes into effect at the very moment when Gama discloses the Indian Ocean.

Adamastor's authoritative status as the collector of disparate spatial and temporal frameworks is enhanced through his exceptionally powerful ontological status, one that Gama himself seems to lack. Despite (or perhaps because of) his monstrosity, his spanning the spectrum of form and formlessness, Adamastor is one of only two characters who introduces himself with the first-person-present form of "to be": "Eu sou" (I am).[101] Positioned in a liminal space, Adamastor is responsible for the impossible task of keeping "together that which does not hold together, and the disparate itself," that which "can be thought . . . only in a dis-located time of the present,"[102] as Jacques Derrida (albeit in a different context) has defined the powerful presence of specters. As he brings together the past and the future into an assemblage of fragmented geographic parts and frameworks of knowledge, as he relocates and rearranges space as well as time, Adamastor becomes the very manifestation of the monstrous present under the aegis of cartography. With Barros in mind, Adamastor exposes and embodies the monstrosity of the map.

The story of Adamastor's monstrous and misshapen body itself is reminiscent of the two liminal rocks Abyla and Calpe, demarcating the limits of the Mediterranean. One might argue that Adamastor himself is a failed copy of Hercules: we learn, in fact, that before being transformed into a rock, Adamastor was, not unlike Gama, "a Captain of the Sea" (*capitão do mar*),[103] as he proudly declares upon his first appearance. Having navigated the seas just like Gama, Adamastor already possesses the theoretical and practical knowledge of the seas that Gama is only now about to gain. From a captain, Adamastor was transformed into an immobile rock—and a telluric marker—just when he hastened to embrace his object of desire, the maritime nymph Tethys:

> Que, crendo ter nos braços quem amava,
> Abraçado me achei c'um duro monte
> De áspero mato e de espessura brava.
> Estando c'um penedo fronte a fronte,
> Que eu pelo rosto angélico apertava,
> Não fiquei homem, não; mas mudo e quedo
> E, junto dum penedo, outro penedo!
>
> (Convinced my beloved was in my arms,
> I found myself hugging a hard mountain
> Of undergrowth and rough bush;
> I was cheek to cheek with a boulder
> I had seized as her angelic face,
> Unmanned utterly, dumb and numb with shock,
> A rock on an escarpment, kissing rock.)[104]

From an aquatic nymph, Thetis was thus transformed into solid rock. As he hastened toward his beloved, Adamastor sealed off with his own body a waterway which had previously been open. His embrace and metamorphosis both symbolizes and provides the mythological background for the landlocked nature of the Indian Ocean, now breached by the cartography-changing naval activities performed by the Portuguese sailors. For Adamastor, Gama's daring attempt to forcefully bypass the Promontory and cut into the Indian Ocean conjures up the memory of Hercules's separation of Abyla and Calpe.

Yet Adamastor's commanding and colossal presence preserves its ambivalent status. It threatens to collapse once his ontological affirmation, "eu sou," is revealed as an echo of another voice uttering "eu sou": the anthropomorphized "famous Ganges," who had appeared in King Manuel's prophetic dream in canto four:[105]

> Eu sou o ilustre Ganges, que na terra
> Celeste tenho o berço verdadeiro;
> Estoutro é o Indo, Rei, que, nesta serra
> Que vês, seu nascimento tem primeiro.
> Custar-te-emos contudo dura guerra;

Mas, insistindo tu, por derradeiro,
Com não vistas vitórias, sem receio
A quantas gentes vês porás o freio.

(I am the famous Ganges whose waters
Have their source in the earthly paradise;
This other is the Indus, which springs
In this mountain which you behold.
We shall cost you unremitting war,
But persevering, you will become
Peerless in victory, knowing no defeat,
Conquering as many peoples as you meet.)[106]

The affirmation "eu sou," uttered only by Ganges and Adamastor, establishes an intercontinental network which conjoins Asia and Africa in a poetic and performative embrace: the juncture which verbally binds the two continents together serves as a poetic *pendant* to the cartographic embrace of Asia and Africa on a Ptolemaic map, where the two continents meet in a gesture enclosing the Indian Ocean. The affirmative "eu sou" which shuttles between India and the Cape of Good Hope positions Ganges and Adamastor as those who connect the two continents like tectonic plates—before they are disjoined by the Portuguese captain major.

However, unlike the "misshapen" Adamastor reminiscent of the underworld and struggling with the overwhelming confluence of cartographic models, Ganges is born in earthly Paradise. As the original (river) flowing out of Paradise, he thus exclaims: "na terra / Celeste tenho o berço verdadeiro" (Heavenly earth is my true birthplace).[107] If Camões foregrounds Ganges as the first to reclaim ontological status ("eu sou"), Ganges nevertheless lacks Adamastor's cartographic vision to create models of reality, that is, to draw, represent, and foreshadow territorial outlines. Unlike Ganges, Adamastor is endowed with the commanding power of cartography. If Adamastor is first introduced as "uma figura" (a figure), it is because of the word's strong cartographic underpinning.

PTOLEMAIC POETICS

While scholars have recognized the powerful presence of cartographic imagery in *Os Lusíadas*, they have directed their attention to one type of map: the nautical chart, and thus to a route-enhancing map associated with the practice of navigation and the exploration of new lands and seas. Bernhard

Klein surmised that the epistemology of Camões's poem is in fact closely allied with that of the sea chart used in deep-sea navigation. For Klein, Camões's poetic rendering of Gama's itineraries across hitherto unexplored maritime spaces is akin to the functioning of nautical charts:

> To make such charts, a mariner would first set down a number of compass stars on a sheep or goatskin, then draw rhumb lines from these compass stars that extended in all directions across the map. Only as a final step would the coastlines be drawn in. The sequence is important: the network of rhumb lines—not yet a system of graduated coordinates such as meridians or parallels—worked as a frame of reference or orientation that was more central to navigation than the depiction of a particular portion of the earth's surface.[108]

Klein contends that sea charts were first "frames of spatial reference that held no content, that circumscribed an emptiness: a blank surface without landmarks or points of orientation."[109] If for Klein "the only true sea chart ever printed" is perhaps the "perfect and absolute blank"[110] of Lewis Carroll's map in *The Hunting of the Snark*, the reduction of the map to its frame is certainly more akin to a Ptolemaic *tabula*. Navigational charts tend to gesture toward openness and continuous itineraries to be plotted beyond the material edges of the map (often a vellum, or animal hide). In contrast, the most salient feature of a Ptolemaic map is its rigorous frame. Gama is sent out, as it were, to discover a maritime space that has already been mapped by both the cartographer (Ptolemy) and the poet (Camões). The second half of *Os Lusíadas* is crystalized as a cartographic surface upon which the poetic stanzas are inscribed with the astuteness of a cartographer mapping the geographic space of an already established framework. Rather than *dis*covering new territories, Camões's cartographic poetics *covers* unfamiliar spaces with familiar cartographic and poetic images, stock images of the Mediterranean. The cartographic medium becomes here a potent epistemological tool that manipulates the reader into experiencing the unknown world through its mediating interface, evoking the imprint of Ptolemy's familiar geography.

Camões's cartographic imagery of the Indian Ocean as an isolated sea dovetails with the poem's self-contained stanza form, isolated units inserted within the defined space of the page: the *ottava rima*, a poetic form of Italian origin first developed by Boccaccio. Eagerly taken up by fifteenth- and sixteenth-century Italian epic poets such as Matteo Maria Boiardo, Luigi Pulci, Ludovico Ariosto, and Torquato Tasso, the *ottava rima* represented for these poets a "little world."¹¹¹ But the *ottava rima* was not only characteristic of epic poems. It was also popular, from the early fifteenth century on, in cosmographic writings such as *La sfera*,¹¹² a cartographic poem attributed to Leonardo Dati (or his brother Gregorio),¹¹³ written not only during the impressive rise of the cartographic genre of the *isolario*, the *Island Book*, but also when the first Ptolemaic maps started to circulate in Florence, before spreading across Europe. *La sfera* puts the question about the chiastic relationship between poetic form and cartographic isolation—or the cartographic form as poetic isolation—center stage.¹¹⁴ Just like Dati's *La sfera*, Camões's *Os Lusíadas* is a unique example of the desire to correlate poetic form and cartographic framing by displaying self-contained poetic stanzas on the page as isolated geographic units—be it a body of land or water. While Dati's *ottava rima* describes the three insular parts of the *oikoumene*—Africa, Europe, and Asia—as self-contained geographic entities encircled, in the tradition of the T-O map, by an ocean, Camões's Ptolemaic imagery inverts the relationship between land and water and uses the *ottava rima* as a poetic device to point to the Ptolemaic isolation of the seas. The choice of *ottava rima*, in contrast to the concatenated *terza rima*, for instance, is more than a nod to Italian epic writers such as Boiardo and Ariosto—who are explicitly referenced in the opening stanzas of canto one.¹¹⁵ It is an attempt to align geographic space with a poetic form that best conveys the idea of isolation and containment.

The paradox of Camonian poetics is, as Hélio Alves observes (quoting Georges Bataille's *Accursed Share*), the strong sense of an "impossibility of continuing growth."¹¹⁶ Alves notes that once Gama reaches the coast of India "the capacity of absorption and evolution has reached its maximum limit."¹¹⁷ Gama's odeporic goal and his subsequent captivity on the Malabar coast are matched by the exhaustion of his poetics, increasingly exacerbated in canto eight: in his lengthy discourse to the ambassador of the Samorin, Gama demonstrates a predilection for contorted syntax and reverts to obscure vocabulary, to words, that is, used exclusively in this very monologue, such as "delitos" (crimes), "vaso da nequícia" (vessel of iniquity), "prisca" (early), and "undívago" (roaming the waves).¹¹⁸ Compressing his obscure lexicon into two stanzas, Gama starts spinning his rhetoric wheels, revealing the limits of his poetic energy.¹¹⁹ Tarrying on the Indian shore sets in motion a discursive "expenditure," as Bataille puts it: "Only the impossibility of continuing growth makes way for squander. Hence the real excess does not begin until the growth of the individual or group has reached its limits."¹²⁰ Gama's communicative exhaustion is mirrored not only in the geographic limits he reaches, but in Camões's own voice as an author. Stepping out of his capacity as the narrator of Gama's deeds, Camões exposes the very limits of poetic expression when, at the end of the seventh canto, he abruptly discontinues a verse in order to question the efficacy of his own poetic voice by pointing to its limits:

> ... Mas, ó cego,
> Eu, que cometo, insano e temerário,
> Sem vós, ninfas do Tejo e do Mondego,
> Por caminho tão árduo, longo e vário!
> Vosso favor invoco, que navego
> Por alto mar, com vento tão contrário
> Que, se não me ajudais, hei grande medo
> Que o meu fraco batel se alague cedo.
>
> (... But what
> Blind folly is this that I embark,
> On a voyage so hard, so long and so varied
> Without you, nymphs of Tagus and Mondego?
> I implore your help, for I am sailing
> The open sea with a wind so contrary
> That, if you cease to inspire and maintain me,
> My slight craft will no longer sustain me.)[121]

The image of poetic plenitude, achieved once the Indian shore is reached and heightened by Camões's powerful metapoetic intervention, dovetails with a cartographic model which aims at distributing its energy equally and uniformly as it fills the available space of the Southern Hemisphere: the equipollent form of a Ptolemaic map. Once Gama accomplishes his voyage across the Indian Ocean, he completes the design of the map and Camões fulfills his poetic goal. Gama's arrival in India and his subsequent arrest (and thus the impossibility of any further movement or expansion) gesture toward the limits of a predetermined cartographic frame beyond which no further poetico-cartographic movement or expansion is possible. If there is still a poetic surplus, it has no poetic or spatial outlet—it spills over into Camões's own authorial voice and tarries as rhetorical expenditure at the geographic limits of the expedition. Gama's unheroic departure demonstrates that his energy has been exhausted—staying longer in Calicut might even result in a further arrest:

Se mais que obrigação, que mando e rogo,
No peito vil o prémio pode e val,
Bem o mostra o Gentio a quem o entenda,
Pois o Gama soltou pela fazenda.

Por ela o solta, crendo que ali tinha
Penhor bastante donde recebesse
Interesse maior do que lhe vinha
Se o Capitão mais tempo detivesse.
Ele, vendo que já lhe não convinha
Tornar a terra, por que não pudesse
Ser mais retido, sendo às naus chegado
Nelas estar se deixa descansado.

(But that profit mattered to him far more
Than honour, orders, or obligation,
The Catual showed to any who had eyes,
Releasing da Gama for the merchandise.

He freed him, convinced he had in hand
Sufficient stake to guarantee him
Greater gain than would accrue
By detaining the captain longer.
Da Gama accepted it would be foolish
To return ashore and risk being
Re-imprisoned, so having boarded ship,
He rested and reflected on his trip.)[122]

Contrary to what one might expect from a travel narrative (and the navigational chart as its cartographic *pendant*), the frequently used ekphrastic descriptions—be it of the banners which, inspired by Ariosto's *Orlando Furioso*, narrate Portuguese history, or the architectural features of Indian palaces and places of worship[123]—operate within a closed and predetermined system that offers only limited outlet for poetic expansion. Gama's conversation with the Samorin's minister, the Catual, is "particularly devoid of any pragmatic sense," while the captain major's interaction with the Samorin turns into a "simple ostentatious display of oratorical proficiency and talent."[124] Ironically, Gama's limited interactive space along the Malabar coast contrasts with the vastness of the Indian territory over which the Samorin presides: the Samorin's very name, first attested in Ibn Battuta's

Travels (1342), is a Portuguese corruption of the Malayalam which signifies, according to two different traditions, either "August Emperor" (from *Svami Tiri* and *Tirumulapad*) or "one who has the sea for his borders" or "lord of the sea" (from *Samudrî Raja* or *Samudhrāthiri Raja*).[125]

EKPHRASTIC TECHNIQUES

Camões's *Os Lusíadas* is the poetic equivalent of Mercator's projection. Both trace, with different means, the process by which Europe emerged as a powerful continent through the use of cartographic projections. No instance in the Camonian poem condenses Eurocentric projection more vigorously than the ekphrastic image Camões creates in canto nine: a detailed description of the Isle of Love in which the lush vegetation of the Indian Ocean, already noticed by Humboldt, is translated into familiar Mediterranean flora. The narrator's reference to the island as a "tapestry," seemingly a poetic detail, exposes Camões's description of the idyllic insular scenery as a reflection upon cartographic projection:

> Pois a tapeçaria bela e fina
> Com que se cobre o rústico terreno,
> Faz ser a de Aqueménia menos dina,
> Mas o sombrio vale mais ameno.
> Ali a cabeça a flor Cifísia inclina
> Sôbolo tanque lúcido e sereno;
> Florece o filho e neto de Ciniras,
> Por quem tu, Deusa Páfia, inda suspiras.
>
> (As for the bright, flowery meadow
> Carpeting the untended ground,
> It exceeded Persia's finest
> As it graced the valley shade.
> Here Narcissus drooped his head,
> Above his fated, flawless pool;
> And springing from Adonis' wound, the flower
> The goddess sighs for, to this very hour.)[126]

While scholars insist that Camões's firsthand experience of crossing the Indian Ocean on his way to India allows him to create a "mental cartographic framework" filled "with empirical detail,"[127] the ekphrastic image of the tapestry that occupies a central position in the latter half of *Os*

Lusíadas calls for a substantial revision of the hitherto insufficiently questioned status of Camões as a "great sea painter."[128] Here, the image of the tapestry extending across the Indian Ocean quite literally covers (up) the sea, superimposing a layer of Mediterranean vegetation upon the tropical climate of the Indian Ocean. The reader is reminded, then, that a tapestry, as a "piece of cloth," is etymologically akin to *mappa* (as in *mappa mundi*), to the very materiality of a map.

The rhetorical use of ekphrasis, which in Barros's *Of Asia* is limited to the initial and final margins,[129] is transformed in *Os Lusíadas* into a central cartographic tool that directs and informs the understanding of the latter half of the poem.[130] The textile element of the tapestry is a preferred ekphrastic intervention—a technique inserted into *each* canto that relates to India and the Indian Ocean (and three times in canto eight). Ekphrasis guides the reader across the stanzas—and across the Southern Hemisphere—not to enhance empirical accuracy, but to showcase the power of the map that informs Gama's travel. The textile elements in their articulations as maps conceal the aquatic element and present the Indian Ocean as a series of maps instead. Rhetorically transformed into an almost uninterrupted sequence of ekphrastic images, these interventions function as cartographic devices prepared by the poem's characters with "singular artífice" (singular artifice).[131] Ekphrastic tools are used by Gama to narrate the Portuguese history to the Samorin and to learn about Indian mythology and the Catual's palace; by Bacchus to describe the architecture of Neptune's underwater palace; by the goddess Thetis in the poem's concluding canto to present the *máquina do mundo* to the captain major. Yet one of the most stunning ekphrases is the "tapestry" which takes the shape of the Isle of Love.

In her study of the history of tapestry as an "age-old and extremely rich textile tradition" in India,[132] Barbara Karl has shown that the international textile trade and "embroidered textiles similar to the colchas [quilts] were produced in India long before the arrival of the Portuguese."[133] Indeed, Camões's contemporary Gaspar Correia points out in his historical work, *Lendas da India*, that it was the king of Melinde on the East African coast who offered Vasco da Gama rich presents in the form of embroidered fabrics: "The King of Melinde gave each of them rich fabrics and to the captain [Vasco da Gama] a valuable necklace of jewels for the King of Portugal. . . . He gave the captain other rich pieces . . . and a chest full of beautiful fabrics for the Queen and a kind of quilt or bed canopy, worked [stitched] in white, the finest embroidery ever seen, which was made in Bengal, a country in which marvellous things are made with needles."[134]

While textiles were used "as diplomatic gifts within the Indo-Oceanic court milieu," their production across the Indian Ocean was so common and so richly elaborated that Barros estimated, in his *Of Asia*, that "there is so much cotton harvested and there are so many artisans who weave the finest textiles that one could dress all of Europe with them."[135] Yet Camões's poem strategically inverts the directionality of textile exchange. By appropriating the production and trade of the Southern Hemisphere, he turns the textile's "singular artifice" into a specifically European product ushered in by the Mediterranean Goddess Venus in the form of the Isle of Love—not to "dress all of Europe," but to cover the Indian Ocean for the pleasure of the European sailors.

Humboldt rightly emphasized the central ekphrastic role of the Isle of Love for the understanding of the complex relationship between the Indian Ocean and the Mediterranean Sea. The "tapestry" of the Isle of Love highlights central cartographic questions such as the slippage between reality and representation; the tension between what is a true and what a false picture; and the process of writing as mapping. Fama, who serves as Venus's "famosa e célebre terceira" (notorious go-between), announces the advent of the isle's artful tapestry in the midst of the ocean. As "[a] Deusa Giganteia, temerária, / Jactante, mentirosa, e verdadeira, / Que com cem olhos vê, e, por onde voa, / O que vê, com mil bocas apregoa" (Hot-blooded, boasting, lying, truthful [Goddess], / Who sees, as she goes, with a hundred eyes, / Bringing a thousand mouths to propagandize),[136] Fama embodies cartography's very tension between the production of truthful and deceiving images. Like Argus, Fama is equipped with a hundred eyes and extensively outfitted to survey multiple lands at a time. At the same time, Fama makes sure to maintain a veil of secrecy over Venus's cartographic design: she does not reveal whether the newly projected Isle of Love will be true or false—it will remain what scholars have termed a "phantom island."[137]

The Isle of Love, an "allegorical topos" whose history can be traced back to René d'Anjou's *Livre du cuer d'amours espris* and Francesco Colonna's *Hypnerotomachia Polyphili*,[138] condenses an imagined mobility, extensively explored over the course of the fifteenth and sixteenth centuries in the "flexible geography" at the center of the *isolarii*. Here, Mediterranean islands were conceived as mobile and "indefinitely reconstructible"[139] units, consistent with a modular mode of thinking. The Isle of Love is conceived to be one such "unit" in Venus's ingenious scheming. In *Os Lusíadas*, she takes on the role of a master cartographer planning out the design of the Isle of Love in great detail. Venus not only "paints" and "maps" (*pintou*)[140] the island, but takes advantage of its mobility, a feature the island shares with Delos, the birthplace of Venus's siblings Apollo and Diana, to make it visible to the sailors:

> De longe a Ilha viram, fresca e bela,
> Que Vénus pelas ondas lhe levava
> . . .
> Que, por que não passassem, sem que nela
> Tomassem porto, como desejava,
> Para onde as naus navegam a movia
> A Acidália, que tudo, enfim, podia.
>
> Mas firme a fez e imóvel, como viu
> Que era dos Nautas vista e demandada,
> Qual ficou Delos, tanto que pariu
> Latona Febo e a Deusa à caça usada.
>
> (The lovely, verdant island hovered
> As Venus wafted it over the waves
> . . .
> For to prevent their sailing past
> Without making port, as she desired,
> Wherever they [the ships] went, she kept it full in view,
> Shifting it, as she had the power to do.
>
> But she anchored it on the instant
> She saw the mariners speeding towards it,
> As Delos paused when Latona gave birth
> To Apollo and Diana, the huntress.)[141]

Once the tapestry is completed, the poetic work of ekphrasis unfolds: with the help of Venus's machinations, the "second Argonauts" (*os segundos Argonautas*)[142] are let loose to pursue their colonizing aggression, each man "like a hunting dog" (*Qual cão de caçador*) practiced in retrieving the "local" maidens like "shot birds from the water" (*a ave ferida [na água]*).[143] Yet one sailor, Leonard, a slower "hunter" than the others, laments his faltering speed (which prevents him from getting hold of a nymph) by citing a line verbatim from Petrarch's *Rime Sparse*, sonnet 56:

> E notarás, no fim deste sucesso,
> "Tra la spica & la man, qual muro he messo."

(And you will notice, at the end of this event,
Tra la spiga e la man, qual muro è messo.)[144]

The line from Petrarch, inserted in *Os Lusíadas* in the original Italian as the poem's sole citation from a foreign language, is an ekphrasis in its own right. Amidst the lush landscape and appealing vegetation forming the "tapestry" of the Isle, the quotation becomes a meta-ekphrastic device introduced with the scope to thematize the very problem of representation, manifested as the fervent desire to touch the untouchable—a blatantly cartographic paradox. Leonardo's poetic line takes on the shape of a line of demarcation. As João R. Figueiredo contends, this specific sonnet by Petrarch marks the "moment in which life experience is co-extensive with the experience of reading: Petrarch understands that his life can only be described by a set of models he has learned."[145] Leonard's unabashed reference to Petrarch, devoid of any poetic veil, linguistic adjustment, or attempt at translation, aggressively exposes the slippage between reality and representation, between truth and deception. Presented in its nude simplicity, Leonard's line creates the very space to think about the power of ekphrasis and mapping. As Figueiredo contends, "if there is anything pornographic about the Isle of Love, it is not the nymphs' 'naked skin' . . . but the absolute and explicit nudity with which the line, *Tra la spica e la man, qual muro he messo*, is exhibited."[146] The isolated Italian line emerges as a true "pornotopia,"[147] a brazenly unadorned *mise-en-abyme* epitomizing the raw presence of cartographic textuality.

From poetic ekphrasis Camões moves to an ekphrastic image in which the oscillation between reality and representation takes on a particularly cartographic valence. In the middle of the island, Camões inserts a lagoon:

Num vale ameno, que os outeiros fende,
Vinham as claras águas ajuntar-se,
Onde uma mesa fazem, que se estende
Tão bela quanto pode imaginar-se.

(Between the hills, in a pleasant valley,
The translucent rivers came together
To form a lagoon which stretched and brimmed
With a beauty beyond imagining.)[148]

The Portuguese word Camões uses here for "lagoon" is not "(a)lagoa," as one might expect, but "mesa," from the Latin "mensa" (table). The waters of the lagoon, situated in the center of the isle, are so still that they are easily confused, Camões suggests, with an immobile blank surface, a table. Inserted within the ekphrastic tapestry of the Isle of Love, the table functions as a meta-ekphrasis challenging the limits of cartographic imagination and geographic mapping. At the same time, the *mesa*—both liquid and solid—is a *mise-en-abyme* of the larger "tabula," the Indian Ocean, described in canto eight as "liquido estanho"—liquid metal or pool. The image of a "table" boldly states that what might appear as an aquatic element (the Indian Ocean) is actually a "tabula," a cartographic table representing the ocean. "Mesa" thus powerfully condenses the imaginative power invested not only by Venus in designing the island, but also by Camões in depicting the Indian Ocean as a process of cartographic projections.

The Camonian "mesa" epitomizes what Farinelli has termed the progressive table-making process characteristic of European thought from the early modern period on. With the rise of cartography, the table took on the function of a "model of the mind." The Latin "mens" (mind) and "mensa" (table) are intimately related, Farinelli argues, and as such, in their intricate entanglement and ekphrastic form, they constitute the Isle of Love's central element. The map is "a logical extension of the table," so that, Farinelli continues, "the passage to the map as a model of the mind only reflects a further degree of the colonization of the world by the Table itself."[149] The appearance of the table on the Isle of Love turns the island into what Dünne has termed a "mobile operating room"[150] that brings into sharp focus the colonizing processes at work: European cartographic models are vigorously imposed at the expense of non-European modes of imagining, representing—and being.[151]

It is perhaps not a coincidence that Camões crafted his poem in the very years that Gerhard Mercator, an attentive reader of Pedro Nunes's *Tratado sobre certas dúvidas da navegação*,[152] invented his cylindrical projection method, which he used, for the first time, to make his 1569 world map titled *Nova et aucta orbis terrae descriptio*.[153] Here, Mercator proposes the impossible task of his new method, as he explicitly contends, to "[project] the sphere onto the plane by a new and most convenient invention, which corresponds to the squaring of the circle" ("inuentione nova & conuenientissima spheram in plano extendendo, quae sic quadraturae circuli respondet").[154] While commonly hailed as a milestone in the history of cartographic projection and still immensely influential, Mercator's effort to literally "square the circle" has met with stark criticism, despite its evident success in imposing itself as *the* model of projection.

If for Dipesh Chakrabarty Europe is an elusive "imaginary figure,"[155] it is because poets such as Camões (alongside cartographers such as Mercator) set out to aggrandize and standardize the continent's contours by producing and promoting the equipollent space of a "mensa," increasingly transportable as a metaphor and "idea," across continents. It is perhaps appropriate to term Camões's "mensa," which epitomizes the darker side of cartography, a model not only of the mind, but of the colonizing mind.[156] If Europe, as Frantz Fanon succinctly put it, "is literally the creation of the Third World,"[157] this creation originated with and was propelled by the rapidly rising authority of the map, a *tabula rasa* upon which regions and continents were projected and divided up even before the process of colonization proper began.

CONCLUSION

In 2017, Nicholas de Genova opens his volume on *The Borders of "Europe": Autonomy of Migration, Tactics of Bordering* with the following introductory remarks:

> This book has arisen in the midst of what has been ubiquitously and virtually unanimously declared in mass-mediated public discourse and the dominant political debate to be a "crisis" of migration in Europe. The first intimations of a European crisis arose amid the unsightly accumulation of dead black and brown bodies awash on the halcyon shores of the Mediterranean Sea. When a ship transporting as many as 850 migrants and refugees capsized on April 19, 2015, all but 28 of the vessel's passengers were sent to their deaths in what appears to have been the worst border-crossing shipwreck in the Mediterranean on record. . . . These human catastrophes at sea have indisputably transformed the maritime borders of Europe into a macabre deathscape.[1]

From its central, but permeable, place on the medieval Hereford map (ca. 1300) upon which the *Mare Nostrum*, the Mediterranean Sea, studded with circles and triangles representing its numerous islands, holds together the *oikoumene* as a single world-continent, the Mediterranean has turned, seven centuries later, into "a macabre deathscape" in which the presence of the human body resurfaces in its most cynical and cruel form: as a late biometric marker of the bounds of a "Fortress Europe" which, its borders increasingly impermeable, is either confused with the European Union or reduced—the specter of the "Iron Curtain" still looming large—to its western (and northern) part. Be it Europe's manifold strategies to address the so-called "migration crisis" (by highlighting Europe's traditional openness toward the newly arrived or, conversely, by evoking Europe's historic Christian faith); be it the partial disintegration of the European Union with the imminent question of Brexit hinging almost exclusively upon the border that separates Northern Ireland and the Republic of Ireland; or be it the Russian-Ukrainian war over the Crimean peninsula: pressing questions concerning Europe and its borders are deeply rooted, one easily forgets, in geography and cartography.

Constantly in flux, continuously reshaping what we mean when we say "Europe," borders—not only national, but also continental—are as powerful as they are fragile. Boundaries are also the consequence of political and economic decisions. But they are foremost the result of cartographic and poetic processes which first imagine them into being. Borders are intrinsically poetic, if with the Greek word ποίησις (poiesis) we understand the active making of something that did not exist before. Before they take on an eco-political meaning, borders are imagined and deployed on maps as a performative, geopoetic, act.[2] For Kenneth White, the founder of the *Institut international de géopoétique* (International Institut of Geopoetics), geopoetics is nothing less than the "*mouvement* majeur qui concerne les fondements mêmes de l'existence de l'homme sur la terre" (major *movement* concerning the very foundations of human existence on earth).[3] Casting ever-changing toponyms on maps (Byzantium, Constantinople, Istanbul, for instance), translating boundaries from one place to another (imagined dividing lines separating the Eurasian land mass convey the idea of Europe as an independent continent), creating previously non-existing—and registering, conversely, the disappearance of formerly existing—locales, regions, and nations (North Macedonia, the USSR) are performative cartographic acts: powerful acts, that is, that are not merely descriptive, but profoundly normative. At once controlling and mobile, a map, intrinsically geopoetic in nature, is a powerful (and not an innocent) epistemic tool with which the world is at once apprehended and produced. I call *europoiesis* the convoluted making of Europe as a continent through the commanding, and at the same time fragile, lens of carto-poetics.

Cartographic Humanism takes its point of departure from the paradoxical observation that while we tend to use the term "Europe" as a fixed category, the continent itself was in the making in the early modern period, in particular from the late fifteenth to the late sixteenth century. While Europe had been in flux since antiquity, it was on the brink of modernity that it was transformed from an ill-defined part of the world, its borders permeable, to a sharply delineated, hegemonic, and metaphysically charged continent. *Cartographic Humanism* unearths the geopoetically and cartographically ambiguous nature of Europe at a particularly momentous turning point in the continent's formation, prior to the creation of nation-states, when the rise of the continent was propelled by the emergence of a novel humanistic discipline: cartography.

Cartographic Humanism is a long-overdue critical intervention into the rise of Europe as a surprisingly unexplored continent. Pushing at once against smooth narratives of progress and all-too-dark scenarios, it mobilizes cartography as a *pharmakon*, present and poison alike, able to offer unprecedented insights into the transformation of Europe from a geographic to a metaphysical idea, interchangeable with globalization. The modern "idea of Europe" is a particularly poignant and pressing case that calls for meticulous examination: the continent's rise as an idea standing for a universal and secular vision of the human and for a modern world-system has proven to be extraordinarily influential—as well as devastating on a global scale, where colonialism, slavery, and exploitation are but a few of its dire outcomes. From the Treaty of Tordesillas (1494) to the Berlin Congo Conference (1884–85) and beyond, cartography has propelled the arbitrary division of land and its inhabitants. Cartographic models elaborated in the course of the fifteenth and sixteenth century (from Ptolemy's coordinate system to the central role of the Mercator map projection) still dominate our twenty-first-century methods of representing space.

Cartographic Humanism discusses Renaissance humanists as "borderlanders"—if by that term one understands, with Oscar J. Martínez, a population living in different linguistic and cultural ecosystems characterized by "the ability to be multilingual and multicultural." The cartographic writers discussed in this book are all driven by a fundamental linguistic, poetic, cartographic, and philological versatility that characterizes Renaissance humanism. In contrast to Martínez, for whom "individuals from interior zones who live in homogeneous environments have no need to develop such multifaceted human proficiencies, or to be knowledgeable and sensitive to the perspectives of other peoples,"[4] *Cartographic Humanism* unearths the heterogeneity and complexity of so-called "homogeneous environments," showing that Renaissance humanists often desired, but ultimately failed to establish Europe's center. The German humanist Conrad Celtis, for instance, for whom Nuremberg was not only the center of the German lands but, as Willibald Pirckheimer put it, the "navel of all of Europe" (*totius Europae umbilico*),[5] described the impossibility to represent Europe's center through the poetic figure of a "liquid vertex." In Celtis's work, the messiness typically assigned to the border is, as Etienne Balibar surmised, "no longer *at the border*," but has multiplied to form *"zones, regions, or countries."*[6] This sense of multiple connectivity and convoluted multilingualism, at the center of a cartographically inflected humanism, unfolded in tandem with Europe's multiplying borders, propelled by Ptolemaic cartography, which drew the continent's increasingly sharp contours as arbitrary borderlines. Since its early beginnings, Europe concomitantly pulled in these two opposing directions, the multifaceted border zone and the clear-cut borderline.

When discussed against the backdrop of world affairs, Europe has often been reduced to a monolithic bloc interchangeable with the hegemonic "West." This equation has led to a practice among scholars of taking shortcuts in their investigations of early modern Europe: conflating "Renaissance Europe" (or "Renaissance humanism") with western Europe is an all-too-common practice. By extending the analysis to regions beyond, but also joint with "the West," this study offers an urgent corrective by dislodging commonly accepted ideas about Europe and its borders. It serves as a reminder that for Aristotle Greece—later typically associated with the origins of quintessential Europeanness—assumed a "middle position" between two continents, Europe and Asia; that one of the most eminent Italian humanists, Enea Silvio Piccolomini, dedicated the opening chapters of his influential book *De Europa* (1458) to eastern and central Europe; that Ptolemy's *Geography* located Italy in Europe's east; and that in the fifteenth and sixteenth centuries the layout of the popular Macrobian zonal map privileged a division of Europe into north and south, not east and west. *Cartographic Humanism* shows that despite our tendency to see, live in, and treat continents as stable and immobile units in which each region or country is permanently anchored, the idea of a continent needs to be continuously historicized and subject to careful scrutiny. Committed to offering an unprecedented insight into early modern Europe and to dispelling geographical and historical commonplaces about the continent, *Cartographic Humanism* tarries on those interstitial zones we now identify as the limits of Europe.

This book is inspired by recent work in the humanities that binds cartography and literature to one another as a single, interdisciplinary, endeavor, capable of questioning and redirecting categories previously considered in isolation. "I foreshadow the possibility of new a-wherenesses." With a cunning pun, Nigel Thrift captures the power of the "spatial turn" of the past decades to radically change awareness, with an eye to the twenty-first century, by redirecting critical thinking and creative production to "what seems like a constantly expanding universe of spaces and territories, each of which provides different kinds of inhabitation—from the bordering provided by the womb . . . through the corporeal traces of buildings and landscapes."[7] Thrift's promise, tinted with a touch of metaphoricity and geared toward a future world which, in Amir Eshel's words, opens up new spaces "by expanding our vocabularies, by probing the human ability to act,"[8] could not be more fitting for the early modern period, when perhaps the first "spatial turn," indeed the first "cartographic turn," at the center of *Cartographic Humanism*, occurred. J. B. Harley has long called for a turn in cartography which would "break the assumed link between reality and representation which has dominated cartographic thinking [and] . . . has led it in the pathway of 'normal science' since the Enlightenment."[9] Advocating an "epistemological shift in the way we interpret the nature of cartography," Harley urges scholars and historians of cartography "to read between the lines of the map . . . to discover the silences and contradictions that challenge the apparent honesty of the image."[10] This book proposes to read between the (poetic) lines of the map to unearth the drive with which poets, historians, and mapmakers—cartographic writers, that is, whose writing aimed at striking a balance between discourse and spatial plotting—started thinking, for the first time, with the map in mind as they embarked on the new endeavor to draw lines between continents.

ACKNOWLEDGMENTS

"A book has neither object nor subject; it is made of variously formed matters, and very different dates and speeds. To attribute the book to a subject is to overlook this" (Gilles Deleuze and Félix Guattari, *A Thousand Plateaus: Capitalism and Schizophrenia*, trans. Brian Massumi [Minneapolis: University of Minnesota Press, 1987], 3). Gilles Deleuze and Félix Guattari's words serve as a reminder that books never originate without the help from others who intervene in these "variously formed matters": teachers, colleagues, institutions, and friends. I feel very honored to acknowledge the support I have received over many years—from my former teachers at New York University, to my colleagues at Harvard University, to several other institutions, and, last but not least, to my family and friends on both sides of the Atlantic. My warmest thanks are due to Jane Tylus, without whose generosity, brilliance, and wisdom this book would simply not exist. Jane's outstanding expertise in early modern literature and her exquisite personality taught me how to read and navigate the Renaissance. I am immensely grateful to Tim Reiss, from whose critical eye and keen thought I have benefited over the years and whose work on a transoceanically imagined Renaissance has significantly affected the course of this book. Emily Apter suggested the book's title years before I understood what it could possibly mean—I thank her for all her invaluable inspiration. Furthermore, I am grateful to Emily and to Jacques Lezra for having given me the opportunity to develop and test my ideas for this project during an engaging NYU Summer Seminar on "The Problem of Translation," sponsored by the Andrew Mellon Foundation in 2011.

I have the good fortune of being surrounded by colleagues from whose intelligence, expertise, and humor I benefit on a daily basis. I thank David Damrosch for his incessant support and contagious cheerfulness. David's encouraging words prompted me to embark on a project that became more comparative than I would ever have imagined. John Hamilton's encyclopedic knowledge and love of philology have been crucial in shaping my own thought. I am immensely thankful for his steady support and ever-present intellectual exchange. I wish to express my heartfelt gratitude to Kay Shelemay for her fine friendship and for being such an inspiring role model—her knowledge and wisdom, paired with generosity and care, have been meaningful to me beyond measure. Karen Thornber has been a most important source of inspiration since my first days at Harvard, and I am grateful for her fantastic collegiality. For an extraordinary *entretien infini* I wish to extend my warmest thanks to Tom Conley, whose rare wit, knowledge, and friendship have redirected not only my work on maps, but also the way I look at the world. I thank, in this regard, Homi Bhabha and Steven Biel for including our "Cartography Seminar" in the program of the Mahindra Humanities Center. Marc Shell's interest in this project and our conversations on an infinite range of topics (but perhaps ultimately on islands) have significantly influenced my writing process. I am, furthermore, grateful to Mariano Siskind for our numerous exhilarating conversations. His unmatched generosity, paired with astute thought, provided a marvelously warm welcome as soon as I had set foot on campus.

I am especially indebted to friends and colleagues who carefully read and wisely commented on drafts of various chapters: Hélio Alves, Vincent Barletta, Joe Blackmore, Tom Conley, David Damrosch, John Hamilton, Jakub Niedźwiedź, Elizabeth Mellyn, Tim Reiss, Kay Shelemay, and Phillip Usher. From the time I was pondering what shape *Cartographic Humanism* would take to its last stages, graduate students who took my seminar on "Cartography and Early Modern Literature" (in 2013, 2014, and 2018) played a most formative role. I thank, in particular, Manuel Azuaje-Alamo, Nina Beguš, Daniel Bergmann, Corrado Confalonieri, Valentina Frasisti, Alex Lambrow, Natalie Nogueira, William Porter, Chiara Trebaiocchi, Gideon Unkeless, Hudson Vincent, and May Wang for pushing my thinking.

Many of this book's insights were triggered by exquisite questions and generous comments I have received from colleagues in informal conversations—over lunches, dinners, teas, coffees, in hallways, at conferences, or during walks. I thank, in particular, Daniel Aguirre Oteiza, Albert Ascoli, Jean-Marc Besse, Ann Blair, Svetlana Boym (whose absence is felt), Julie Buckler, Ted Cachey, Giancarlo Casale, Verena Conley, Joseph Connors, Virginia Cox, Surekha Davies, Emma Dench, Jörg Dünne, David Elmer, Anders Engberg-Pederson, Jim Engell, Francesco Erspamer, Franco Farinelli, Valeria Finucci, Michael Flier, Grzegorz Franczak, David Frick, Jay Garcia, Luís

Girón-Negrón, Bill Granara, Stephen Greenblatt, Virginie Greene, Sylvaine Guyot, Timothy Hampton, Eric Hayot, Tamar Herzog, Cemal Kafadar, Annabel Kim, Valerie Kivelson, Michael Kunichika, Valentina Lepri, Frank Lestringant, Annette Lienau, Françoise Lionnet, Shannon McHugh, Sandra Naddaff, Donald Ostrowski, Katharine Park, Simone Pinet, Serhii Plokhii, Martin Puchner, Ayesha Ramachandran, Eugenio Refini, Panagiotis Roilos, Judith Ryan, Dana Sajdi, Stephanie Sandler, Jeffrey Schnapp, Deanna Shemek, Daniel Lord Smail, Doris Sommer, Diana Sorensen, David Stern, Michael Stolz, Ramie Targoff, Gordon Teskey, Martin Thiering, Bill Todd, Teresa Ulewiczowa, Kate Van Orden, Jean-Claude Vuillemin, Nicholas Watson, Justin Weir, Naomi Weiss, Leah Whittington, Kären Wigen, Toby Wikström, Larry Wolff, and Saul Zaritt.

In 2015–16, I had the good fortune of spending a year as a Distinguished Junior External Fellow at the Stanford Humanities Center. This book would not have been completed (in a timely manner and in its current shape) without the time I spent there, surrounded by new friends and congenial colleagues with whom I was able to discuss my ideas while inhaling the scent of eucalyptus trees. I wish to express my gratitude, in particular, to Caroline Winterer, Andrea Davies, Bob Barrick, Patricia Terrazas, and the SHC team for their extraordinary and hard work; Jenna Gibbs and Peter Reill, as well as Kay and Jack Shelemay, for their friendship; Ruth Ahnert, Barney Bates (dearly missed), Günter Blamberger, Scott Bukatman, Niloofar Haeri, Stefan Hoffmann, Jennifer Iverson, and Nancy Kollmann for always engaging conversations; and colleagues at Stanford, who have enriched my year in California in manifold ways: Vincent Barletta, Ximena Briceño, Margaret Cohen, Ewa Domańska, Paula Findlen, Roland Greene, Monika Greenleaf, Sepp Gumbrecht, Robert Pogue Harrison, Héctor Hoyos, Ivan Lupić, Patricia Parker, Nancy Ruttenburg, Kathryn Starkey, and Laura Wittman. My special thanks to Anna Ranieri and Stephen Boyd for making my year at Stanford more homey and welcoming than I could have possibly hoped for.

It was a great privilege to spend the spring of 2017 as a research fellow at the Internationales Forschungszentrum Kulturwissenschaften in Vienna. Thomas Macho and the dedicated IFK team worked hard to create a most stimulating intellectual environment which turned my stay at the Center into a most inspiring research semester—my heartfelt thanks to them and to all the fellows for most thought-provoking conversations, in particular to Eve Blau, Corina Caduff, Sabina Folnović-Jaitner, Golan Gur, Ross Lipton, Winfried Meyer, Thilo Neidhöfer, Mareike Peschl, and Jens Schröter. My family and friends in Vienna made my time in Austria particularly memorable.

This book project was supported by numerous fellowships and grants from Harvard University. I am particularly grateful to acknowledge the John F. Cogan Junior Faculty Leave Fellowship from the Davis Center for Russian and Eurasian Studies, the William F. Milton Endowment Research Grant, the Clark Research Award, the Anne and Jim Rothenberg Humanities Research Grant, the GSAS Junior Faculty Research Assistance Grant, and the Radcliffe Institute for Advanced Study Exploratory Workshop Grant. The book has also been aided immeasurably by Harvard's FAS and divisional deans—Michael Smith, Claudine Gay, Diana Sorenson, and Robin Kelsey—who granted me both funding and leave time, necessary to complete the book. I spent substantial time poring over early modern maps and texts in US and European libraries and benefited greatly from the generous help and vast knowledge of librarians and curators. I wish to thank, in particular, David Weimer from the Harvard Map Collection and Susan Halpert from Harvard's Houghton Library for their unwavering support.

For their generous invitations to give talks and test my ideas on cartography, in different stages, in front of a broader audience I thank, in particular, Albrecht Koschorke, Kirsten Mahlke, and Nicolas Detering from the Europa-Zentrum at Konstanz University; Roland Greene and Nicholas Fenech from the Renaissance Worldmaking workshop at the Stanford Humanities Center; Jakub Niedźwiedź from the Katedra Literatury Staropolskiej (Department of Old Polish Literature) at the Jagiellonian University in Kraków; Ellen Boucher from the European Studies Program at Amherst College; Monika Raič from the Department of Comparative Literature at Innsbruck University; Matts Hallenberg and Charlotta Forss from the Department of History at Stockholm University; Magalí Armillas-Tiseyra and Jonathan Eburne from the Pennsylvania State University (Comparative Literature Luncheon); Kathryn Starkey and Elaine Treharne from the Center for Medieval and Early Modern Studies at Stanford University; Fernando Loffredo from Stony Brook (for a repeated invitation, Art Department and Center for Italian Studies); Tanja Michalsky and Ted Cachey from the Bibliotheca Hertziana and the University of Notre Dame Rome; Tanja Michalsky and Martin Thiering from the Bibliotheca Hertziana/Max Planck Institut in Rome; Donald Ostrowski from the Early Slavists' Seminar at Harvard; Rawi Abdelal from the Davis Center for Russian and Eurasian Studies at Harvard; and Marta Wojtkowska-Maksymik from the Instytut Literatury Polskiej (Institute of Polish Literature) at Warsaw University. I also wish to acknowledge the support of Abbey Berg and James Stenerson from Pace University for including me in the Faculty Research Forum during my time as assistant professor at Pace (2012–13). The Forum was a crucial sounding board for my book ideas at an early stage.

I have been immensely privileged to work on this book with Randolph Petilos from the University of Chicago Press: his enthusiasm for the topic has been inspirational for me since we first were in touch, when I was transforming my ideas on Europe and cartographic humanism into a coherent argument. I could not have hoped for a more congenial, warm, and generous interlocutor. Great thanks are due to Jo Ann Kiser for copyediting my manuscript. Hudson Vincent and Marco Romani-Mistretta have been absolutely fabulous research assistants at Harvard, while Penelope Burt and Troy Tower worked wonders with my prose. This publication was subsidized in part by Harvard Studies in Comparative Literature (Schofield Publication Fund) and by Harvard's FAS Tenure-Track Faculty Publication Fund.

Numerous friends have contributed, over many years and in manifold ways, to this book, and it is a great honor and a pleasure to acknowledge them: Sage Anderson, Magalí Armillas-Tiseyra, Marlen Bidwell-Steiner, Guido Caniglia, Giada Ceri, Luciana Ceri, Kahlil Chaar, Yali Dekel, Catherine Deutsch, Monika Dockendorff, Javier Domingo, Guadalupe Gonzalez Dieguez, Nikolaus Hink, Adrián Izquierdo, Eva Kernbauer, Racha Kirakosian, Maarten Knechtelsdorfer, Sergius Kodera, Aleksander Krzyżowski, Ewa Krzyżowska, Katarzyna Kurkiewicz, Justine Landau, Gabriella Lerario, Fernando Loffredo, Daniel Lukes, Dane Matthews, Elizabeth Mellyn, Ali Mostashari, Martin Müller, Jonathan Mullins, Stefano Nencioni, Barbara Nestola, Maria Pia Paoli, Joanna Partyka, Fernando Pérez Villalón, Joseph Perna, Beata Potocki, Michaela Reimers, Nicole Sütterlin, Roberto Teichner, Eran Tzelgov, Javier Uriarte, Phillip Usher, Chris Van Ginhoven, Caren Van Houwelingen, Almudena Vidorreta Torres, Vissia Viglietta, Erica Weitzman, Daniel Winkler, Jalal Zaïm, and Simona Zecchi.

My boundless gratitude goes to my family—this book is in many ways also theirs: my parents, Małgorzata and Wiesław Piechocki, who have supported me unconditionally and nurtured my love for languages, arts, and the humanities in communist Poland, through rocky times as political refugees and, at first, stateless dwellers in Austria, and smoother periods across continents; my sisters Claudia and Julia, who are not only the most extraordinary siblings but also the most precious friends one can possibly hope for; my wonderful brothers-in-law Werner Steiner and Pedro Serra, whom I wish I could see more often for merry times; Miruna and Dante—fortunate enough to have not only "il mezzo del cammin," but their entire lives ahead of them; and Piotr Paczulla (more brother than cousin) and Beata Stebnicka for reminding me, with great humor, excellent stories, and immense affection, of my real, imagined, and possible roots. Finally, I owe more than words can say to Anthony Sepulveda, the only one who knows how to make lemonade out of rocks: thank you for being in my life.

Parts of chapter 4 have previously appeared as "Syphilologies: Fracastoro's Cure and the Creation of Immunopoetics" in *Comparative Literature* 68, no. 1 (2016) and are republished by permission of the publisher, Duke University Press.

NOTES

INTRODUCTION

1. Marshall G. S. Hodgson, *Rethinking World History: Essays on Europe, Islam, and World History*, ed. Edmund Burke, vol. 3 (Cambridge: Cambridge University Press, 1993), 4.

2. For a description of the Hereford Map, see David Woodward, "Medieval *Mappaemundi*," in *The History of Cartography*, vol. 1: *Cartography in Prehistoric, Ancient, and Medieval Europe and the Mediterranean*, ed. J. B. Harley and David Woodward (Chicago: University of Chicago Press, 1987), 286–370; and Evelyn Edson: *The World Map, 1300–1492: The Persistence of Tradition and Transformation* (Baltimore: Johns Hopkins University Press, 2007), in particular 11–30.

3. *The New Oxford Annotated Bible* (Oxford: Oxford University Press, 2001), Genesis 9:19.

4. See Woodward, "Medieval *Mappaemundi*," 345. Marcia Kupfer has recently argued that what is commonly considered a limner's error must, instead, be embedded within the larger internal logic of the map. The toponymic inversion can then be understood as an astute and self-conscious play with mirror images and speculation; see Marcia Kupfer, *Art and Optics in the Hereford Map: An English Mappa Mundi, c. 1300* (New Haven, CT: Yale University Press, 2016).

5. Michael Eitzinger, *De Europae Virginis, Tauro insidentis, topographica atque historica descriptione, liber* (Cologne: Gottfried van Kempen, 1588). On Eitzinger's "Europa" map see also Michael Wintle, *The Image of Europe: Visualizing Europe in Cartography and Iconography throughout the Ages* (Cambridge: Cambridge University Press, 2009), 250–51.

6. See Peter Meurer, "Europa Regina. Sixteenth-Century Maps of Europe in the Form of a Queen," in *Belgeo* 3–4 (2008): 1–16: 12. See also Peter Meurer, "Cartography in the German Lands, 1450–1650," in *The History of Cartography*, ed. David Woodward, vol. 3: *Cartography in the European Renaissance*, part 2 (Chicago: University of Chicago Press, 2007), 1172–1245: 1235.

7. Meurer, "Europa Regina," 6.

8. Camões repeatedly refers to Europe as "soberba," although the adjective is sometimes lost in translation; see Luís de Camões, *Os Lusíadas*, trans. Hayden White (Oxford: Oxford University Press, 2001), 2.80, 41 and 3.6, 49.

9. Dipesh Chakrabarty, *Provincializing Europe: Postcolonial Thought and Historical Difference* (Princeton, NJ: Princeton University Press, 2000), 27–28.

10. The "cartographic turn" of the 1980s and 1990s has produced numerous important books and essays underscoring the significance of cartography across all the disciplines of the humanities. These include the six-volume *The History of Cartography* project, published by the University of Chicago Press (vol. 1 was published in 1987, while vol. 5, the last volume, is forthcoming); J. B. Harley, "Deconstructing the Map," *Cartographica* 26, no. 2 (1989): 1–20; Doreen B. Massey, *Space, Place, and Gender* (Minneapolis: University of Minnesota Press, 1994); Denis Cosgrove, ed., *Mappings* (London: Reaktion, 1999); Christian Jacob, *The Sovereign Map: Theoretical Approaches in Cartography throughout History*, trans. Tom Conley, ed. Edward H. Dahl (Chicago: University of Chicago Press, 2006); Karen Elizabeth Bishop, ed., *Cartographies of Exile: A New Spatial Literacy* (New York: Routledge, 2016); Anders Engberg-Pederson, ed., *Literature and Cartography* (Cambridge, MA: MIT Press, 2017); and Franco Farinelli, *Blinding Polyphemus: Geography and the Models of the World*, trans. Christina Chalmers (York, PA: Seagull Books, 2018).

11. Attempts at defining (the idea of) Europe are, of course, countless. They range from Denys Hay's pioneering study on "the idea of Europe" to Roberto Esposito's recent repositioning of philosophical centers in (Western) Europe. See Denys Hay, *Europe: The Emergence of an Idea* (Edinburgh: Edinburgh University Press, 1957); J. R. Hale, *Renaissance Europe: Individual and Society, 1480–1520* (New York: Harper & Row, 1971); Robert Bartlett, *The Making of Europe: Conquest, Colonization and Cultural Change 950–1350* (Princeton, NJ: Princeton University Press, 1993); Hodgson, *Rethinking World History*; Roberto Esposito, *A Philosophy for Europe: From the Outside*, trans. Zakiya Hanafi (Cambridge: Polity Press, 2018).

12. Walter Mignolo, *Local Histories/Global Designs* (Princeton, NJ: Princeton University Press, 2012), 236.

13. Chakrabarty, *Provincializing Europe*, 4.

14. Immanuel Wallerstein, *The Modern World-System: Capitalist Agriculture and the Origins of the European World-Economy in the Sixteenth Century* (New York: Academic Press, 1976).

15. Walter Mignolo, *The Darker Side of the Renaissance: Literacy, Territoriality, & Colonization* (Ann Arbor: University of Michigan Press, 2003), xi.

16. The "spatial turn" (which tentatively took shape in the 1950s and unfolded, first in the social sciences, from the 1970s and 1980s on) has productively carried over to the humanities. Important works include Gaston Bachelard, *The Poetics of Space: The Classic Look at How We Experience Intimate Places*, trans. Maria Jolas (Boston: Beacon Press, 1969); Michel de Certeau, *The Practice of Everyday Life*, trans. Steven Rendall (Berkeley: University of California Press, 1984); Michel Foucault, "Of Other Spaces," trans. Jay Miskowiec, *Diacritics* 16, no. 1 (1986): 22–27; Gilles Deleuze and Félix Guattari, *A Thousand Plateaus: Capitalism and Schizophrenia*, trans. Brian Massumi (Minneapolis: University of Minnesota Press, 1987); Gilles Deleuze, *The Logic of Sense*, trans. Mark Lester (New York: Columbia University Press, 1990); Henri Lefebvre, *The Production of Space*, trans. Donald Nicholson-Smith (Cambridge: Blackwell, 1991); David Harvey, *Justice, Nature, and the Geography of Difference* (Malden, MA: Blackwell Publishers, 1996); Frederic

Jameson, "Cognitive Mapping," in *Marxism and the Interpretation of Culture*, ed. Cary Nelson and Lawrence Grossberg (Urbana: University of Illinois Press, 1988), 347–57.

17. For the English translation, see Ptolemy, *Geography*, ed. Luther Stevenson, trans. Joseph Fischer (New York: Cosimo Classics, 2011).

18. Recent interdisciplinary studies, often (but not always) spearheaded by historians, classicists, and literary scholars, show a greater awareness of this geographically reduced understanding of Europe and offer more inclusive perspectives on Europe. Such in-depth studies include Almut-Barbara Renger and Roland Alexander Ißler, eds., *Europa—Stier und Sternenkranz. Von der Union mit Zeus zum Staatenverbund* (Bonn: Bonn University Press, 2009); Peter Hanenberg, ed., *Der literarische Europa-Diskurs: Festschrift für Paul Michael Lützeler zum 70. Geburtstag* (Würzburg: Königshausen & Neumann, 2013); and Nicolas Detering, *Krise und Kontinent: Die Entstehung der deutschen Europa-Literatur in der Frühen Neuzeit* (Cologne: Böhlau Verlag, 2017).

19. For a detailed and insightful history of continents, see Martin W. Lewis and Kären E. Wigen, *The Myth of Continents: A Critique of Metageography* (Berkeley: University of California Press, 1997). The number of continents today is still not consolidated across the globe. If children in twenty-first-century US elementary schools learn about the existence of seven continents [Africa, Antarctica, Asia, Australia (Australasia or Oceania), Europe, North America, and South America], their European counterparts, for instance, often leave school with a fivefold continental scheme in mind (Africa, Asia, Europe, America, and Australia)—a scheme that persists, among others, in the iconic image of the five interlacing Olympic rings invented at the beginning of the twentieth century.

20. To investigate Renaissance humanism and early modern literature through the lens of cartography is a relatively recent development. While several scholars have brought these two strands together in the past twenty years (following the "cartographic turn" in literary studies with the pioneering work, among others, of Tom Conley, Frank Lestringant, and Theodore J. Cachey Jr. since the 1990s), *Cartographic Humanism* is the first study to take cartography as a major catalyst for the rise of Renaissance humanism itself, by broaching the topic from a broad comparative and geographically inclusive perspective. Important studies on cartography and literature of the early modern period include Frank Lestringant, *L'atelier du cosmographe: ou l'image du monde à la Renaissance* (Paris: A. Michel, 1991); Tom Conley, *The Self-Made Map: Cartographic Writing in Early Modern France* (Minneapolis: University of Minnesota Press, 1996); Theodore J. Cachey Jr., "Petrarch, Boccaccio, and the New World Encounter," *Stanford Italian Review* 10, no. 1 (1991): 45–59, and "Cartographic Dante," *Italica* 87, no. 3 (2010): 325–54; Jean-Marc Besse, *Les grandeurs de la terre: Aspects du savoir géographique à la Renaissance* (Lyon: ENS, 2003); Ricardo Padrón, *The Spacious Word: Cartography, Literature, and Empire in Early Modern Spain* (Chicago: University of Chicago Press, 2004); Phillip Usher, *Errance et cohérence. Essai sur la littérature transfrontalière à la Renaissance* (Paris: Editions Classiques Garnier, Zolo); Ayesha Ramachandran, *The Worldmakers: Global Imagining in Early Modern Europe* (Chicago: University of Chicago Press, 2015).

21. Roberto Esposito, *Da fuori: Una filosofia per l'Europa* (Turin: Einaudi, 2016), 3. While he references geography, Esposito does not engage with questions of space which he deems to flatten, rather than enrich, a "philosophy for Europe."

22. See Christy Constantakopoulou, *The Dance of the Islands: Insularity, Networks, the Athenian Empire and the Aegean World* (Oxford: Oxford University Press, 2007), 23.

23. See Lewis and Wigen, *The Myth of Continents*, 21.

24. Constantakopoulou, *The Dance of the Islands*, 229. Miletus used its insular possessions, as Hecataeus of Miletus suggested, "as a refuge in case Miletus fell into Persian hands." Constantakopoulou, *The Dance of the Islands*, 229.

25. Constantakopoulou, *The Dance of the Islands*, 19.

26. Strabo, *Geography*, trans. Horace Leonard Jones (Cambridge, MA: Harvard University Press, 2014), 3.5.3, 128–29.

27. Homer, *Iliad*, trans. A. T. Murray (Cambridge, MA: Harvard University Press, 2014), 2.635, 108–9. See also Constantakopoulou, *The Dance of the Islands*, 234.

28. Farinelli, *Blinding Polyphemus*, 34. This book is a translation of Franco Farinelli, *Geografia: Un'introduzione ai modelli del mondo* (Turin: Einaudi, 2003).

29. The word απειρος is the Doric form of ἤπειρος: in the *Women of Trachis* Sophocles refers to Europe and Asia as "twin continents" (δισσαῖσιν ἀπείροις) by using απειρος, while in his poem "Europa" Moschus, defining Europe as "the land opposite (ἀντιπέρην)" of Asia, designates both as the "two continents" (ἠπείρους δοιὰς) (subsuming Africa under Asia). Sophocles, *The Women of Trachis*, trans. Hugh Lloyd-Jones (Cambridge, MA: Harvard University Press, 1994), 101, 140–41. I slightly changed the English translation. Moschus, "Europa," in *Theocritus, Moschus, Bion*, trans. Neil Hopkinson (Cambridge, MA: Harvard University Press, 2015), 2.9–10; 2.8, 450–51. Marc Shell has recently argued that "the logical definition of *island* is linked with the logical circumscription of definition in a way that cannot avoid the linguistics and natural history of islands." Marc Shell, *Islandology: Geography, Rhetoric, Politics* (Stanford, CA: Stanford University Press, 2014), 8.

30. Strabo, *Geography*, 1.1., 3.

31. See George Tolias, "*Isolarii*, Fifteenth to Seventeenth Century," in *The History of Cartography*, ed. David Woodward, vol. 3: *Cartography in the European Renaissance*, part 1 (Chicago: University of Chicago Press, 2007), 263–84: 265, and *Îles et Insulaires (XVIe-VIIIe siècle)*, ed. Frank Lestringant and Alexandre Tarrête (Paris: Presses de l'Université Paris-Sorbonne, 2017). See also Shell, *Islandology*, 21.

32. Herodotus, *The Histories*, 1.17.3. Quoted by Constantakopoulou, *The Dance of the Islands*, 229.

33. Thucydides, *Peloponnesian War*, 1.15.2. Quoted by Constantakopoulou, *The Dance of the Islands*, 85. My emphasis.

34. Fernand Braudel forged the expression of "almost islands" to describe areas of experienced, not necessarily geographic, isolation. Thriving in the tension between connectivity and separation (safety), Athens was a complex case of such an "almost island." See Fernand Braudel, *The Mediterranean and the Mediterranean World in the Age of Philip II* (London: HarperCollins, 1972), 160–61, discussed in Constantakopoulou, *The Dance of the Islands*, 16.

35. See Benedetto Bordone, *Libro . . . nel quale si ragiona de tutte l'isole del mondo con li lor nomi antichi, moderni, historie, favole, modi del loro vivere, in qual parte del mare stanno, in qual parallelo & clima giacciono* (Venice: Zoppino, 1528).

36. See Constantakopoulou, *The Dance of the Islands*, 18.

37. See Lewis and Wigen, *The Myth of Continents*, 22.

38. Strabo, *Geography*, 1.4.7., 243.
39. Strabo, *Geography*, 1.4.7., 244–47.
40. Strabo, *Geography*, 1.4.7., 244–45.
41. Ptolemy, *Cosmographia*, trans. Jacopo Angeli da Scarperia (Rome: Arnoldus Buckinck, 1478), 1.5, n.p./J. Lennart Berggren and Alexander Jones, eds., *Ptolemy's Geography: An Annotated Translation of the Theoretical Chapters* (Princeton, NJ: Princeton University Press, 2000), 1.5, 63. I slightly altered the English translation.
42. Patrick Gautier Dalché, "The Reception of Ptolemy's *Geography* (End of the Fourteenth to Beginning of the Sixteenth Century)," in *The History of Cartography*, ed. David Woodward, vol. 3: *Cartography in the European Renaissance*, part 1 (Chicago: University of Chicago Press, 2007), 285–364: 304.
43. *Oxford Latin Dictionary*, ed. P. G. W. Glare, vol. 1 (Oxford: Oxford University Press, 2012), entry "continuō," 474.
44. Isidore of Seville, *Liber ethymologiarum* (Basel: Michael Furter, 1489), 14.8.9., fol. 75. For the English translation, see Isidore of Seville, *Etymologies*, trans. Stephen A. Barney, W. J. Lewis, J. A. Beach, and Oliver Berghof (Cambridge: Cambridge University Press, 2006), 14.8.9., 300.
45. *Oxford Latin Dictionary*, entry "contineō," 472.
46. John A. Marino, "The Invention of Europe," in *The Renaissance World*, ed. John Jeffries Martin (New York: Routledge, 2007), 140–65: 141.
47. Marino, "The Invention of Europe," 141.
48. Hay, *Europe*, 96.
49. Christopher Columbus, *The 'Libro de las profecías'*, trans. Delno C. West and August King (Gainesville, FL: University of Florida Press, 1991); and Christopher Columbus, *The Book of Prophecies*, ed. Roberto Rusconi (Berkeley: University of California Press, Repertorium Columbianum, 1997). See also Stephen Greenblatt, *Marvelous Possessions: The Wonder of the New World* (Chicago: University of Chicago Press, 1991), 26–51; and Carol Delaney, *Columbus and the Quest for Jerusalem* (New York: Free Press, 2011).
50. Marino, "The Invention of Europe," 141.
51. Ptolemy describes Italy in the third book of his *Geography*, dedicated to "partis Europae orientalis iuxta subiectas provincias aut satrapas" (the provinces and prefectures found in that part of Europe which lies toward the east). Ptolemy, *Geographia*, ed. Marco Beneventano and Giovanni Cotta (Rome: Bernardino Vitali for Evangelista Tosino, 1507), n.p. For the English translation, see Ptolemy, *Geography*, trans. Joseph Fischer, 71.
52. This did not prevent Pope Pius II from assimilating Christians and Europeans in his other writings, in particular since it was during his pontificate (in 1461) that the last Christian stronghold in Asia Minor, Trebizond, fell to the Ottomans. Even so, when Hay writes that "Europe and Christendom are identified by the pope: or rather, Christendom is seen as *radiating out from* a European base," he adds a significantly more cautious qualification. See Hay, *Europe*, 84–85. My emphasis.
53. Giovanni Boccaccio, *Il Comento alla Divina Commedia e gli altri scritti intorno a Dante*, ed. Domenico Guerri, vol. 3 (Bari: Laterza, 1918), 180; Marino, "The Invention of Europe," 141. The expression "Boccaccio the geographer" was coined in Roberta Morosini, *Boccaccio geografo: Un viaggio nel Mediterraneo tra le città, i giardini e il 'mondo' di Giovanni Boccaccio* (Florence: Polistampa, 2010).

54. Maciej Miechowita, *Tractatus de duabus Sarmatiis Asiana et Europiana et de contentis in eis* (Kraków: Johannes Haller, 1517).

55. Lewis and Wigen, *The Myth of Continents*, 2.

56. Even though isolated historical studies of the cartographic underpinnings of early modern Europe and its formation as a continent are slowly emerging, such as Valerie Kivelson's synoptic "The Cartographic Emergence of Europe" and Michael Wintle's sweeping *The Image of Europe* (a comprehensive overview of visual representations of Europe across the ages), studies by (comparative) literary scholars do not exist. See Valerie A. Kivelson, "The Cartographic Emergence of Europe," in *The Oxford Handbook of Early Modern European History, 1350–1750*, ed. Hamish Scott, vol. 1: *People and Place* (Oxford: Oxford University Press, 2015), 37–69; and Wintle, *The Image of Europe*.

57. My coinage is inspired by Giuseppe Mazzotta's *cosmopoiesis*. See Giuseppe Mazzotta, *Cosmopoiesis: The Renaissance Experiment* (Toronto: University of Toronto Press, 2001).

58. The *oikoumene* was depicted as a single continent on many fifteenth-century *mappae mundi*, including those of Andreas Walsperger (from 1448), Giovanni Leardo (1452–53), and Fra Mauro (1457–59). See Denis Cosgrove, "Images of Renaissance Cosmography, 1450–1650," in *The History of Cartography*, ed. David Woodward, vol. 3: *Cartography in the European Renaissance*, part 1 (Chicago: University of Chicago Press, 2007), 55–98: 59.

59. The early modern rise of continental thinking is different from "thinking continental," an expression recently employed by Susan Naramore Maher, who investigate the figuring of Deep Time against the backdrop of environmental threats. See Susan Naramore Maher et al., *Thinking Continental: Writing the Planet One Place at a Time* (Lincoln: University of Nebraska Press, 2017).

60. Hay, *Europe*, xxii–xxiii. This statement appears only in the second edition. Quoted by Benjamin Braude, "The Sons of Noah and the Construction of Ethnic and Geographical Identities in the Medieval and Early Modern Periods," *William and Mary Quarterly* 54, no. 1 (1997): 103–42: 110.

61. See Hans Baron, *The Crisis of the Early Italian Renaissance: Civic Humanism and Republican Liberty in an Age of Classicism and Tyranny* (Princeton, NJ: Princeton University Press, 1955). Nicolas Mann defines humanism's concern with the legacy of antiquity as involving "above all the rediscovery and study of ancient Greek and Roman texts, the restoration and interpretation of them and the assimilation of ideas and values that they contain. . . . And in this way it was to become the embodiment of, and vehicle for, that very classical tradition that is the most fundamental aspect of the continuity of European cultural and intellectual history." Nicolas Mann, "The Origins of Humanism," in *The Cambridge Companion to Renaissance Humanism*, ed. Jill Kraye (Cambridge: Cambridge University Press, 2011), 1–19: 2.

62. Mignolo, *The Darker Side of the Renaissance*, 1.

63. Mignolo, *The Darker Side of the Renaissance*, 256–58.

64. Doreen Massey has long called for a move to "decentre" Europe, to recognize it "as merely one (though most certainly in military and other terms the most powerful) of the histories being made at that time [the early modern period]." Her intention is not, however, to look at the complexity and instability of Europe itself. Doreen Massey, *For Space* (London: Sage Publications, 2005), 63.

65. Divided into eight books, Ptolemy's *Geography* is framed by a theoretical introductory chapter (book one), which provides an insight, among others, into different projection methods, and a concluding chapter (book eight), an atlas *avant la lettre* which contains a world map and a series of twenty-six regional maps (none of which are extant): ten maps dedicated to Europe, four to Africa, and twelve to Asia. Ptolemy conceives two books on Europe (books two and three), one on Africa (book four), and three on Asia (books five to seven).

66. Having reached northwestern Europe via Islamic Spain as early as the eleventh century, the concept of terrestrial longitudes and latitudes was "sufficiently known in the thirteenth century." We have evidence of single scholars such as Roger Bacon using the coordinate system for geographic purposes. But the true breakthrough of the concept occurred once Ptolemy's *Geography* started circulating. See David Woodward, "Roger Bacon's Terrestrial Coordinate System," *Annals of the Association of American Geographers* 80, no. 1 (1990): 109–22: 118.

67. David Woodward developed the three types of imagining and plotting cartographic space in Woodward, "Roger Bacon's Terrestrial Coordinate System," 119.

68. Farinelli, *Blinding Polyphemus*, 167.

69. David Woodward, "Cartography and the Renaissance: Continuity and Change," in *The History of Cartography*, ed. David Woodward, vol. 3: *Cartography in the European Renaissance*, part 1 (Chicago: University of Chicago Press, 2007), 3–24: 12.

70. The *Compact between Spain and Portugal*, signed by the Catholic Sovereigns at Madrid, May 7, 1495, demanded that "the line of the said division be placed on all navigational charts [*cartas de marear*] made hereafter in our kingdoms and territories by those journeying in the said ocean sea." Frances G. Davenport, ed., *European Treaties Bearing on the History of the United States and its Dependencies to 1648* (Washington, DC: Carnegie Institution of Washington, 1917), 103–5. I have slightly changed the English translation of the Spanish original.

71. See Alison Sandman, "Spanish Nautical Cartography in the Renaissance," in *The History of Cartography*, ed. David Woodward, vol. 3: *Cartography in the European Renaissance*, part 1 (Chicago: University of Chicago Press, 2007), 1095–1142: 1108; and Jesús Varela Marcos et al., eds., *El Tratado de Tordesillas en la Cartografía Histórica* (Valladolid: Sociedad V Centenario del Tratado de Tordesillas, 1994).

72. Farinelli offers as an example of the embodied experience of space the case of the medieval traveler Marco Polo, who measured space with time, the uneven experience of days and months in which "every day the things of the world reveal[ed] their proper duration to him." Farinelli, *Blinding Polyphemus*, 18.

73. Farinelli, *Blinding Polyphemus*, 30.

74. Robert D. Sack, *Homo Geographicus: A Framework for Action, Awareness, and Moral Concern* (Baltimore: Johns Hopkins University Press, 1997), 1.

75. Ramachandran, *The Worldmakers*, 22.

76. Franco Farinelli, *La crisi della ragione cartografica* (Turin: Einaudi, 2009), 13. For Heidegger's influential essay, see Martin Heidegger, "The Age of the World-Picture," in *The Question Concerning Technology and Other Essays*, trans. William Lovitt (New York: Harper Torchbooks, 1977).

77. Woodward, "Roger Bacon's Terrestrial Coordinate System," 119.

78. "The cartographic partition of Africa," Thomas J. Bassett reminds us, "inextricably linked mapmaking and empire building." Thomas J. Bassett, "Cartography and Empire Building in Nineteenth-Century West Africa," *Geographical Review* 84, no. 3 (1994): 316–35: 316. The act of drawing arbitrary territorial lines across the African continent at the Berlin Congo Conference uprooted entire peoples and led to (still ongoing) violent struggles, civil wars, and mass migration—all this along with economic and environmental disasters.

79. Gerard González Germain, "The Copy of Ptolemy's *Geography* (1478) owned by the Vespuccis," in *Shores of Vespucci: A Historical Research of Amerigo Vespucci's Life and Contexts*, ed. Angelo Cattaneo (Berlin: Peter Lang, 2018), 87–100: 87. See also Gautier Dalché, "The Reception of Ptolemy's *Geography*," 361.

80. The tripartite type of a *mappa mundi* is commonly known as the T-O map, showing the three continents as segments of a circle (typically associated with the *okeanos*), the "O," divided by three bodies of water in the shape of a "T": the Mediterranean Sea, the Nile, and the Don (Tanais) River. See Woodward, "Medieval *Mappaemundi*," 297.

81. Isidore of Seville, *Etymologiae* (Augsburg: Günther Zainer, 1472).

82. As Walter Mignolo has pointed out, the oft-forgotten name of fifth-century CE Cosmas Indicopleustes needs to be brought into the discussion: "His *Christian Topography* is one of the first descriptions of a tripartite and Christian conception of the earth." Mignolo, *The Darker Side of the Renaissance*, 232.

83. Braude, "The Sons of Noah," 111. What is more, in Jewish exegesis, in the midrashic commentary on Genesis 9–10, Japheth's sons "are placed in the Euphrates Valley, Thrace, and Africa." Braude, "The Sons of Noah," 111.

84. Braude, "The Sons of Noah," 112–13.

85. Cosgrove, "Images of Renaissance Cosmography, 1450–1650," 59.

86. Braude, "The Sons of Noah," 107. A particularly striking example is the tradition of pseudoepigraphical medieval Christian commentators, who invented a fourth son of Noah, Jonathan, "who did not long survive the introduction of the printed Bible." While the *Nuremberg Chronicle*, a masterpiece of fifteenth-century printing from 1493, included this tradition, "a century later, [Jonathan's] existence was dismissed." Braude, "The Sons of Noah," 108 [note 9].

87. Farinelli, *Blinding Polyphemus*, 13.

88. Strabo, *Geography*, 2.4.7., 415.

89. Pliny the Elder, *Natural History*, trans. Harris Rackham (Cambridge, MA: Harvard University Press, 1962), 3.1.5., 5; 7.

90. Jerry Brotton, "A 'Devious Course': Projecting Toleration on Mercator's 'Map of the World', 1569," *Cartographic Journal* 49, no. 2 (2012): 101–6: 101.

91. Brotton, "A 'Devious Course'," 101.

92. Hodgson, *Rethinking World History*, 4–5.

93. Mark Monmonier, *Drawing the Line: Tales of Maps and Cartocontroversy* (New York: Henry Holt, 1995), 1.

94. J. B. Harley, "The Map and the Development of the History of Cartography," in *The History of Cartography*, vol. 1, 1–42: 1.

95. Conley, *The Self-Made Map*, 1.

96. Conley, *The Self-Made Map*, 5.

97. For the impact of the mathematical *quadrivium* upon the language-centered *trivium* in sixteenth-century Europe, see Timothy J. Reiss, *Knowledge, Discovery and Imagination in Early Modern Europe: The Rise of Aesthetic Rationalism* (Cambridge: Cambridge University Press, 1997).

98. Mann, "The Origins of Humanism," 2–3; and Cachey, "Petrarch, Boccaccio, and the New World Encounter," 45–59. While the interest of Italy's *tre corone* in geography and, to a lesser extent, cartography is increasingly recognized, the impact of cartography as a driving force of their work—and humanism writ large—has not yet been fully explored.

99. Ramachandran, *The Worldmakers*, 17.

CHAPTER ONE

1. For the Latin edition, see Hartmann Schedel, *Liber Chronicarum* (Nuremberg: Anton Koberger, 1493). In the same year, the *Chronicle* was translated into German by the city secretary Georg Alt; see Hartmann Schedel, *Buch der Chroniken und Geschichten* (Nuremberg: Anton Koberger, 1493). Celtis later considered Alt as translator for his own *Norimberga* in 1495; see William Hammer, *Latin and German Encomia of Cities* (PhD dissertation, University of Chicago, 1937), 22. Both editions featured appendices of selected passages from Enea Silvio Piccolomini's *De Europa* (268r in the Latin, 268v in the German edition).

2. Jeffrey Ashcroft, "Black Arts Renaissance and Printing Press in Nuremberg, 1493–1528," *Forum for Modern Language Studies* 45, no. 1 (2009): 3–18: 6.

3. "Ich Cunnradus Celtis . . . das werck der Cronica, so jetzo gedruckt ist, sol vnd will . . . mit allem vleyß von newem corigieren vnd in ainen anndern form prynngen mit sampt ainer Newen Europa vnd anderm darczu gehorig vnd notturfftig." Quoted by Dieter Wuttke, "Humanismus als integrative Kraft. Die *Philosophia* des deutschen 'Erzhumanisten' Conrad Celtis. Eine ikonologische Studie zu programmatischer Graphik Dürers und Burgkmairs," *Artibus et historiae*, 6, no. 11 (1985): 65–99: 68. For the transcription of the entire contract, see Hans Bösch, "Eine projektiert gewesene zweite Ausgabe der sogen. Schedel'schen Chronik," *Mitteilungen aus dem germanischen Museum* 1 (1884): 37–39. For an abridged English translation, see Ashcroft, "Black Arts Renaissance," 8–9.

4. Ashcroft, "Black Arts Renaissance," 8.

5. Ayesha Ramachandran, *The Worldmakers: Global Imagining in Early Modern Europe* (Chicago: University of Chicago Press, 2015), 67.

6. The work as a whole was not completed due to Celtis's untimely death in 1508. The *Germania illustrata* is included, however, as a completed volume on the memorial woodcut portrait of Celtis that was produced, a year before Celtis's death, by the famous artist Hans Burgkmair. Furthermore, in 1505, Burgkmair created the first broadsheet illustrating Amerigo Vespucci's *Mundus novus* and the peoples of the New World and, in 1508, a woodcut representing *The King of Cochin*. See Stephanie Leitch, "Burgkmair's *Peoples of Africa and India* (1508) and the Origins of Ethnography in Print," *Art Bulletin* 91, no. 2 (2009): 134–59: 134.

7. Conrad Celtis, *Quatuor Libri Amorum secundum Quatuor Latera Germanie* (Nuremberg: Sodalitas Celtica, 1502). For a modern edition, see Conrad Celtis, *Quatuor*

Libri Amorum secundum Quattuor Latera Germanie: Germania Generalis, ed. Felicitas Pindter (Leipzig: Teubner, 1934). The most insightful discussions of Celtis's work and cartography are Gernot Michael Müller, *Die Germania generalis des Conrad Celtis: Studien mit Edition, Übersetzung und Kommentar* (Tübingen: Max Niemeyer Verlag, 2001); Jörg Robert, "Celtis' *Amores* und die Tradition der Liebeselegie," in *Amor als Topograph: 500 Jahre Amores des Conrad Celtis. Ein Manifest des deutschen Humanismus. Kabinettausstellung der Bibliothek Otto Schäfer, 7. April–30. Juni 2002* (Schweinfurt: Bibliothek Otto Schäfer Ausstellungskatalog, 2002), 9–17; and Peter Meurer, "Cartography in the German Lands, 1450–1650," in *The History of Cartography*, ed. David Woodward, vol. 3: *Cartography in the European Renaissance*, part 2 (Chicago: University of Chicago Press, 2007), 1172–1245, esp. 1189–91. I quote from the 1502 edition of the *Quatuor Libri Amorum*, unless otherwise indicated. The *Quatuor Libri Amorum* has not yet been translated into English.

8. Jörg Robert, *Konrad Celtis und das Projekt der deutschen Dichtung: Studien zur humanistischen Konstitution von Poetik, Philosophie, Nation und Ich* (Tübingen: Max Niemeyer Verlag, 2003), 133.

9. Celtis, *Quatuor Libri Amorum*, Praefatio, n.p.

10. Robert, "Celtis' *Amores* und die Tradition der Liebeselegie," 9.

11. Celtis, *Quatuor Libri Amorum*, [fol. 6v]. In a passage preserved solely in Aulus Gellius's *Noctes Atticae* (*Attic Nights*), the first-century BC Roman poet Lucius Afranius wrote: "Usús me genuit, máter peperit Memoria, / Sophiám vocant me Grái, vos Sapientiam." Aulus Gellius, *Attic Nights*, trans. John C. Rolfe (Cambridge, MA: Harvard University Press, 1967–84), 13.8. On the importance of this passage for Celtis see Wuttke, "Humanismus als integrative Kraft," 80.

12. Celtis, *Quatuor Libri Amorum*, [fol. 6v].

13. Raymond Klibansky, Erwin Panofsky, and Fritz Saxl, *Saturn and Melancholy: Studies in the History of Natural Philosophy, Religion, and Art* (London: Nelson, 1964), 279.

14. Wuttke has argued that, without being new at all, the Philosophy of Celtis corresponds to the definition of "sapientia" already advanced by the ancients. Wuttke, "Humanismus als integrative Kraft," 74.

15. See József Babicz, "Die mathematisch-geographischen und kartographischen Ideen von Albertus Magnus und ihre Stelle in der Geschichte der Geographie," in Albert Zimmermann, ed., *Die Kölner Universität im Mittelalter: Geistige Wurzeln und soziale Wirklichkeit* (Berlin: De Gruyter, 1989), 97–110.

16. Albertus Magnus, *De natura locorum* (Vienna: Hieronymus Vietor, 1513).

17. On the *translatio studii* from Rome to France see Karlheinz Stierle, "*Translatio Studii* and Renaissance: From Vertical to Horizontal Translation," in *The Translatability of Cultures: Figurations of the Space Between*, ed. Sanford Budick and Wolfgang Iser (Stanford, CA: Stanford University Press, 1996), 55–67.

18. The Strasbourg-based printer Johannes Grüninger was among the first to include woodcuts in editions of Latin classics and, among others, in Sebastian Brant's *Narrenschiff* (1496). Early woodcuts were often reused in different contexts. See Peter Luh, *Kaiser Maximilian gewidmet: Die unvollendete Werkausgabe des Conrad Celtis und ihre Holzschnitte* (Frankfurt am Main: Peter Lang, 2001), 11–12.

19. Theodore J. Cachey Jr., "Maps and Literature in Renaissance Italy," in *The History of Cartography*, ed. David Woodward, vol. 3: *Cartography in the European Renaissance*, part 1 (Chicago: University of Chicago Press, 2007), 450–60: 454. An important exception to this rule is Francesco Berlinghieri's (rhymed) *Le Septe Giornate della Geographia* (Florence: Nicolo Todesco, ca. 1480).

20. Wuttke, "Humanismus als integrative Kraft," 65, 66.

21. Stuchs's *Buchlein yn die kunst Cosmographia* was reprinted, in 1910, as the "German Ptolemy" (Deutscher Ptolemäus). Franz Machilek, "Kartographie, Welt- und Landesbeschreibungen in Nürnberg um 1500," in *Landesbeschreibungen Mitteleuropas vom 15. bis 17. Jahrhundert*, ed. Hans-Bernd Harder (Cologne: Böhlau, 1983), 4. See also Erwin Rosenthal, *The German Ptolemy and Its World Map* (New York: New York Public Library, 1944), 9 [note 13], and Meurer, "Cartography in the German Lands, 1450–1650," 1193.

22. See Machilek, "Kartographie, Welt- und Landesbeschreibungen," 1–12; and John W. Hessler, *A Renaissance Globemaker's Toolbox. Johannes Schöner and the Revolution of Modern Science 1475–1550* (Washington, DC: Library of Congress, 2013). As Patrick Gautier Dalché points out, the French Jean Fusoris, author of a treatise on the sphere, "had produced the oldest known non-classical version of a terrestrial globe (1432)," unfortunately not extant. Patrick Gautier Dalché, "The Reception of Ptolemy's *Geography* (End of the Fourteenth to Beginning of the Sixteenth Century)," in *The History of Cartography*, ed. David Woodward, vol. 3: *Cartography in the European Renaissance*, part 1 (Chicago: University of Chicago Press, 2007), 285–364: 306.

23. Whether Behaim indeed participated in the early Portuguese travels along Africa's western coast is still a question of debate. See Johannes Willers, "Leben und Werk des Martin Behaim," in *Focus Behaim Globus*, ed. Gerhard Bott, vol. 1 (Nuremberg: Verlag des Germanischen Nationalmuseums, 1992), 173–88: 181.

24. A document drafted by the City Council of Nuremberg describes Behaim's creation of the globe in the following words: "diese figur des apffels [ist] gebrackticirt und gemacht worden aus kunst angebung Fleys durch den gestrengen und erbar herrn Martin Behaim Ritter der sich dann in diser kunst kosmographia vil erfahren hot und bey einen dritten der welt umfahren solches alles mit fleiss ausgezogen aus den püchern ptolomei plinii strabonis und Marko Polo" (this figure of an apple was achieved and made by strictly following the skill [art], directions, and industry of the rigorous and honorable knight, Martin Behaim, who is much experienced in the art of Cosmography and has circumnavigated one third of the world. All this has been extracted industriously from the books of Ptolemy, Pliny, Strabo, and Marco Polo). Quoted in Ernst Georg Ravenstein, *Martin Behaim: His Life and His Globe* (London: George Philip, 1908), 71. English translation slightly adapted.

25. Hernán Cortés, *Praeclara Ferdina[n]di Cortesii de nova maris oceani Hyspania narratio* (Nuremberg: Fridericus Peypus, 1524). By 1524, the Aztec capital had, of course, already been destroyed by the Spanish conquerors, so that the map aims at capturing, cynically as well as tragically, Tenochtitlan's past glory. See Barbara E. Mundy, "Mapping the Aztec Capital: The 1524 Nuremberg Map of Tenochtitlan, Its Sources and Meanings," *Imago Mundi* 50 (1998), 11–33: 11; and, more recently, Elizabeth Horodowich, *The Venetian Discovery*

of America: Geographic Imagination and Print Culture in the Age of Encounters (Cambridge: Cambridge University Press, 2018), 180.

26. See Kirsten A. Seaver, "Norumbega and *Harmonia Mundi* in Sixteenth-Century Cartography," *Imago Mundi* 50 (1998): 34–58.

27. Stephanie Leitch, *Mapping Ethnography in Early Modern Germany: New Worlds in Print Culture* (New York: Palgrave Macmillan, 2010), 33.

28. The term "umbilication," used by Guy Rosolato (*La relation d'inconnu*, 1978, 257), is taken up by Tom Conley, *The Self-Made Map: Cartographic Writing in Early Modern France* (Minneapolis: University of Minnesota Press, 1996), 9.

29. See Brigitte English, "Erhard Etzlaub's Projection and Methods of Mapping," *Imago Mundi* 48 (1996): 103–23: 108.

30. Etzlaub was probably personally involved in the creation of Nicolaus Claudianus's first map of Bohemia in 1518. Claudianus, a member of the Bohemian Brethren, collaborated closely with Etzlaub as well as with the Nuremberg printer Hieronymus Höltzel. Together, they produced a Czech almanac in 1517. See Machilek, "Kartographie, Welt- und Landesbeschreibungen," 3.

31. The city is indexed by its lesser municipal coats of arms, "a split shield with a half-eagle on the right and five diagonal bars ("bends") on the left"; see *Gothic and Renaissance Art in Nuremberg, 1300–1550*, ed. Rainer Kahsnitz and William D. Wixom (New York: Metropolitan Museum of Art, 1986), 230.

32. See Adrian Wilson, *The Making of the Nuremberg Chronicle* (Amsterdam: Nico Israel, 1976), 21; Christopher Wood, *Forgery, Replica, Fiction: Temporalities of German Renaissance Art* (Chicago: University of Chicago Press, 2008), 241; Wuttke, "Humanismus als integrative Kraft," 67.

33. The *Chronicle* was equally unprecedented in its linguistic scope: no other work of this caliber was published in two different languages within only a few months—in scholarly Latin in July and in German translation in December 1493.

34. Celtis, often called an "itinerant humanist" (*Wanderhumanist*), spent many years traveling through Europe as a scholar and researcher, studying astronomy and mathematics in different cities in Italy, Poland, Bohemia, Hungary, and Austria. His interest in and knowledge of geography and cartography had earned him the sobriquet "poet-cosmographer" (*Dichterkosmograph*). Wuttke, "Humanismus als integrative Kraft," 67. Celtis is also remembered as the discoverer of the alleged Roman "Tabula Peutingeriana," named after Celtis's friend Konrad Peutinger, to whom he bequeathed the map after his death.

35. Denis Cosgrove, "Images of Renaissance Cosmography, 1450–1650," in *The History of Cartography*, ed. David Woodward, vol. 3: *Cartography in the European Renaissance*, part 1 (Chicago: University of Chicago Press, 2007), 55–98: 89.

36. Schedel, *Liber Chronicarum* (Latin edition), fol. 1r. Schedel, *Buch der Chroniken und Geschichten* (German edition), fol. 1r.

37. Schedel, *Liber Chronicarum*, fol. 2v. "Darnach hat got die andern örter nemlich des mittags und mitternacht in derselben gestalt außgemessen. Die auch den vordern zwayen örtern mit verwantschaft zugesellet werden. Dann das ort das vo[m] wirm und sunne[n] heißer ist das hangt den anfang allernechst an. Aber das ort das in kelt un[d] ewigen gefrüst qualet ist des tails des letzten nidergangs. Wann als die finsternus den liecht, also ist auch die kelt der wirm widerwertig. Darümb als die wirm den liecht, also ist der mittag den

auffgang und die kelt der finsternus. Also auch die mitternacht dem nidergang aller nehst." Schedel, *Buch der Chroniken und Geschichten* (German edition), fol. 2v.

38. See also Lactantius, *Divinarum institutionum* (Subiaco: Sweynheym and Pannartz, 1465), n.p./Lactantius, *Divine Institutes*, trans. Anthony Bowen and Peter Garnsey (Liverpool: Liverpool University Press, 2003), 9.7–9, 148.

39. For a recent discussion of iconographic representations of God as a geometer, see Ramachandran, *The Worldmakers*, 63–64.

40. Friedrich Hölderlin, "Urtheil und Seyn," in *Sämtliche Werke*, vol. 4 (Stuttgart: Cotta, 1943–1977). Quoted by Roberto Esposito, *Da fuori: Una filosofia per l'Europa* (Turin: Einaudi, 2016), 30.

41. *The New Oxford Annotated Bible with the Apocryphal/Deuterocanonical Books* (Oxford: Oxford University Press, 2007), Jeremiah 1:14, 1076.

42. See Karen C. Pinto, *Medieval Islamic Maps: An Exploration* (Chicago: University of Chicago Press, 2016).

43. The woodcut artist is probably Hans Süss von Kulmbach, a student of Albrecht Dürer. See Jeffrey Chipps Smith, *Nuremberg, A Renaissance City, 1500–1618* (Austin: University of Texas Press, 1983), 91.

44. Tom Conley, "Inklines and Lifelines: About *La coche* (1547) by Marguerite de Navarre," *Parallax* 6, no. 1 (2000): 92–110: 94.

45. Conley, "Inklines and Lifelines," 94.

46. Celtis, *Quatuor Libri Amorum*, 1.1, vv. 7–22, fol. 9r.

47. For the English translation, see Áron Orbán, *Born for Phoebus: Solar-Astral Symbolism and Poetical Self-Representation in Conrad Celtis and his Humanist Circles* (PhD dissertation, Central European University, Budapest, 2017), 120.

48. Hesiod, *Theogony*, trans. Glenn W. Most (Cambridge, MA: Harvard University Press, 2007), vv. 123–25, 13.

49. Conley, *The Self-Made Map*, 9.

50. Celtis was actually born in the lesser-known Wipfeld, close to Würzburg. See Robert, *Konrad Celtis und das Projekt der deutschen Dichtung*, 369.

51. Celtis, *Quatuor Libri Amorum*, 1.12, v. 334, fol. 19v.

52. See Robert, *Konrad Celtis und das Projekt der deutschen Dichtung*, 370. With a nod to the *Pharsalia* of Lucan, who mentions both "Erebos" and "Dis" with the alleged priests of the Greeks, the Druids, Celtis establishes a continuity between the culture of ancient Greece and his own birthplace in Franconia.

53. Theodore J. Cachey Jr., "Petrarchan Cartographic Writing," in *Medieval and Renaissance Humanism: Rhetoric, Representation and Reform*, ed. Stephen Gersh and Bert Roest (Leiden: Brill, 2003), 73–91: 78.

54. Robert H. F. Carver, *The Protean Ass: The Metamorphoses of Apuleius from Antiquity to the Renaissance* (Oxford: Oxford University Press, 2007), 162. Leon Battista Alberti, who probably visited the monastery of Subiaco, described the printing press as "the new German invention that enables three men to produce two hundred volumes in one hundred days." Leon Battista Alberti, "Preface," *Dello scrivere in cifra*, ed. D. Kahn (Turin: Galimberti, 1994), 27–28, quoted by Carver, *The Protean Ass*, 162 [note 7].

55. The 1468 edition of the *Divine Institutes* was the earliest printed book found in Schedel's private library and one of five early editions of Sweynheym and Pannartz he

possessed. See Martin Lowry, "Venetian Capital, German Technology and Renaissance Culture in the Later Fifteenth Century," *Renaissance Studies* 2, no. 1 (1988): 1–13: 5.

56. Carver, *The Protean Ass*, 162.

57. Ilse Slot, "Michelangelo's 'Q': A Contribution to the Interpretation of the Sistine Chapel's Sibyls," *Fragmenta* 5 (2011): 13–21: 16.

58. "Cogitate dignitatibus vestris splendorem virtute, doctrina et eruditione quaerentes sanctisque moribus titulos addere, ut vos dignos illis homines existiment et vos illae, non vos illas tanquam aucupes gregem avium persequamini." Conrad Celtis, *Public Oration Delivered in the University of Ingolstadt*, in *Selections from Conrad Celtis: 1459–1508*, ed. and trans. Leonard Forster (Cambridge: Cambridge University Press, 1948), 36–64: 43; 42. On the *Public Oration* see also Leitch, *Mapping Ethnography in Early Modern Germany*, 44–45.

59. Celtis, *Public Oration Delivered in the University of Ingolstadt*, 43; 42.

60. Celtis, *Public Oration Delivered in the University of Ingolstadt*, 43; 42. Translation slightly changed.

61. Celtis, *Public Oration Delivered in the University of Ingolstadt*, 47; 46

62. Tacitus, *Germania*, trans. M. Hutton, revised by Robert Maxwell Ogilvie (Cambridge, MA: Harvard University Press, 1970), 2.2, 130; 131.

63. Celtis invites his students to "do away with that old disrepute of the Germans in Greek, Latin and Hebrew writers, who ascribe to us drunkenness, cruelty, savagery and every other vice bordering on bestiality and excess." Celtis, *Public Oration Delivered in the University of Ingolstadt*, 43; 42.

64. "Magno vobis pudori ducite Graecorum et Latinorum nescire historias et super omnem impudentiam regionis nostrae et terrae nescire situm, sidera, flumina, montes, antiquitates, nationes, denique quae peregrine homines de nobis ita scite collegere, ut apud me magnum miraculum sit." Celtis, *Public Oration Delivered in the University of Ingolstadt*, 43; 42.

65. Celtis, *Public Oration Delivered in the University of Ingolstadt*, 43–45; 42–44.

66. "Induite veteres illos animos, viri Germanii, quibus totiens Romanis terrori et formidini fuistis, et ad angulos Germaniae oculos convertite limitesque eius laceros et distractos colligate!" Celtis, *Public Oration Delivered in the University of Ingolstadt*, 47; 46.

67. Terence Cave, *Pré-histoires: textes troublés au seuil de la modernité* (Geneva: Droz, 1999), 16.

68. Ptolemy, *Geography: An Annotated Translation of the Theoretical Chapters*, ed. J. Lennart Berggren and Alexander Jones (Princeton, NJ: Princeton University Press, 2002), 1, 57. I slightly modified the translation. "Finis corographie est partem totius sigillatim animadvertere: ut si quis aurem tantum aut oculum pingat." Ptolemy, *Cosmographia* (Ulm: Lienhart Holle, 1482), 1, n.p.

69. Ptolemy, *Geography*, 1, 57. "Cosmographie proprium est unam eandemque habitabilem terram nobis cognitam ostendere quo modo natura situque se habeat." Ptolemy, *Cosmographia*, 1, n.p.

70. Ptolemy, *Cosmographia*, 1, n.p. Ptolemy, *Geography*, 1, 57.

71. See Franco Farinelli, *Blinding Polyphemus: Geography and the Models of the World*, trans. Christina Chalmers (London: Seagull Books, 2018), 10–11.

72. See Franco Farinelli, *La crisi della ragione cartografica* (Turin: Einaudi, 2009), 88–89.

73. See Carl Kerényi, *Dionysos: Archetypal Image of Indestructible Life*, trans. Ralph Manheim (Princeton, NJ: Princeton University Press, 1976), 66.

74. Farinelli, *Blinding Polyphemus*, 15.

75. Farinelli, *La crisi della ragione cartografica*, esp. 95–99.

76. Farinelli, *La crisi della ragione cartografica*, 89.

77. Farinelli, *La crisi della ragione cartografica*, 88. Suddenly, the act of lying down becomes visible behind the terminology used for measurement, as in the English word "cubit," from the Latin "cubitus" (elbow), the result of "lying down" (*cubare*).

78. Farinelli, *La crisi della ragione cartografica*, 89.

79. Apollo permeates Celtis's four books of poems from the beginning to the end. "Eight of the first ten poems have significant references to Apollo, and this percentage holds, roughly, for the entire collection." David Hotchkiss Price, *Albrecht Dürer's Renaissance: Humanism, Reformation, and the Art of Faith* (Ann Arbor: University of Michigan Press, 2003), 80.

80. The motif of Apollo as the subjugator of the python was predominantly conveyed through the Homeric *Hymn to Apollo* and, briefly, in Seneca's *Hercules furens*, a tragedy Celtis edited and published in 1487, one year after publishing his first extant book, a short treatise on the poetic meter, *Ars versificandi et carminum*. See Conrad Celtis, *Selections*, 5. His first work, titled *Tractatus de preceptis rhetoris*, is lost. It is the *Ars versificandi et carminum* that afforded Celtis, early in his career, the prestigious title of poet laureate. For the second edition of the *Ars versificandi*, see Conrad Celtis, *Ars versificandi et carminum* (n.p. [Leipzig]: n.p. [Mauritius Brandis], n.d. [1487]). For the lost *Tractatus*, see Gustav Bauch, *Geschichte des Leipziger Frühhumanismus mit besonderer Rücksicht auf die Streitigkeiten zwischen Konrad Wimpina und Martin Mellerstadt* (Leipzig: Otto Harrassowitz, 1899), 350.

81. Ovid, *Metamorphoses*, trans. Frank Justus Miller (Cambridge, MA: Harvard University Press, 1977), 1. 434–44, 32; 33, with my slight amendations.

82. As Stuart Elden points out, Roman surveyors were aware that the common etymology of "territory" and "terror" was disputed. See Stuart Elden, *The Birth of Territory* (Chicago: University of Chicago Press, 2013), 85.

83. "Territis fugatisque inde civibus, territoria dixerunt." Siculus Flaccus, *De conditionibus agrorum*, in *De Agrorum conditionibus, & constitutionibus limitum* (Paris: Adrianus Turnebus, 1554), 3.

84. Frontinus, *De limitibus agrorum*, in *De Agrorum conditionibus, & constitutionibus limitum*, 49.

85. Between 1498 and 1500, Celtis published Tacitus's *Germania* with the title *De origine et situ Germanorum*. It was the second edition of Tacitus's *Germania* in Germany, but the first that featured an addition: Celtis's own treatise on Germany, *De situ et moribus Germanie additiones*, later republished, with the title *Germania generalis*, as part of the *Quatuor Libri Amorum*.

86. Tacitus, *Germania*, trans. M. Hutton and W. Peterson (Cambridge, MA: Harvard University Press, 1970), 1.1, 128–29, with my slight amendations.

87. Ptolemy, *Geographicae Enarrationis Libri Octo*, trans. Willibald Pirckheimer, notes by Johannes Regiomontanus (Strasbourg: Johannes Grüninger, 1525), book 2, fol. 21r.

88. Ptolemy, *Geographicae Enarrationis Libri Octo*, book 8, Table 4, "Germania magna," n.p.

89. "Cetera [Germaniae] Oceanus ambit, latos sinus et insularum immensa spatia complectens." Tacitus, *Germania*, 1.1, 128–29.

90. Celtis, *Quatuor Libri Amorum*, "Ad Divum Maximylianum Invictissimum & sereniss. Rhom. Regem . . . praefatio," n.p.

91. What seemingly aligns itself into an unproblematic "natural" boundary through woodcut artistry is, in fact, a highly manipulated boundary. Kraków, for instance, the "German" city with which Celtis's elegies open, was not a part of Germany—in fact, it was then the capital of the Polish kingdom. In his description of German peoples, Piccolomini, for instance, described Kraków as a Polish city: "Cracovia quamvis polonici iuris est trans viscellam condita fluvium. In solo sarmatico eum tamen presulem obtinuit." Enea Silvio Piccolomini, *De ritu, situ, moribus, et condicione theutonie descriptio* (Leipzig: Wolfgang Stockel, 1496), fol. 8v. David Frick points out that it was only in 1772, "in the so-called First Partition of Poland . . . [that] the Vistula River was made into a border . . . and Kraków, for the first time in its history, became a border town." David A. Frick, "Franklin's Free Will; or, Optimism in Cracow, 1798," *Austrian History Yearbook* 28 (1997): 59–94: 62.

92. Meurer, "Cartography in the German Lands," 1177–78.

93. See Meurer, "Cartography in the German Lands," 1181.

94. Translated into French and published in Paris (1488–89) and Lyon (1491) under the title *Mer des Hystoires* (*Sea of Histories*), the anonymous *Rudimentum novitiorum*, originally published in Latin, became widely popular. *Mer des Hystoires* (Paris: Pierre Le Rouge, 1488, and Lyon: Jean de Pré, 1491). Andrea Worm, "Rudimentum Novitiorum," in *Encyclopedia of the Medieval Chronicle*, ed. Dunphy Graeme and Cristian Bratu, last updated in 2016, http://dx.doi.org/10.1163/2213-2139_emc_SIM_000333. For the *Rudimentum novitiorum*, see also Robert Karrow, "Centers of Map Publishing in Europe, 1472–1600," in *The History of Cartography*, ed. David Woodward, vol. 3: *Cartography in the European Renaissance*, part 1 (Chicago: University of Chicago Press, 2007), 611–21: 615.

95. Meurer, "Cartography in the German Lands," 1181.

96. The scale is playfully supplanted by a division into nine sections corresponding to the following criteria: season, life cycle, direction, wind, humor, position of the sun, element, temperature, and color.

97. Nigel Thrift, "Space," *Theory, Culture & Society* 23, nos. 2–3 (2006): 139–55: 141.

98. It was in 1530 that the Bavarian humanist Jacob Ziegler completed his map of "Schondia" (Scandinavia), published in 1532 in *Quae intus continentur: Syria, ad Ptolomaici operis rationem, praeterea Strabone, Plinio, & Antonio auctoribus locupletata . . .* (Strasbourg: Peter Schöffer, 1532). In 1539, the Swedish historian and cartographer Olaus Magnus published his *Carta Marina* (Venice, 1539) depicting Europe's north and Scandinavia in particular. William R. Mead, "Scandinavian Renaissance Cartography," in *The History of Cartography*, vol. 3: *Cartography in the European Renaissance*, part 2 (Chicago: University of Chicago Press, 2007), 1781–1805: 1786.

99. Celtis, *Quatuor Libri Amorum*, 4.9, v. 14, fol. 65v.

100. Also, the English word "(h)ora" means "limit." Marc Shell points out that in John Trevisa's translation of Ranulph Higden's *Polychronicon* (1387) "orisoun (bourn) is akin

to the borderline that divides living persons from ghosts in *Hamlet*. Likewise, *orisoun* recalls two kinds of horizons: the 'cut' that separates the sea from the sky and the 'coast', or horizontal limit (*horos*), that surrounds *terra firma*." Marc Shell, *Islandology: Geography, Rhetoric, Politics* (Stanford, CA: Stanford University Press, 2014), 301 [notes 37 and 39].

101. On a physiological level, from antiquity on the shape of one's feet (*pedes*) was considered one of the markers of civilization or lack thereof. Already Scylax of Caryndia, a fifth-century sea captain, told of monstrous races inhabiting the eastern edges of the world, among whom he found the *Skiapodes*, "Shadow-feet," who "have extremely flat feet, and at high noon they fall on the ground and stretch out their feet so as to make shade." In Johannes Tzetzes, *Historiarum variarum Chiliades* (Leipzig: Theophilus Kiessling, 1826), 7.629–36. Quoted by James S. Romm, *The Edges of the Earth in Ancient Thought: Geography, Exploration, and Fiction* (Princeton, NJ: Princeton University Press, 1992), 85.

102. Celtis, *Quatuor Libri Amorum*, 4.9, v. 18, fol. 65v.

103. Quintilian, *The Orator's Education*, trans. Donald A. Russell (Cambridge, MA: Harvard University Press, 2001), 10.1, 19: 260; 261.

104. See Surekha Davies, *Renaissance Ethnography and the Invention of the Human* (Cambridge: Cambridge University Press, 2016).

105. Tacitus, *Germania*, 45.1–2, 206; 207.

106. Tacitus. *Germania*, 46.3, 212; 213.

107. Tacitus. *Germania*, 46.6, 214; 215.

108. Celtis, *Quatuor Libri Amorum*, 4.9, vv. 21–22, fol. 65v.

109. Celtis, *Quatuor Libri Amorum*, Praefatio, n.p. While elegies commonly stressed the relation between love and youth, Ovid's *Tristia* and *Epistolae ex Ponto* as well as the popular corpus of love elegies by the sixth-century elegiac poet Maximianus thematized the equally trendy topic of the "senex amans."

110. Gilles Deleuze, "Ecrivain non: un nouveau cartographe," *Critique* 343 (1975): 1207–27: 1217. Tom Conley translates "quadrillage" as "gridding" in Tom Conley, "Mapping in the Folds: Deleuze *Cartographe*," *Discourse* 20, no. 3 (1998): 123–38: 129.

111. Conley, "Mapping in the Folds," 129.

112. Conley, "Mapping in the Folds," 130.

113. Gilles Deleuze, *Logic of Sense*, trans. Mark Lester (New York: Columbia University Press, 1990), 9.

114. Deleuze, *Logic of Sense*, 9.

115. Quintilian writes that "some call clear and vivid descriptions of places . . . *topographia*" ("Locorum quoque dilucida et significans descriptio . . . alii τοπογραφίαν dicunt"). In Quintilian, *The Orator's Education*, 9.2, 58; 59. For Walter of Châtillon, see Alfred Hiatt, "Geography in Walter of Châtillon's *Alexandreis* and its Medieval Reception," *Journal of Medieval Latin* 23 (2013), 255–94: 258.

116. See Martin Korenjak, "Deutschland als Landschaft. Konrad Celtis und der Herkynische Wald," in Thomas Schultheiß and Jochen Baier, eds., *Würzburger Humanismus* (Tübingen: Narr Verlag, 2015), 19–36: 29.

117. Korenjak, "Deutschland als Landschaft," 29.

118. Korenjak, "Deutschland als Landschaft," 29. Celtis's reference to "Germania . . . illustrata" is ambiguous: while he planned on publishing a major work with this title

(never completed), Celtis wrote a poetic description of Germany titled *Germania generalis*. Based on this fact, I chose to consider the *Germania generalis* as the point of reference in this quotation.

119. Quoted by Robert, *Konrad Celtis und das Projekt*, 178 [note 126].

120. Celtis, *De situ & moribus Germaniae additiones*, in Müller, *Die* Germania generalis *des Conrad Celtis*, vv. 187–196, 102.

121. These four rivers evocative of Paradise are also included on the woodcut in the *Quatuor Libri Amorum* that precedes the description of southern Germany and Regensburg.

122. Müller, *Die* Germania generalis *des Conrad Celtis*, vv. 197–99, 102. While the north (Arcton), east (Eurus), and west (Zephir) are listed, Celtis does not explicitly mention the south (wind) here.

123. Celtis, *De situ*, in Müller, *Die* Germania generalis *des Conrad Celtis*, vv. 215–225, 104.

124. Pomponius Mela contended that the Hercynian Forest "covers a distance of sixty days' march." Pomponius Mela, *Description of the World*, trans. F. E. Romer (Ann Arbor: University of Michigan Press, 1998), 3.29, 109.

125. Celtis, *De situ*, in Müller, *Die* Germania generalis *des Conrad Celtis*, vv. 237–239, 104.

126. See Cosmas of Prague, *Chronicle of the Czechs*, trans. Lisa Wolverton (Washington, DC: Catholic University of America Press, 2012), and Enea Silvio Piccolomini, *Historia Bohemorum* (Helmstadt: Melchior Sustermann, 1699). Piccolomini's work on Bohemia was "first published in Czech in 1510 as *Kronika česka*." Derek Sayer, *The Coasts of Bohemia: A Czech History* (Princeton, NJ: Princeton University Press, 1998), 41.

127. Celtis, *De situ*, in Müller, *Die* Germania generalis *des Conrad Celtis*, vv. 243–247; 250–253, 106.

128. Martin Korenjak surmises that the fact that for Celtis the Hercynian Forest reaches beyond the German land's actual territory has to do with Celtis's desire to put Germany into a wider "historical, cultural historical, cosmological and geographic context" in order to prevent "an isolation of Germany from the rest of the world." Korenjak, "Deutschland als Landschaft," 28.

129. Ancient geographical writers such as Pomponius Mela and Strabo mentioned Germany's territory as one difficult to measure. In his *Description of the World*, Mela writes that the land "is not easily passable, because of its many rivers; it is rugged on account of its numerous mountains; and to a large extent it is impassable with its forests and swamps." Mela, *Description of the World*, 109.

130. Celtis, *Quatuor Libri Amorum*, 2.10, vv. 21–30, fol. 37r.

131. Müller, *Die* Germania generalis *des Conrad Celtis*, 78.

132. Celtis, *De situ*, in Müller, *Die* Germania generalis *des Conrad Celtis*, v. 189, 80.

133. Celtis, *De situ*, in Müller, *Die* Germania generalis *des Conrad Celtis*, v. 221, 80.

134. Celtis, *De situ*, in Müller, *Die* Germania generalis *des Conrad Celtis*, vv. 190–191, 102.

135. On poetics and speech-fluency disorder see Marc Shell, *Stutter* (Cambridge, MA: Harvard University Press, 2005), 7.

136. Robert, *Konrad Celtis und das Projekt der deutschen Dichtung*, 405. Robert misreads Celtis's "quadrifluvium" as "quadrifluus."

137. Gerard O'Daly, *Days Linked by Song: Prudentius' Cathemerinon* (Oxford: Oxford University Press, 2012), "Before Taking Food," 3, vv. 101–105, 87–88. These lines are also quoted in Latin by Robert, *Konrad Celtis und das Projekt der deutschen Dichtung*, 405.

138. O'Daly, *Days Linked by Song*, 3, vv. 26–30, 83–84.

139. O'Daly, *Days Linked by Song*, 99.

140. The "dactylus" serves the self-referential purpose of pointing to the poem's own meter. Prudentius chooses here the catalectic dactylic tetrameter, a meter divided into four feet, which, in turn, consist of dactyls and spondees. Dag Norberg, *An Introduction to the Study of Medieval Latin Versification*, ed. Jan Ziolkowski, trans. Grant C. Roti and Jacqueline de La Chapelle Skubly (Washington, DC: Catholic University of America Press, 2004), 77.

141. This reading is rejected by Gerard O'Daly, who emphasizes instead "the Pauline echoes in the stanza." O'Daly, *Days Linked by Song*, 98.

142. O'Daly, *Days Linked by Song*, 3, vv. 16–20, 82.

143. Celtis, *Quatuor Libri Amorum*, Praefatio, n.p.

144. "The central boss on military shields was called the *umbilicus* or navel." David Leatherbarrow, *Topographical Stories: Studies in Landscape and Architecture* (Philadelphia: University of Pennsylvania Press, 2004), 124, referencing James Ackerman, *The Architecture of Michelangelo* (London: Zwemmer, 1961), 167–69.

145. See Wilhelm Heinrich Roscher, *Omphalos* (Leipzig: Teubner, 1913), 10.

146. The Latin "ferculum" not only means "dish of food," but also the support on which the dish is prepared and eaten: "a tray for food." As "a frame or stretcher for carrying things," it was used especially in triumphal processions, but also for images of Gods. *Oxford Latin Dictionary*, ed. P. G W. Glare, vol. 1 (Oxford: Oxford University Press, 2012), entry "ferculum," 752.

147. Scholars have long pointed out that the Ebsdorf Map, like other maps of the thirteenth century, emerged during a period of dispute over the doctrine of transubstantiation. Thus, the rise of Christological *mappae mundi* in the first half of the thirteenth century signifies, Farinelli argues, a "transfer": "just as the host becomes, by virtue of the Eucharist, the true body of Christ, so the map, host of the Earth, is transformed into its [the Earth's] true body." Farinelli, *La crisi della ragione cartografica*, 22.

CHAPTER TWO

1. Ludovico Ariosto, *Orlando Furioso secondo l'editio princeps del 1516*, ed. Tina Matarrese and Marco Praloran (Turin: Einaudi, 2016), 9.60, 260; 9.20, 248; 9.58, 260. The English translation is based on the 1532 edition: Ludovico Ariosto, *Orlando Furioso*, trans. Guido Waldman (Oxford: Oxford University Press, 2008), 10.72; 10.69, 101.

2. Ariosto, *Orlando Furioso*, ed. Matarrese and Praloran, 9.59, 260. This stanza appeared, slightly changed, in canto 10, stanza 71, in the definitive 1532 edition of *Orlando Furioso*. Ludovico Ariosto, *Orlando Furioso* (Lyon: Bastiano di Bartholomeo Honorati, 1556), 10.71, 80/Ariosto, *Orlando Furioso*, trans. Guido Waldman, 10.71, 101. I slightly changed the English translation.

3. While Waldman translates Ariosto's "Russi" as "Russians," in the sixteenth century the ethnonym "Russi" denoted the Ruthenians, an Eastern Slavic people constituting

the main population of the Grand Duchy of Lithuania and occupying a vast territory from Kiev in the south to Novgorod in the north (while the "Russians" would have corresponded to the sixteenth-century Muscovites). Today, Ruthenia (or Kievan Rus', as it was known from the ninth to the twelfth century), to a large extent overlaps with Ukraine, Belarus, and the western part of Russia. For the question of mapping Sarmatia in Ariosto's *Orlando Furioso*, see Alexandre Doroszlaï, *Ptolémée et l'hyppogriffe: la géographie de l'Arioste soumise à l'épreuve des cartes* (Turin: Edizioni dell'Orso, 1998), 62–65.

4. Thibaut Maus de Rolley, "Le globe e le chevalier: variations sur la méditation cosmographique dans la fiction chevaleresque de la Renaissance," in *Les Méditations cosmographiques à la Renaissance*, ed. Frank Lestringant, Jean-Marc Besse, and Marie-Dominique Couzinet (Paris: Presses de l'Université Paris-Sorbonne, 2009), 129.

5. Gaston Bachelard, *The Poetics of Space*, trans. Maria Jolas (Boston: Beacon, 1969), 155.

6. The "definitive" edition of *Orlando Furioso* was published—after numerous unauthorized reprints, several stylistic changes, and the addition of six cantos—in Ferrara in 1532. See Conor Fahy, "Some Observations on the 1532 Edition of Ludovico Ariosto's *Orlando Furioso*," *Studies in Bibliography* 40 (1987): 72–85: 72.

7. Maciej Miechowita, *Tractatus de duabus Sarmatiis Asiana et Europiana et de contentis in eis* (Kraków: Johannes Haller, 1517), n.p. I quote from this edition, if not otherwise indicated. On Miechowita and his reception of Ptolemy see Katharina N. Piechocki, "Erroneous Mappings: Ptolemy and the Visualization of Europe's East," in *Early Modern Cultures of Translation*, ed. Karen Newman and Jane Tylus (Philadelphia: University of Pennsylvania Press, 2015), 76–96.

8. The term was used by Mariano Cuesta Domingo, "La cartografía en 'prosa' durante la época de los grandes descubrimientos Americanos," in *Actas del Congreso de Historia del Descubrimiento, 1492–1556* (Madrid: Real Academia de la Historia, 1992), 299. Quoted in Ricardo Padrón, *The Spacious Word: Cartography, Literature, and Empire in Early Modern Spain* (Chicago: University of Chicago Press, 2004), 89.

9. Teresa Jaroszewska, "A la découverte de l'Europe de l'Est: *Tractatus de duabus Sarmatiis Asiana et Europiana* de Mathias de Miechow," in *Les représentations de l'autre: du Moyen Age au XVIIe siècle. Mélanges en l'honneur de Kazimierz Kupisz*, ed. Evelyne Berriot-Salvadore (Saint-Etienne: Publications de l'Université de Saint-Etienne, 1995), 17–30: 17. For Miechowita's impact on subsequent travel writers and politicians such as Herberstein, see Larry Wolff, *Inventing Eastern Europe: The Map of Civilization on the Mind of the Enlightenment* (Stanford, CA: Stanford University Press, 1994), 152.

10. Maciej Miechowita, *Tractat von baiden Sarmatien aund anderen anstossenden landen in Asia und Europa von sitten und gepräuchen der völcker so darinnen wonen*, trans. Johannes Eck (Augsburg: Marx Wirsung, 1518). For a comparison of the different translations of the *Tractatus*, see Saskia Metan, *Wissen über das östliche Europa im Transfer: Edition, Übersetzung und Rezeption des "Tractatus de duabus Sarmatiis" (1517)* (Cologne: Böhlau, 2019).

11. Maciej Miechowita, *Descriptio Sarmatiarum Asianae et Europianae et eorum quae in eis continentur* (Kraków: Johannes Haller, 1521).

12. Simon Grynaeus, *Novus orbis regionum ac insularum veteribus incognitarum* (Basel: Johannes Hervagius, 1532).

13. One year later, Grynaeus's volume appeared in a German translation under the

title *Die new welt, der landschaften unnd Insulen, so bis hie her allen Altweltbeschrybern unbekant*, trans. Michael Herr (Strasbourg: Georg Ulrich, 1534).

14. Giovanni Battista Ramusio, *Navigazioni e viaggi*, ed. Marica Milanesi (Turin: Einaudi, 1978–88). To this day, this groundbreaking travel anthology has not been translated into English.

15. The edition is dedicated to Severino Ciceri. In 1562, Porcacchi dedicated to Ciceri the first reprint of Ariosto's comedies. See Ludovico Ariosto, *Comedie* (Venice: Gabriel Giolito de' Ferrari, 1562).

16. Maciej Miechowita, *Historia delle due Sarmazie*, trans. Annibal Maggi (Venice: Gabriel Giolito de' Ferrari, 1561).

17. Giovanni Battista Ramusio, *Secondo volume delle navigationi et viaggi* (Venice: Giunti, 1583). This is the second, substantially augmented, edition of the *Secondo volume delle navigationi et viaggi*. The first edition, published in 1559, offers a much more modest choice of travel narratives pertaining to Europe's east and Asia (and does not contain Miechowita's *Tractatus*). Scholars have counted more than twenty editions of the *Tractatus*. See David Thomas and John Chesworth, eds., *Christian-Muslim Relations*, vol. 7: *Central and Eastern Europe, Asia, Africa and South America (1500–1600)* (Leiden: Brill, 2015), 66.

18. Michael Herkenhoff, *Die Darstellung Außereuropäischer Welten in Drucken deutscher Offizinen des 15. Jahrhunderts* (Berlin: Akademie Verlag, 1996), 125–26. Copernicus was born in the royal Prussian city of Toruń (Thorn), a part of the Polish kingdom since 1466, and studied, among other places, in Kraków, Bologna, and Padua.

19. Miechowita left an inventory of the books and manuscripts he owned, and the *Commentariolus* is recorded on May 1, 1514. See Leszek Hajdukiewicz, *Biblioteka Macieja z Miechowa* (Wrocław: Zakład Narodowy im. Ossolińskich, 1960), 70.

20. Wapowski collaborated with the Italian cartographer Marco Beneventano on the 1507 edition of Ptolemy's *Geography* and supplied him with updated information about the geography of Europe's east, in particular Poland. See Ptolemy, *Geographia* (Rome: Bernardino Vitali, 1507); Zsolt G. Török, "Renaissance Cartography in East-Central Europe, ca. 1450–1650," in *The History of Cartography*, ed. David Woodward, vol. 3: *Cartography in the European Renaissance*, part 1 (Chicago: University of Chicago Press, 2007), 1806–51: 1816; Karol Buczek, "Bernard Wapowski, der Gründer der polnischen Kartographie," in *Comptes rendus du Congrès International de Géographie, Varsovie 1934* (Warsaw: Kasa imienia Mianowskiego, 1935; reprinted Nendeln, Liechtenstein: Kraus Reprint, 1972), 4: 61–63; Karol Buczek, *The History of Polish Cartography from the 15th to the 18th Century*, trans. Andrzej Potocki (Wrocław: Ossolineum, 1966), in particular 32–40; and Bożena Modelska-Strzelecka, "Bernard Wapowski," *Zeszyty naukowe Uniwersytetu Jagiellońskiego. Prace Historyczne* 42 (1973): 23–59: 57.

21. See Jakub Niedźwiedź, "Wielopiśmienność Wielkiego Księstwa Litewskiego: nowe perspektywy badawcze," *Wielogłos* 20, no. 2 (2014): 11–21: 11.

22. Niedźwiedź, "Wielopiśmienność," 11.

23. Oscar J. Martínez, *Border People: Life and Society in the U.S.–Mexico Borderlands* (Tucson: University of Arizona Press, 1994), 20.

24. Maciej Miechowita, *Chronica Polonorum* (Kraków: Hieronymus Vietor, 1519). Jan Długosz's *Annales seu cronici incliti regni Poloniae* (*Annals or Chronicles of the*

famous Kingdom of Poland) was partially printed in 1615, in Ruthenia, and as a whole in 1701–3. The *Annales* widely circulated in manuscript form during Miechowita's lifetime. See Jan Długosz, *Historia polonica Ioannis Długossi seu Longini Canonici Cracoviensis*, ed. Jan Szczęsny Herburt (Dobromyl: Jan Szeliga, 1615). This edition also includes Miechowita's *Chronica Polonorum*. For an abridged English translation, see Jan Długosz, *The Annals of Jan Długosz: An English Abridgement = Annales seu cronicae incliti regni Poloniae*, trans. Maurice Michael (Charlton, West Sussex: IM Publications, 1997).

25. After Miechowita's death, his map collection became the first publicly accessible collection as a part of the cartographic holdings of the University of Kraków. See Hajdukiewicz, *Biblioteka*, 154.

26. Valerie Kivelson, "The Cartographic Emergence of Europe?" in *The Oxford Handbook of Early Modern European History, 1350–1750*, vol. 1: *Peoples and Place*, ed. Hamish Scott (Oxford: Oxford University Press, 2015), 37–69: 54.

27. Kivelson, "The Cartographic Emergence of Europe?" 37.

28. Pomponius Mela, *Description of the World*, trans. F. E. Romer (Ann Arbor: University of Michigan Press, 1998), 1.7, 35. Translation slightly adapted.

29. Pliny the Elder, *Natural History*, trans. H. Rackham (Cambridge, MA: Harvard University Press, 1962), 3.5, 5; 7.

30. *An Eleventh-Century Egyptian Guide to the Universe: The Book of Curiosities*, ed. and trans. Yossef Rapoport and Emilie Savage-Smith (Leiden: Brill, 2014).

31. When he uses the toponym Hiberia, Strabo refers at once to Spain (in book three) and to the Caucasus (book eleven); in his *Life of Antony*, Plutarch uses "Iberia" for Spain and, in the *Life of Pompey*, for the Caucasus; Pliny the Elder refers in book thirty-seven to Spain, in book four to the Caucasus, when using the toponym.

32. See Johannes Honter, *Rudimentorum cosmographiae libri duo: Quorum prior astronomiae, posterior geographiae primapia complectitur* (Kraków: Maciej Szarfenberg, 1530).

33. Johannes Honter, *Rudimentorum cosmographicorum Libri III* (Zurich: Christoph Froschauer, 1581), book 2, n.p. See also Christine R. Johnson, *The German Discovery of the World: Renaissance Encounters with the Strange and Marvelous* (Charlottesville: University of Virginia Press, 2008), 229 [note 143].

34. Miechowita, *Tractatus*, 2.3.2, n.p.

35. Lucan, *The Civil War*, trans. J. D. Duff (Cambridge, MA: Harvard University Press, 1957), 3, vv. 272–79, 134–35.

36. Leonid S. Chekin, *Northern Eurasia in Medieval Cartography: Inventory, Text, Translation, and Commentary* (Turnhout: Brepols, 2006), 46.

37. Marc Shell, *Islandology* (Stanford, CA: Stanford University Press, 2014), 3–4.

38. Franco Farinelli, *La crisi della ragione cartografica* (Turin: Einaudi, 2009), 12.

39. Shell, *Islandology*, 110.

40. The ancient Greek "nēsos" can mean both "island" and "peninsula," and several Greek peninsulas shifted back and forth between a peninsular and insular status. While the Greek toponym *Peloponnesus* contains the term "nēsos," the peninsular Mount Athos had actually been separated from the mainland for some time; see Shell, *Islandology*, 21.

41. Giovanni Battista Ramusio would take up the image of Crimea as an island in the second volume of *Navigazioni e viaggi*, where he writes that "Taurica Chersonesso"

is "l'Isola attaccata con la terra ferma" (the Island attached to the mainland). Ramusio, *Navigationi et viaggi*, vol. 2 (Venice: Giunti, 1583), 4v.

42. Miechowita, *Tractatus*, 2.3.2.

43. Cristoforo Buondelmonti, *Liber insularum archipelagi* (1420), Biblioteca Nacional de España, Madrid, MS 18246.

44. Tommaso Porcacchi, *L'isole più famose del mondo descritte da Thomaso Porcacchi da Castiglione* (Venice: Simon Galignani, 1572), 110.

45. Martin W. Lewis and Kären E. Wigen, *The Myth of Continents: A Critique of Metageography* (Berkeley: University of California Press, 1997), 26, my emphasis.

46. On these two conceptualizations of border in the twentieth century, see David Newman, "Borders and Bordering: Towards an Interdisciplinary Dialogue," *European Journal of Social Theory* 9, no. 2 (2006): 171–86: 175.

47. Nigel Thrift, "Space," *Theory, Culture & Society* 23, nos. 2–3 (2006): 139–55: 144.

48. Paweł Bukowiec, "O potrzebie ujęć subwersywnych w badaniach nad pograniczem," *Wielogłos* 20, no. 2 (2014): 81–90: 89.

49. Cédric Parizot et al., "The antiAtlas of Borders, A Manifesto," *Journal of Borderlands Studies* 29, no. 4 (2014): 503–12: 504.

50. Abraham Ortelius' 1570 *Theatrum Orbis Terrarum* is commonly considered the first atlas, while Gerhard Mercator's 1595 *Atlas* is the first to use the word "atlas" in the title. See Abraham Ortelius, *Theatrum Orbis Terrarum* (Antwerp: Aegidius Coppenius Diesth, 1570); and Gerhard Mercator, *Atlas sive Cosmographicae Meditationes de Fabrica Mundi et Fabricati Figura* (Duisburg: Albertus Busius, 1595).

51. Chris Rumford, "Introduction: Theorizing Borders," *European Journal of Social Theory* 9, no. 2 (2006): 155–69: 161.

52. Larry Wolff, *Inventing Eastern Europe*, 148.

53. Valerie Kivelson, "The Cartographic Emergence of Europe?" 54.

54. The use of Alexander's Altars as a boundary separating Europe from Asia was curiously grounded in a geographic error. Pliny the Elder recalls that "ultra Sogdiani, oppidum Panda et in ultimis eorum finibus Alexandria ab Alexandro Magno conditum. arae ibi sunt ab Hercule ac Libero Patre constitutae, item Cyro et Samiramide atque Alexandro: finis omnium eorum ductus ab illa parte terrarum, includente flumine Iaxarte, quod Scythae Silim vocant, Alexander militesque eius Tanain putavere esse" (Beyond [the sources of the Indus] are the Sogdiani and the town of Panda, and on the farthest confines of their territory Alexandria, founded by Alexander the Great. At this place there are altars set up by Hercules and Father Liber [Dionysus], and also by Cyrus and Samiramis and by Alexander, all of whom found their limit in this region of the world, where they were shut in by the river Syr Darya, which the Scythians call the Silis and which Alexander and his soldiers supposed to be the Don)." Pliny the Elder, *Natural History*, 6.49, 374–75. The early modern cartographic conflation of the sources of the Tanais River with the Altars of Alexander stems from the geographic confusion of two river names: the Iaxartes (nowadays Syr Daria in Turkmenistan) and the Don (Tanais) River. The Iaxartes River, east of the Aral Sea, was the historical northeastern border of the reign of Alexander the Great. See Arrian, *Anabasis Alexandrou. The Landmark Arrian: The Campaigns of Alexander*, ed. James Romm, trans. Pamela Mensch (New York: Pantheon, 2010), 3.30, 7–8, 149.

55. See Jaroszewska, "A la découverte de l'Europe de l'Est," 24.

56. Miechowita, *Tractatus*, dedicatory epistle to Stanisław Turzo, n.p.

57. In an attempt to track Miechowita's rejection of the Riphean Mountains, Emperor Maximilian I sent his delegate Sigismund von Herberstein to the Grand Duchy of Muscovy, who returned with a confirmation of Miechowita's findings and abundant new material for what became the first great description of Muscovy. See Sigismund von Herberstein, *Rerum moscoviticarum comentarii* (Vienna: n.p., 1549).

58. "I grandi fiumi che sfociano sulla riva settentrionale del Mar Nero dovevano [per i geografi Rinascimentali] avere come origine grandi laghi o, preferibilmente, grandi e lontane montagne." Marica Milanesi, "Il confine degli Urali: un'invenzione geopolitica," *Limes: Rivista italiana di geopolitica* 1 (1994): 109–18: 110.

59. Milanesi, "Il confine degli Urali," 110.

60. It is with Miechowita in mind that Abraham Ortelius opened, many years later, his description of Poland in the *Theatrum Orbis Terrarum*: "Polonia, quae a planitie terrae, (quam ipsi vernacule etiamnum Pole vocant) nomen habet, vasta Regio est" (Poland, which derives its name from the flatness of the earth—called "pole" in the vernacular—is a vast region). Ortelius, *Theatrum Orbis Terrarum*, 44.

61. Herodotus, *The Histories*, trans. Aubrey de Sélincourt (New York: Penguin, 2003), 258.

62. Ovid, *Tristia*, trans. Arthur Leslie Wheeler (Cambridge, MA: Harvard University Press, 1996), 3.12.29, 148–49.

63. Tacitus, *Germania*, trans. Maurice Hutton, revised by Robert Maxwell Ogilvie (Cambridge, MA: Harvard University Press, 1970), 1.1., 128–29. I slightly altered the translation.

64. Ralph Tuchtenhagen, "Antikerezeption und Herrschaftslegitimation in der frühen Neuzeit am Beispiel der Theorien über den Ursprung der Völker Europas," *Chloe* 41 (2010): 134–35.

65. The most detailed study of the importance of "Sarmatia" for early modern Polish history, literature, and culture is Tadeusz Ulewicz, *Sarmacja: Studium z problematyki słowiańskiej XV i XVIw. Zagadnienie sarmatyzmu w kulturze i literaturze polskiej*, intro. Teresa Bałuk-Ulewiczowa (Kraków: Collegium Columbinum, 2006). See also Hans-Jürgen Bömelburg, "Sarmatismus—Zur Begriffsgeschichte und den Chancen und Grenzen als forschungsleitender Begriff," *Jahrbücher für Geschichte Osteuropas* 57, no. 3 (2009): 402–8.

66. Ptolemy, *Geography*, ed. Luther Stevenson, trans. Joseph Fischer (New York: Cosimo Classics, 2011), 3.5, 79.

67. Ptolemy, *Geography*, 5.8, 120.

68. Ptolemy, *Geography*, "Tabula Octava," Nancy, BM 441, annotated by Cardinal Guillaume Fillastre. Quoted by Christiane Deluz, "L'Europe selon Pierre d'Ailly ou selon Guillaume Fillastre? De l'*Ymago Mundi* aux Légendes de la Carte de Nancy," in *Humanisme et culture géographique à l'époque du concile de Constance: Autour de Guillaume Fillastre, Actes du Colloque de l'Université de Reims, 18–19 November 1999*, ed. Didier Marcotte (Turnhout: Brepols, 2001), 151–59: 153.

69. Gervase of Tilbury, *Otia Imperialia: Recreation for an Emperor*, ed. and trans. S. E. Banks and J. W. Binns (Oxford: Clarendon, 2002), 2.7., 238–39.

70. Jan Długosz, *Historiae Polonicae* (= *Polish Histories*), ed. Aleksander Przeździecki (Kraków: Typographia Kirchmayeriana, 1873), 1, 28.

71. Długosz, *Historiae Polonicae*, 40.

72. In a different passage, Długosz equates the Poles with the Sarmatians.

73. The rare name Saruth seems to have been introduced by the Benedictine monk Rupert of Deutz in his twelfth-century commentary on Genesis, *De Victoria verbi*, which circulated in several print editions since 1487. See Rupert of Deutz, *De Victoria Verbi Dei* (Augsburg: Anton Sorg, 1487), fol. 20v-21r.

74. Denis Guénoun, *Hypothèses sur l'Europe: Un essai de philosophie* (Belfort: Circé, 2000), 42. Quoted by Rodolphe Gasché, *Europe, or the Infinite Task: A Study of a Philosophical Concept* (Stanford, CA: Stanford University Press, 2008), 11.

75. Bukowiec, "O potrzebie," 89.

76. Gasché, *Europe, or the Infinite Task*, 11.

77. Gunnar Olsson, *Lines of Power/Limits of Language* (Minneapolis: University of Minnesota Press, 1991), 181.

78. Bernhard Siegert, *Cultural Techniques: Grids, Filters, Doors, and other Articulations of the Real* (New York: Fordham University Press, 2015), 97.

79. See Charles S. Peirce, *The Art of Reasoning* (1893). Quoted in Robert Stockhammer, "'An dieser Stelle': Kartographie und die Literatur der Moderne," *Poetica: Zeitschrift für Sprach- und Literaturwissenschaft* 33, nos. 3–4 (2001): 272–306: 279. With Peirce in mind, Robert Stockhammer and Jörg Dünne refer to the map as a "deictic surface" (*Zeigefläche*). For the German term "Zeigefläche," see Stockhammer, "'An dieser Stelle,'" 280; for the English term "deictic surface," see Jörg Dünne, "Map Line Narratives," in *Literature and Cartography*, ed. Anders Engberg-Pederson (Cambridge, MA: MIT Press, 2017), 361–90: 383 [note 12].

80. Siegert, *Cultural Techniques*, 98.

81. Siegert, *Cultural Techniques*, 98.

82. Amerigo Vespucci, *Mundus Novus* (Antwerp?: W. Vorsterman?, 1504), n.p.

83. Miechowita, *Tractatus*, 1.1.1. My emphasis.

84. Marcus Tullius Cicero, *Timaeus. De universitate/Timaeus. Über das Weltall (Latin-German)*, ed. and trans. Karl and Gertrud Bayer (Düsseldorf: Artemis & Winkler, 2006), 6.18, 32.

85. Cicero, *Timaeus*, 3.13, 24.

86. Fluent in German, Latin, and Greek, Andrzej Glaber was the chair of Astrology at the University of Kraków. See Jolanta Migdał, "Glaberowskie korektywy gramatyczne w Żołtarzu Walentego Wróbla," *Studia Polonistyczne* 16–17 (1988): 71–91: 72.

87. Maciej Miechowita, *Polskie wypisanie dwoiey krainy swiata: ktorą po łacinie Sarmatią, takież y lud tam przebywaiąci zową Sarmate, iakoby zawsze gotowi a zbroyni. Gdzież też obiawione są niektore dawne dzieie polskie. Z wypisania doctora Macieia Miechowity dopiro wyłożone (Polish Description of a twofold world region: which is called in Latin Sarmatia, and the people living there are called Sarmatians, meaning always ready to take up arms. Where also some older Polish deeds are explained. Written by dr. Maciej Miechowita, just translated)*, trans. Andrzej Glaber (Kraków: Florian Ungler, 1535).

88. "The Sarmatians rode armed (*armatus*) over the open fields before Lentulus restrained them at the Danube, and from their enthusiasm for weaponry (*arma*) they are thought to have received the name Sarmatians." Isidore of Seville, *Etymologies*, trans. Stephen A. Barney, W. J. Lewis, J. A. Beach, and Oliver Berghof (Cambridge: Cambridge University Press, 2006), 9.2, 93, 197.

89. According to Herodotus, the Tanais River "divides the Royal Scythians from the Sauromatae." East of the Tanais, he continues (in direct opposition to Miechowita), "is no longer Scythia; the first of the divisions belongs to the Sauromatae, whose country begins at the inner end of the Maeotian lake and stretches fifteen days' journey to the north, and is all bare of both forest and garden trees." Herodotus, *The Persian Wars*, trans. A. D. Godley (Cambridge, MA: Harvard University Press, 1995), 4.57, 257; 4.21, 221. For Lucan, both the "nomad peoples of Scythia . . . dipp[ing] their arrows in poison" (*tinxere sagittas/Errantes Scythiae populi*) and the "Sarmatians, akin to the savage Moschi" (*saevisque adfinis Sarmata Moschis*), were equally "dangerous people" (*gens aspera*). Lucan, *The Civil War*, 3, vv. 266–67, 270, 269; 134–35.

90. Isidore of Seville, *Etymologies*, 16.8, 3, 363.

91. Aethicus Ister, *Cosmography*. Quoted by Chekin, *Northern Eurasia*, 24.

92. Sandro Mezzadra and Brett Neilson, *Border as Method, or, the Multiplication of Labor* (Durham, NC: Duke University Press, 2013), 54–55.

93. Mezzadra and Neilson, *Border as Method*, 55.

94. See Oscar Halecki, *Borderlands of Western Civilization: A History of East Central Europe* (New York: Ronald Press Company, 1952), 118.

95. *Carmina de memorabili cede Scismaticorum Moscoviorum per Serenissim[um] ac Invictissim[um] D. Sigismundu[m] Rege[m] Polonie* (Rome: ?, 1515). See Johannes Dantiscus, *Corpus Epistularum*, vol. 1: *Ioannes Dantiscus' Correspondence with Sigmund von Herberstein*, ed. Jerzy Axer et al. (Warsaw: OBTA, 2008).

96. *Carmina*, 1515.

97. From Herodotus on, the people dwelling beyond Alexander's Altars were known as the Scythians: "On the opposite bank of the river [Iaxartes] camped a Skythian army, which Alexander intended to attack. These Skythians had never before been subdued. . . . Alexander was aware of the magnitude of the situation. . . . As usual, the crossing was successful and the Skythian force was routed. During the chase Alexander found that he had outstripped someone greater than the Great Kings." Ory Amitay, *From Alexander to Jesus* (Berkeley: University of California Press, 2010), 31.

98. Jan Łaski was the uncle of John a Lasco (Jan Łaski), the itinerant Reformer to whom Erasmus would sell his private library. On Jan Łaski the Reformer, see Jacqueline Glomski, *Patronage and Humanist Literature in the Age of the Jagiellons: Court and Career in the Writings of Rudolf Agricola Junior, Valentin Eck, and Leonard Cox* (Toronto: University of Toronto Press, 2007), 42.

99. Łaski, "Epistola," in *Carmina*, n.p.

100. Łaski, "Epistola," in *Carmina*, n.p.

101. Miechowita, *Tractatus*, 1.2.3.

102. "Scythia," Miechowita contends, "nunc Thartaria vocatur" (Scythia is now called Tartaria). Miechowita, *Tractatus*, 2.1.1.

103. Miechowita describes several customs of the Tatars, whom he divides into different peoples such as the Trans-Volga (Zavolhenses), Crimean (or Perekopian) (Przecopenses), Kazan (Cosanenses), and Noghay (Nohacenses) Tatars. Scholars have noted that while Miechowita "knows virtually nothing of the Kazan khanate," his account of the Crimean Tatars, to whom he equally refers as Ulans, "is by far the most complete, suggesting that his sources were most familiar with Crimean affairs." Devin DeWeese, *Islam-

ization and Native Religion in the Golden Horde: Baba Tükles and Conversion to Islam in Historical and Epic Tradition (University Park: Pennsylvania State University Press, 1994), 349.

104. The Lithuanian Tatars used a version of Belorussian written in Arabic alphabet. See Niedźwiedź, "Wielopiśmienność," 11.

105. Ptolemy, *Geography* (Rome: 1507), 47, n.p.

106. Kivelson, "The Cartographic Emergence of Europe?" 46–47.

107. Ludovico Ariosto, *Orland Szalony*, trans. Piotr Kochanowski, vol. 1 (Kraków: Wydawnictwo Akademii Umiejętności, 1905), 10.71, 214, my emphasis. Piotr Kochanowski was the nephew of the renowned Polish Renaissance poet and translator Jan Kochanowski.

108. Niccolò Machiavelli, *Discorsi della prima Deca di Tito Livio* (Firenze: Sansoni, 1971), 2.8, 180. For the English translation, see Niccolò Machiavelli, *Discourses on the First Decade of Titus Livius*, trans. Ninian Hill Thomson (London: Kegan Paul, 1883), 2.8, 224–25.

109. Łaski, "Epistola," in *Carmina*, n.p.

110. See Wiktor Weintraub, "Renaissance Poland and 'Antemurale Christianitatis'," *Harvard Ukrainian Studies* 3 (1979): 920–30: 921; and Janusz Tazbir, *Poland as the Rampart of Christian Europe: Myth and Historical Reality*, trans. Chester A. Kisiel (Warsaw: Interpress, 1983).

111. Francesco Filelfo, "Letter to Ladislaus" ("Franciscus Philelfus Vladislao, Hungariae regi, salute plurimam dicit," November 5, 1444), in Filelfo, *Collected Letters: Epistolarum Libri XLVIII*, vol. 1, ed. Jeroen De Keyser (Alessandria: Edizioni dell'Orso, 2015), VI, PhE.06.01, 327.

112. Filelfo, "Letter to Ladislaus," 325.

113. Mezzadra and Neilson, *Border as Method*, vii. The emphasis is theirs.

114. Stanisław Turzo was bishop of the Bohemian town of Olomouc and member of a powerful banking family related to leading political figures in Kraków as well as the Fugger family in Augsburg. From ca. 1494, Miechowita lived in Stanisław Turzo's family house in Kraków. Stanisław's father was one of Kraków's richest inhabitants, who promoted the rise of humanism, art, and book print. Stanisław's brother, Johannes, became one of the most important entrepreneurs in the newly emerging field of metallurgic technology.

115. Miechowita, *Tractatus*, dedicatory epistle to Stanisław Turzo. I discuss this passage also in Piechocki, "Erroneous Mappings," 90–91.

116. In the above-quoted passage, Miechowita uses three other words, besides "aperta," that denote disclosure and openness: "patefacta," "pateat," and "clarescat." "Patefacere" means "to make visible, reveal, uncover, lay bare," "to make or lay open, to open," and, more specifically, "to open the way as a discoverer or pioneer; to be the first to find." "Pateo" denotes "to stand open, lie open, be open," especially in the context of doors, gates, and buildings. It further means "to stretch out, extend; to be accessible, attainable." "Clarescere" signifies "to be illuminated, become bright, shine," but also "to become famous." *Oxford Latin Dictionary*, ed. P. G W. Glare, vols. 2 and 1 (Oxford: Oxford University Press, 2012), entries "patefaciō" and "clarescō," and *Harpers' Latin Dictionary* (New York: American Book Company, 1907), entries "patefaciō" and "pateō."

117. I thank Jakub Niedźwiedź for this information.
118. Witold Wilczyński, "Geografia i metageografia ziem dawnej Sarmacji," *Przegląd Geopolityczny* 9 (2014); 9–30: 26.
119. Miechowita, *Tractatus*, dedicatory letter to Bishop Stanisław Turzo. The dedicatory letter is also reprinted, in Latin, in Martin Rothkegel, ed., *Der lateinische Briefwechsel des Olmützer Bischofs Stanislaus Thurzó* (Hamburg: LIT, 2007), 147–48.
120. See Nancy Bisaha, *Creating East and West: Renaissance Humanists and the Ottoman Turks* (Philadelphia: University of Pennsylvania Press, 2004).
121. Giovanni Battista Ramusio, "Discorso sopra varii viaggi per li quali sono state condotte fino a' tempi nostri le spezierie e altri nuovi che se potriano usare per condurle," in *Navigazioni e viaggi*, ed. Milanesi, vol. 2, 967–90: 976.
122. Together with the Tanais River, Ramusio mentions the ancient Greek city of Tana, located at the mouth of the Tanais. In the late Middle Ages, the Venetians and the Genovese turned the conveniently located city into an emporium for trade between Asia and Europe. However, with the rise of the Ottoman Empire the Venetians and Genovese lost their possessions on the northern shores of the Black Sea and the Sea of Azov. For a discussion of Tana's different names in Slavic, Arabic, and Mongolian languages, see Donald Ostrowski, "City Names of the Western Steppe at the Time of the Mongol Invasion," *Bulletin of the School of Oriental and African Studies* 61, no. 3 (1998): 465–75.
123. Ramusio, *Navigazioni e viaggi*, 978.
124. Ramusio, *Navigazioni e viaggi*, 981–82.
125. Two decades after Miechowita published his *Tractatus*, the Swedish cartographer Olaus Magnus, who spent many years in Danzig putting together his *Carta marina* (1539), would display the "abundantia" of Europe's northern regions on his map, thus promoting a vigorous and thriving image of Europe's North equal to the pulsating network of the Mediterranean. Miechowita corresponded with Olaus's brother, Johannes, on the origin of the Goths. Two letters are included in Maciej Miechowita, *Descriptio duarum Sarmatiarum*.
126. Miechowita often refers to Ruthenia as "Roxolania."
127. Miechowita, *Tractatus*, 2.1.1.
128. Miechowita, *Tractatus*, 2.1.1.
129. Miechowita, *Tractatus*, 2.1.1.
130. Jakub Parkosz, *Traktat o ortografii polskiej*, ed. Marian Kucała (Warsaw: PWN, 1985), 57–58.
131. Mezzadra and Neilson, *Border as Method*, vii.
132. Mezzadra and Neilson, *Border as Method*, ix.
133. Mezzadra and Neilson, *Border as Method*, viii.
134. Miechowita, *Tractatus*, 2.2.1.
135. Valerie Kivelson, *Cartographies of Tsardom. The Land and Its Meanings in Seventeenth-Century Russia* (Ithaca, NY: Cornell University Press, 2006), 136.
136. Miechowita, *Descriptio Sarmatiarum*, dedicatory letter to Johannes Haller, 1521. The adjective "gold(en)" is spelled "złota" in Polish and "zlata" or "zolotaia" in the Russian transliteration.

137. See Grzegorz Franczak, "*Faex Gentium*: Polacy w Moskwie wobec rosyjskiej 'mniejszości' (1606–1612)," in *Etniczność, Tożsamość, Literatura. Zbiór Studiów*, ed. Paweł Bukowiec and Dorota Siwor (Kraków: Universitas, 2010), 45–67: 67.

138. The "Slata Baba" or "Aurea Anus" (or "Vetula") appears on all of Herberstein's maps of Muscovy, albeit in different forms and with altering attributes. Later editions of *Rerum moscoviticarum* (e.g., those from 1556 and 1557) show the *aurea vetula* seated on a throne-like seat with a scepter in her hand. Not mentioned in Miechowita, here the scepter seems to conflate the *aurea vetula* with the Nordic legend of Thor, "who seems to simulate Jupiter with his scepter" (*Thor autem cum sceptro Iovem simulare videtur*) in Adam of Bremen and, in the fifteenth and sixteenth centuries, in Ericus Olai and Johannes Magnus. On Herberstein's map, the scepter perhaps takes on the functions of a line and instrument of territorial demarcation. Ericus Olai, *Chronica regni Gothorum*, ii, 6. Quoted by Maria Elena Ruggerini, "Gli idoli del tempio di Uppsala: Tradizione ed ermeneutica in Johannes e Olaus Magnus," in *I fratelli Giovanni e Olao Magno. Opera e cultura tra due mondi*, ed. Carlo Santini (Rome: Il Calamo, 1999), 261–307: 281.

139. Richard Chancellor was the captain of the *Edward Bonaventure* and "Pilot Maior of the fleete" on the expedition of 1553 that discovered the northern sea route to Russia leading to the opening up of trade between England and Russia.

140. Richard Chancellor, *The Voyages and Discoveries*, in Richard Hakluyt, *The Principal Navigations* (London: George Bishop and Ralph Newberie, 1589), 279.

141. For Augustin Hirschvogel's and Gerhard Mercator's depiction of the *aurea vetula*, see Friedrich Wilhelm Krücken, *Ad Maiorem Gerardi Mercatoris Gloriam: Abhandlungen zum Leben und Werk Gerhard Mercators* (Münster: Monsenstein und Vannerdat, 2009), 287–98; and Roland Cvetkovski and Alexis Hofmeister, eds., *An Empire of Others: Creating Ethnographic Knowledge in Imperial Russia and the USSR* (Budapest: Central European University Press, 2014), 179–81.

142. Abraham Ortelius, "Russiae, Moscoviae et Tartariae descriptio: Auctore Antonio Ienkensono Anglo, edita Londini anno 1562 & dedicata ilustriß. D. Henrico Sydneo Wallie presidi," in Abraham Ortelius, *Theatrum Orbis Terrarum* (Antwerp: Aegidius Coppenius Diesth, 1570), n.p. In his description of Russia and Tartaria, Ortelius explicitly mentions Maciej Miechowita as his source.

143. Jenkinson's map, deemed lost, was rediscovered by Krystyna Szykuła and purchased by the cartographic collection of Wrocław University Library. Krystyna Szykuła, "Anthony Jenkinson's unique wall map of Russia (1562) and its influence on European cartography," *Belgeo* 3–4 (2008): 325–40; Krystyna Szykuła, "Odnaleziona mapa Rosji Jenkinsona z 1562 roku. Pierwsza próba analizy mapy," *Acta Universitatis Wratislaviensis* 1678, no. 2 (1995): 7–31; Krystyna Szykuła, "Mapa Rosji Jenkinsona (1562)— Kolejne podsumowanie wyników badań," *Czasopismo Geograficzne* 71, no. 1 (2000): 67–97.

144. Ortelius, *Theatrum Orbis Terrarum*, n.p.

145. Ortelius, *Theatrum Orbis Terrarum*, n.p.

146. Karen Elizabeth Bishop, "Introduction. The Cartographical Necessity of Exile," in Karen Elizabeth Bishop, ed., *Cartographies of Exile: A New Spatial Literacy* (New York: Routledge, 2016), 1–22: 1.

147. Krystyna Szykuła, "Anthony Jenkinson's unique wall map of Russia (1562)," 328.

CHAPTER THREE

1. Ptolemy, *Cosmographia*, trans. Jacopo Angeli da Scarperia (Naples: Bernardo Sylvano, 1490), 292v-293r, MS, Bibliothèque nationale de France, Lat. 10764. See Henri Michelant, *Catalogue de la Bibliothèque de François I*er *à Blois, en 1518: publié d'après le manuscrit de la Bibliothèque impériale de Vienne* (Paris: A. Franck, 1863), 43.

2. Germaine Aujac notes that "[l]'Europe est la plus mal traitée, qui ne montre que la félicité de la Phénicie" (Europe is the worst treated, showing only the bliss of Phoenicia). Germaine Aujac, *La Géographie de Ptolémée: le manuscrit d'Andrea Matteo Acquaviva et d'Isabella Piccolomini* (Arcueil: Editions Anthèse, 1998), 87 [note 10].

3. Guillaume Postel, *Cosmographiae Disciplinae Compendium* (Basel: Johannes Oporinus, 1561), 15.

4. Postel, *Cosmographiae*, 15.

5. "Nam a Gal, quod diluvii fluctum sonat, Galli sunt nuncupati: ut conservarent soli pro toto mundo tanti miraculi nunquam obliterandam memoriam, quae alioqui ad universos orbis habitatores pertinent." Postel, *Cosmographiae*, 20.

6. "Iapetus, quem sacra vera Iaphetum sive Iephetum dicunt, iustius suae parti, ut in nostro orbe inter tres minimae, ita monumentis nulli omnino inferiori, nomen Iapetie dare aut reddere debet, quam a nefarii nebulonis cum vacca congressu fabulosam Europe nomenclaturam illi permittere." Postel, *Cosmographiae*, 2.

7. See, for instance, Strabo, *Geographia interpretibus Guarino Veronesi et Gregorio Tiphernate*, Österreichische Nationalbibliothek, Vienna, Austria, n.d., Cod. 3 Han Mag, fol. 1r.

8. Tom Conley, "A Cartography of Exile: Du Bellay's *France, mere des arts*," in *Cartographies of Exile: A New Spatial Literacy*, ed. Karen Elizabeth Bishop (New York: Routledge, 2016), 44–66: 62.

9. Allegorical depictions of Europe such as Münster's "Europa Virgo" started circulating in particular from the last quarter of the sixteenth century on. In Münster's 1544 *Cosmographia*, Europe is not yet stylized as a woman. The 1588 German edition of the *Cosmographia* appears to be the first in which Europe is represented as a female allegory.

10. Ján Pravda, "Map Language: A Logical Graphic System," in *Selected Papers on Cartography as Language or Sign System*, ed. C. Grant Head and Hansgeorg Schlichtmann, 1978, typescript, 6 pages. Quoted by C. Grant Head, "Mapping as Language or Semiotic System: Review and Comment," in *Cognitive and Linguistic Aspects of Geographic Space*, ed. David M. Mark and Andrew U. Frank (Dordrecht: Springer Science+Business, 1991), 237–62: 244.

11. Geoffroy Tory, *Champ fleury* (Paris: Gilles de Gourmont, 1529), n.p./Geoffroy Tory, *Champ fleury*, trans. George B. Ives (New York: Grolier Club, 1927), xxiii.

12. For a discussion of the etymological proximity between "pagina" and "pagus," see Cornelius J. Crowley, "Romance Derivatives of Latin 'tangere, pangere, tingere, pingere'," *General Linguistics* 4, no. 2 (1959): 56–69: 59–60. The cognate "paganus" (pagan) references, as Richard Stanley observes, a "dweller within that space." Richard M. Stanley, *Literary Constructions of Youth in the Early Empire: The Case of Nero* (PhD dissertation, University of North Carolina at Chapel Hill, 2003), 94.

13. Tory, *Champ fleury*, fol. 111r/Tory, *Champ fleury*, 29.

14. See Pierre Cordier, "Geoffroy Tory et les leçons de l'Antique," *Anabases* 4 (2006) : 11–32: 27.

15. Tom Conley, *The Self-Made Map: Cartographic Writing in Early Modern France* (Minneapolis: Minnesota University Press, 1996), 87.

16. Tory became a lecturer at the Collège du Plessis and later at the Collège de Coqueret soon upon his arrival in Paris. See Stéphanie Deprouw, "De Bourges à Paris en passant par l'Italie," in *Geoffroy Tory: imprimeur de François Ier, graphiste avant la lettre*, ed. Stéphanie Deprouw et al. (Paris: Editions Rmn-Grand Palais, 2011), 18–29: 20.

17. Timothy J. Reiss, *Knowledge, Discovery and Imagination in Early Modern Europe: The Rise of Aesthetic Rationalism* (Cambridge: Cambridge University Press, 1997), 12.

18. Reiss, *Knowledge, Discovery and Imagination*, 2.

19. Nicholas Jardine, "The Forging of Modern Realism: Clavius and Kepler against the Sceptics," *Studies in History and Philosophy of Science* 10 (1979): 141–73: 147. Quoted by Timothy J. Reiss, *Knowledge, Discovery and Imagination*, 4.

20. Franco Farinelli, *Blinding Polyphemus: Geography and the Models of the World*, trans. Christina Chalmers (London: Seagull Books, 2018), 26–27.

21. Farinelli, *Blinding Polyphemus*, 25.

22. In 1529, the year *Champ fleury* was published, François I promoted the Franciscan friar Jean Thénaud, author of two cabbalistic treatises (*La saincte et très chrestienne cabale métrifiée* [1519] and *Traicté de la cabale* [1521], to the position of abbot of Melinays.

23. Pierre Cordier, "Geoffroy Tory," 29.

24. John L. Austin, *How to Do Things with Words* (Cambridge, MA: Harvard University Press, 1975).

25. Marsilio Ficino, *The Hieroglyphics of Horapollo*, trans. George Boas (Princeton, NJ: Princeton University Press, 1993), 14.

26. Tory, *Champ fleury*, fol. 43r/Tory, *Champ fleury*, 105. Some scholars identify Horapollon Neiloios with the grammarian Flavius Horapollon. See Pedro Germano Leal, "Reassessing Horapollon: A Contemporary View on *Hieroglyphica*," *Emblematica* 21 (2014): 37–75: 41.

27. Behind the discovery of the *Hieroglyphica* lies an astounding cartographic trajectory across the Mediterranean: the oldest extant manuscript was purchased on the Aegean island of Andros in 1419 by Cristoforo Buondelmonti, a Florentine merchant and author of the first extant *isolario* (Island Book). See Leal, "Reassessing Horapollon," 42.

28. Ricardo Padrón, *The Spacious Word: Cartography, Literature, and Empire in Early Modern Spain* (Chicago: University of Chicago Press, 2004).

29. Anthony Grafton, *Defenders of the Text: The Traditions of Scholarship in an Age of Science, 1450–1800* (Cambridge, MA: Harvard University Press, 1991), 82. See Elio Antonio de Nebrija, *Opuscula in hoc voluminie co[n]tenta, . . . Berosus chaldeus de origine & successione regnoru[m] . . .* (Burgos: n.p., 1512).

30. Pierre d'Ailly, *Ymago mundi* (Paris: Buron, 1930), 334. Quoted in Christiane Deluz, "L'Europe selon Pierre d'Ailly ou selon Guillaume Fillastre? De l'*Ymago Mundi* aux légendes de la carte de Nancy," in Didier Marcotte, ed., *Humanisme et culture géographique à l'époque du concile de Constance: Autour de Guillaume Fillastre, Actes du Colloque de l'Université de Reims, 18–19 novembre 1999* (Turnhout: Brepols, 2002), 151–59: 152.

31. See Patrick Gautier Dalché, "The Reception of Ptolemy's *Geography* (End of the Fourteenth to Beginning of the Sixteenth Century), in *The History of Cartography*, ed. David Woodward, vol. 3: *Cartography in the European Renaissance*, part 1 (Chicago: University of Chicago Press, 2007), 285–364: 299. D'Ailly's *Ymago mundi* served as a central touchstone for Christopher Columbus, who owned a heavily annotated manuscript copy of the work (Louvain, ca. 1480–83). See Delno C. West and August Kling, "Intellectual and Cultural Background of Christopher Columbus," in *The* Libro de las profecías *of Christopher Columbus*, ed. and trans. Delno C. West and August Kling (Gainesville: University of Florida Press, 1991), 7–40: 24.

32. Pomponius Mela, *De totius orbis descriptione . . . Nu[n]qua[m] anea citra montes impressus* (Paris: Jehan Petit, 1507). This work is also known as *De situ orbis*.

33. Enea Silvio Piccolomini, *Cosmographia in Asiae & Europae eleganti descriptione* (Paris: Collège de Plessis, 1509).

34. Valerius Probus, *De interpretandis Romanorum literis opusculum* (Paris: Collège de Plessis, 1510).

35. Probus, *De interpretandis Romanorum literis opusculum*, frontispiece.

36. The volume includes Priscian's *De ponderibus & mensuris*, Giorgio Valla's *De expetendis & fugiendis rebus opus* (originally published by Aldo Manuzio in 1501), Tory's own dialogue in praise of his native city, Bourges (*Dialogus per Godofredum Torinum in quo urbs Biturica sub loquente persona describitur*), an *Epitaphium* celebrating the Arch of the Sergii in the Istrian city of Pula (*Epitaphium Sertii Polensis Parasiti histrionisque festivissimi apud Salonas Dalmatie in marmore ex libro reconditarum*), and the sixth book of Columella's popular twelve-book treatise on agriculture, *Res rustica*.

37. Denis J. J. Robichaud, "Competing Claims on the Legacies of Renaissance Humanism in Histories of Philology," *Erudition and the Republic of Letters* 3 (2018): 177–222: 184.

38. Jill Kraye, "Philologists and Philosophers," in *The Cambridge Companion to Renaissance Humanism*, ed. Jill Kraye (Cambridge: Cambridge University Press, 1996), 142–60: 142.

39. Aristotle is a powerful case in point: sensitive to questions of space and conscious of Greece's liminal position between two continents, Aristotle has been "Europeanized" over the centuries, becoming one of the founding figures of so-called European thought. Not only has Aristotle's philosophical thought been disconnected, over the centuries, from its intimate relationship with his geographic background, but his writing has undergone a *translatio* from Asia to Europe: continentally ambiguous at first, Aristotle has become an unambiguously European philosopher. In his recent book on European philosophy, Roberto Esposito thus argues, in a discussion of Aristotle's *De anima* in the context of the importance of territory and earth writ large, that "in philosophy earth has never acquired the same status as the other three elements." Spatial thinking and philosophy are typically disconnected. Roberto Esposito, *A Philosophy for Europe: From the Outside*, trans. Zakiya Hanafi (Cambridge: Polity Press, 2018), 39.

40. Aristotle, *Politics*, trans. H. Rackham (Cambridge, MA: Harvard University Press, 2014), 7.6, 565; 567. See also Martin W. Lewis and Kären E. Wigen, *The Myth of Continents: A Critique of Metageography* (Berkeley: University of California Press, 1997), 22.

41. See Aristotle, *Politicorum libri octo*, trans. Leonardo Bruni, ed. Jean Lefèvre d'Etaples (Paris: Henri Estienne, 1506), 7.6, 105r.

42. Rodolphe Gasché, *Europe, or the Infinite Task: A Study of a Philosophical Concept* (Stanford, CA: Stanford University Press, 2008), 1.

43. See Lewis and Wigen, *The Myth of Continents*, 22.

44. For the cartographic term "equipollent," see David Woodward, "Roger Bacon's Terrestrial Coordinate System," *Annals of the Association of American Geographers* 80, no. 1 (1990): 109–22: 119.

45. Tory, *Champ fleury*, n.p./Tory, *Champ fleury*, ii.

46. Pomponius Mela, *De totius orbis descriptione* (Paris: Jehan Petit, 1507).

47. Tory, *Champ fleury*, fol. 5r/Tory, *Champ fleury*, 13.

48. Tory, *Champ fleury*, fol. 7r/Tory, *Champ fleury*, 19.

49. Tory, *Champ fleury*, fol. 6r/Tory, *Champ fleury*, 15–16.

50. See Tory, *Champ fleury*, fol. 34v/Tory, *Champ fleury*, 85.

51. Tory, *Champ fleury*, fol. 10r/Tory, *Champ fleury*, 26.

52. Gilles Deleuze, *Logic of Sense*, trans. Mark Lester (New York: Columbia University Press, 1990), 132.

53. Hesiod, *Theogony*, trans. Glenn W. Most (Cambridge, MA: Harvard University Press, 2006), 215; 518.

54. Mela, *De totius orbis descriptione* (1507), fol. 5v. Pomponius Mela, *Description of the World*, trans. F. E. Romer (Ann Arbor: University of Michigan Press, 1998) I, 42.

55. Fabio Barry, "The Mouth of Truth and the Forum Boarium: Oceanus, Hercules, and Hadrian," *Art Bulletin* 93, no. 1 (2011): 7–37: 21.

56. "Whether it be that Hercules visited those shores, or because we have agreed to enter all marvels everywhere to his credit" (Sive adiit Hercules, seu quidquid ubique magnificum est, in claritatem eius referre consensimus). Tacitus, *Germania*, trans. M. Hutton, revised by Robert Maxwell Ogilvie (Cambridge, MA: Harvard University Press, 1970–81), 184–85.

57. Barry, "The Mouth of Truth and the Forum Boarium," 21.

58. Erasmus, *De pueris instituendis / De recta pronuntiatione (On Education for Children / The Right Way of Speaking Latin and Greek)*, in *The Collected Works of Erasmus, Literary and Educational Writings*, vol. 4 (Toronto: University of Toronto Press, 1985), 360. Quoted by Katie Chenoweth, "The Beast, the Sovereign, and the Letter: Vernacular Posthumanism," *Symploke* 23, no. 1–2 (2015): 41–56: 48.

59. Katie Chenoweth, "The Beast," 48. Italics hers.

60. Tory, *Champ fleury*, fol. 46v./Tory, *Champ fleury*, 114.

61. Pomponius Mela, *De orbis situ*, frontispiece.

62. Tory, *Champ fleury*, fol. 8v/Tory, *Champ fleury*, 22–23.

63. Ovid, *Metamorphoses*, trans. Frank Justus Miller, revised by G. P. Goold (Cambridge, MA: Harvard University Press, 1994), book I, vv. 568–746, 42–55.

64. Tory, *Champ fleury*, fol. 7v/Tory, *Champ fleury*, 19–20.

65. Tory, *Champ fleury*, fol. 7v/Tory, *Champ fleury*, 20.

66. See Julia Kristeva, *The Severed Head*, trans. Jody Gladding (New York: Columbia University Press, 2012), 75.

67. Farinelli, *Blinding Polyphemus*, 21; 18.

68. Farinelli, *Blinding Polyphemus*, 20.

69. Farinelli, *Blinding Polyphemus*, 22.

70. Farinelli, *Blinding Polyphemus*, 30.
71. Farinelli, *Blinding Polyphemus*, 30–31. Translation slightly adapted.
72. Farinelli, *Blinding Polyphemus*, 31.
73. See Geoffrey Bennington, *Frontiers (Kant, Hegel, Frege, Wittgenstein)* [University of Sussex Seminars 1989–1992] (ebook, www.bennington.zsoft.co.uk, 2003), 82.
74. Gottlob Frege, "Über Sinn und Bedeutung," *Zeitschrift für Philosophie und philosophische Kritik* 100 (1892): 25–50: 26. Frege's argument is succinctly summarized in Farinelli, *Blinding Polyphemus*, 33.
75. Tory, *Champ fleury*, "Salut a tous vrayz & devotz Amateurs de bonnes Lettres," n.p./Tory, *Champ fleury*, "Greeting to all true and devoted Lovers of well-formed Letters," n.p.
76. "Qui vouldroit seullement escrire en long Trois ou Quatre Versetz, il fauldroit que le Raouleau fust plus long quil n y a despace dicy au Isles des Molucques, et principallement qui vouldroit escripre en grosse Lettre. La maniere descripre en Raouleaux est icy tresabusive en beaucoup de facons...." Tory, *Champ fleury*, "Salut," n.p./Tory, *Champ fleury*, "Greeting," n.p.
77. Tory, "Salut," n.p./Tory, *Champ fleury*, "Greeting," n.p. Translation slightly changed.
78. Farinelli, *Blinding Polyphemus*, 32.
79. When Emperor Charles V's cartographers established that the Moluccas "lay within the Spanish sphere of influence," a cash settlement allowed the Portuguese to continue to consider the islands as their property. Lisa Jardine, *Worldly Goods: A New History of the Renaissance* (New York: Norton, 1996), 288.
80. Woodward, "Roger Bacon's Terrestrial Coordinate System," 119.
81. Tory, *Champ fleury*, fol. 26r/Tory, *Champ fleury*, 65.
82. Tory, *Champ fleury*, 30/"Ce Quarre [contiendra] cent petits Quarreaulx lesquelz iappelleray Corps, pource que la largeur de le I, qui sera proportionaire de toutes les autres lettres, sera co[n]tenue en lung des dessusdictz petits Quarreaulx." Tory, *Champ fleury*, fol. 11v.
83. Tory, *Champ fleury*, fol. 20r/Tory, *Champ fleury*, 51.
84. Tory, *Champ fleury*, fol. 72v/Tory, *Champ fleury*, 179.
85. Sigismondo Fanti, *Trattato di scrittura / Theorica et pratica de modo scribendi (Venezia 1514)* [facsimile], ed. Antonio Ciaralli and Paolo Procaccioli (Roma: Salerno, 2013), fol. 2r.
86. Brian Rotman, *Signifying Nothing: The Semiotics of Zero* (Stanford, CA: Stanford University Press, 1987), 2.
87. Paul Saenger, *Space between Words: The Origins of Silent Reading* (Stanford, CA: Stanford University Press, 1997), 143. See also Ronald Green, *Nothing Matters: A Book about Nothing* (Alresford, UK: John Hunt Publishing, 2011), 19.
88. See Leo Jordan, "Materialien zur Geschichte der arabischen Zahlzeichen in Frankreich," *Archiv für Kulturgeschichte* 3, no. 2 (1905): 155–95: 158.
89. Green, *Nothing Matters*, 13.
90. Green, *Nothing Matters*, 20–21.
91. Green, *Nothing Matters*, 4.
92. Karen Pinto, *Medieval Islamic Maps: An Exploration* (Chicago: University of Chicago Press, 2016), 13 and Yossef Rapoport and Emilie Savage-Smith, *Lost Maps of the*

Caliphs: Drawing the World in Eleventh-Century Cairo (Chicago: University of Chicago Press, 2018), 79.

93. Jordan, "Materialien zur Geschichte der arabischen Zahlzeichen in Frankreich," 160 [n. 1]. Quoted by Eugene Ostashevsky, *Quintessence from Nothingness: Zero, Platonism, and the Renaissance* (PhD dissertation, Stanford University, 2000), 174.

94. Charles de Bovelles, "Liber de Nichilo," in *Que hoc volumine contine[n]tur* (Paris: Henri Estienne, 1510), 63v.

95. Charles de Bovelles, *Ars oppositorum* (1511), ed. Pierre Magnard (Paris: Vrin, 1984), 54–56.

96. Jan Miernowski, *Le Dieu Néant: Théologies négatives à l'aube des temps modernes* (Leiden: Brill, 1999), 78. For an in-depth discussion of the cipher zero in early modernity within an "analytico-referential discourse" see Timothy J. Reiss, "Peirce and Frege: In the Matter of Truth," in Timothy J. Reiss, *The Uncertainty of Analysis: Problems in Truth, Meaning, and Culture* (Ithaca, NY: Cornell University Press, 1988), 19–55.

97. Farinelli, *Blinding Polyphemus*, 96.

98. Farinelli, *Blinding Polyphemus*, 96. See also Robert Kaplan, *The Nothing That Is: A Natural History of Zero* (Oxford: Oxford University Press, 1999).

99. Farinelli, *Blinding Polyphemus*, 97.

100. Angelo Michele Piemontese, "Vitruvio tra gli alfabeti proporzionali arabo e latino," *Litterae Caelestes* 2, no. 1 (2007): 71–97: 72.

101. Piemontese, "Vitruvio tra gli alfabeti," 72.

102. Annemarie Schimmel, *Calligraphy and Islamic Culture* (New York: New York University Press, 1990), 106.

103. Tory, *Champ fleury*, fol. 21r-21v/Tory, *Champ fleury*, 53–54.

104. Albrecht Dürer, *Unterweysung der Messung, mit dem Zirckel und Richtscheyt, in Linien Ebnen unnd gantzen Corporen* (Nuremberg: [n.p.], 1525).

105. Alexander Nagel has insightfully analyzed the strong affinity that fifteenth- and sixteenth-century Italian visual artists exhibited for non-European scripts (predominantly Arabic and Hebrew), which they readily transformed into artistic pseudoscripts and ornaments—often with the purpose to inquire into "the border between meaning and non-meaning." Alexander Nagel, "Twenty-Five Notes on Pseudoscript in Italian Art," *RES: Anthropology and Aesthetics* 59/60 (2011): 228–48: 239.

106. Henri Meschonnic, *Des mots et des mondes: dictionnaires, encyclopédies, grammaires, nomenclatures* (Paris: Hatier, 1991), 41.

107. "At the word go we are within the multiplicity of languages and the impurity of the limit." Jacques Derrida, "What Is a 'Relevant' Translation?" trans. Lawrence Venuti, *Critical Inquiry* 27, no. 2 (2001): 174–200: 176.

108. Franco Farinelli, *La crisi della ragione cartografica* (Turin: Einaudi, 2009), 13.

109. For a discussion of *Utopia* and its toponyms, see Farinelli, *La crisi della ragione cartografica*, 12–13. In his *Pantagruel* (1532), François Rabelais explicitly refers to More's city of Amaurotes: "Gargantua, at the age of four hundred four score and forty-four years, begat his son Pantagruel by his wife Badebec, daughter of the king of the Amaurots in Utopia, who died in childbirth." François Rabelais, *Pantagruel*, in *The Complete Works*, trans. Donald M. Frame (Berkeley: University of California Press, 1991), 2.2, 140.

110. Farinelli, *La crisi della ragione cartografica*, 13.

111. Jeanne Fahnestock, *Rhetorical Style: The Uses of Language in Persuasion* (Oxford: Oxford University Press, 2011), "Situation and Occasion," 338–39.

112. Sixteenth-century grammar and rhetoric manuals such as Richard Sherry's define topographia as "the description of a place, as of Carthago in the fyrst of Eneidos" and topothesia as "the faynyng of a place, when a place is descrybed, as peradventure none is. Example of this is the Utopia of Syr Thomas Moore." Richard Sherry, *A Treatise of the Figures of Grammer and Rhetorike* (London: Richard Tottel, 1555), fol. 7r. Quoted by Jack Bernard Oruch, *Topography in the Prose and Poetry of the English Renaissance, 1540–1640* (PhD dissertation, Indiana University, 1964), 8.

113. Erasmus, *De duplici copia verborum ac rerum commentarii duo* (Strasbourg: Matthias Schürer, 1516), vol. 2, fol. 49v. Lactantius is the first known author to have used the term "topothesia," in his fourth-century commentary on Statius's *Thebaid*, where he juxtaposed it with "topographia," the description of an existing place. Lactantius Placidus, *Commentarius in Statii Thebaida* (Leipzig: Teubner, 1898), book 2: 32, 80. See also Heinrich Lausberg, *Handbook of Literary Rhetoric*, trans. Matthew Bliss et al. (Leiden: Brill, 1998), §819, 366.

114. Virgil, *Aeneid*, trans. H. Rushton Fairclough (Cambridge, MA: Harvard University Press, 1999–2000), 1:157–61, 272–73: "Defessi Aeneadae, quae proxima litora, cursu / contendunt petere, et Libyae vertunt ad oras. / est in secessu longo locus: insula portum / efficit obiectu laterum, quibus omnis ab alto / frangitur inque sinus scindit sese unda reductos" ("The wearied followers of Aeneas strive to run for the nearest shore and turn towards the coast of Libya. There in a deep inlet lies a spot, where an island forms a harbour with the barrier of its sides, on which every wave from the main is broken, then parts into receding ripples"). See Patrick Gautier Dalché, "L'enseignement de la géographie dans l'antiquité tardive," *Klio* 96, no. 1 (2014) : 144–82: 151.

115. "*Est in secessu* topothesia est, id est fictus secundum poeticam licentiam locus. Ne autem videatur penitus a veritate discedere, Hispaniensis Carthaginis portum descripsit. Ceterum hunc locum in Africa nusquam esse constat, nec incongrue propter nominis similitudinem posuit. Nam topographia est rei verae descriptio." Servius, *In Vergilii Aeneidos Libros I-III Commentarii*, ed. Georg Thilo (Leipzig: Teubner, 1878), 66.

116. Erasmus lists the description of "a city, a mountain, a region, a river, a port, a villa, gardens, an amphitheatre, a fountain, a cavern, a temple, a grove" ("urbis, montis, regionis, fluminis, portus, villae, hortorum, amphiteatri, fontis, specus, temple, luci") and, more widely, "the whole appearance of a place just as if it were in sight" ("tota loci facies, veluti spectanda depingitur") as examples of "τοπογραφια" (*topographia*). Erasmus, *On copia of Words and Ideas*, trans. D. B. King and D. Rix (Milwaukee: Marquette University Press, 1963), 47.

117. Tory, *Champ fleury*, 58. "Assemblees les unes avec les autres en sorte quelles rendent & font vng oeuvre tresbel & parfaict qui est dict en Latin 'Opus vermiculatum', 'Opus tessellatum', & 'Assarotum', du quel Pline en so[n] Histoire naturelle, & Vitruve en son Liure Darchitecture parlent assez amplement." Tory, *Champ fleury*, fol. 23r.

118. "Quam lepide λέξεις compostae! Ut tesserulae omnes / Arte pavimento atque emblemate vermiculato" (How charmingly he *fait ses phrases*—set in order, like the lines / Of mosaic in a pavement, and his inlaid work he twines). Cicero, *De Oratore*, trans. E. W. Sutton (Cambridge, MA: Harvard University Press, 1948), 3, 136–37. Sutton's emphasis.

119. Franco Farinelli, *L'invenzione della Terra* (Palermo: Sellerio, 2007), 100.

120. Tory, *Champ fleury*, fol. 6r-v/Tory, *Champ fleury*, 15.

121. Simone Pinet, *Archipelagoes: Insular Fictions from Chivalric Romance to the Novel* (Minneapolis: Minnesota University Press, 2011), xxvi.

122. Tory, *Champ fleury*, 16–17.

123. Pinet, *Archipelagoes*, xxvi.

124. See Thomas More, *De optimo reipublicae statu deque nova insula Utopia libellus* (Louvain: Theodore Martin, 1516), 70; and *Utopia*, book 2, trans. George M. Logan (New York: Norton, 2011), 38–39. I modified Logan's translation.

125. More, *Utopia*, trans. George M. Logan, 69.

126. *Oxford Latin Dictionary*, ed. P. G. W. Glare, vol. 2 (Oxford: Oxford University Press, 2012), "seminarium," 1908.

127. Thomas More, *La description de l'isle d'Utopie où est comprins le miroer des republicques du monde et l'exemplaire de vie heureuse*, trans. Jean Le Blond (Paris: C. L'Angelier, 1550), epistle by Guillaume Budé, n.p. The first English translation of *Utopia*, by Ralph Robynson, was published in 1551.

128. Plato, *Timaeus*, trans. Harold North Fowler (Cambridge, MA: Harvard University Press, 1977), 48E, 112–13.

129. Plato, *Timaeus*, 48E, 112–13.

130. Plato, *Timaeus*, 48E, 112–13. Punctuation slightly adapted.

131. Plato, *Timaeus*, 50B-C, 117.

132. Stuart Elden, *The Birth of Territory* (Chicago: University of Chicago Press, 2013), 39.

133. Plato, *Oeuvres complètes*, trans. Albert Rivaud, vol. 10 (Paris: Les Belles Lettres, 1925). Quoted by Anthony Vidler, "Chôra," in *Dictionary of Untranslatables: A Philosophical Lexicon*, ed. Barbara Cassin, trans. Emily Apter, Jacques Lezra, and Michael Wood (Princeton, NJ: Princeton University Press, 2014), 131–35: 134.

134. Plato, *Timaeus*, 33B. See Farinelli, *La crisi della ragione cartografica*, 60.

135. Marsilio Ficino, "Timeus, vel de Natura Compendium," in *Omnia D. Platonis Opera* (Venice: Girolamo Scotto, 1571), book 32, 401.

136. On Berlinghieri's *Le Septe Giornate della Geographia* see Sean Roberts, *Printing a Mediterranean World: Florence, Constantinople, and the Renaissance of Geography* (Cambridge, MA: Harvard University Press, 2013).

137. Ficino, "Timeus, vel de Natura Compendium," book 32, 401.

138. Alice Pechriggl, *Corps transfigurés. Stratifications de l'imaginaire des sexes/genres*, vol. 2: *Critique de la métaphysique des sexes* (Paris: L'Harmattan, 2000), 113.

139. Pechriggl, *Corps transfigurés*, 114: "La chôra, elle, n'est pas marquée: malgré tous les passages des formes sensibles qu'il reçoit, cet 'espace' reste vierge."

140. Pechriggl, *Corps transfigurés*, 114: "N'est-ce pas là un vieux rêve que celui de la vierge à marquer?"

141. In an interesting variation on the myth of Europa, Giles of Viterbo (1469–1532) imagines the high ridge of the Apennines (*dorsum Apennini*) as the back of the (male) bull abducting Europa: "tota autem Italia fabelle est taurus" ("the whole of Italy is the bull"). Giles of Viterbo, "Fragment of a Text," in John Monfasani, "Hermes Trismegistus, Rome, and the Myth of Europa: An Unknown Text of Giles of Viterbo," *Viator* 22 (1991): 311–42: 337.

142. Woodward, "Cartography and the Renaissance," 5.

143. Emanuela Bianchi, *The Feminine Symptom: Aleatory Matter in the Aristotelian Cosmos* (New York: Fordham University Press, 2014), 109.

144. Plato, *Timaeus*, 52D, 125.

145. Plato, *Timaeus*, 51E-52C, 124–25.

146. Farinelli, *La crisi della ragione cartografia*, 63.

147. Bianchi, *The Feminine Symptom*, 106.

CHAPTER FOUR

1. Waldseemüller publishes Vespucci's letters documenting his voyages to the New World together with the map and an illustrated cartographic work titled *Cosmographie Introductio (An Introduction to Cosmography)*. Martin Waldseemüller and Matthias Ringmann, *Cosmographiae Introductio, cum quibusdam geometriae ac astronomiae principiis ad eam rem necessariis. Insuper quatuor Americi Vespucii navigationes* (*Introduction to Cosmography Containing the Requisite Principles of Geometry and Astronomy beside the Four Voyages of Amerigo Vespucci*) (St. Dié: [Gaulthier Ludd], 1507). For the English translation see John W. Hessler, ed., *The Naming of America: Martin Waldseemüller's 1507 World Map and the* Cosmographiae Introductio (Washington, DC: Library of Congress, 2008).

2. "Alia quarta pars [mundi] per Americum Vesputium . . . inventa est, quam non video cur quis iure vetet ab Americo inventore sagacis ingenij viro Amerigen quasi Americi terram sive Americam dicendam: cum & Europa & Asia a mulieribus sua sortita sint nomina." Waldseemüller and Ringmann, *Cosmographiae Introductio*, n.p./Hessler, ed., *The Naming of America*, 101. I slightly altered the English translation. The performative power of Waldseemüller's toponymic choice could not have been more decisive and has long been recognized by scholars of the Renaissance. Edmundo O'Gorman, *The Invention of America: An Inquiry into the Historical Nature of the New World and the Meaning of Its History* (Bloomington: Indiana University Press, 1961).

3. Hessler, ed., *The Naming of America*, 101.

4. In the earliest dated edition of the Latin translation (from a lost Italian original), Vespucci writes: "Satis ample tibi scripsi de reditu meo ab novis illis regionibus quas et classe et impensis et mandato istius serenissimi portugalie regis perquisivimus et invenimus, quasque novum mundum appellare licet. Quando apud maiores nostros nulla de ipsis fuerit habita cognitio et audientibus omnibus sit novissima res. Et enim hec opinionem nostrorum antiquorum excedit, cum illorum maior pars dicat ultra lineam equinoctialem, et versus meridiem non esse continentem, sed mare tantum quod atlanticum vocavere" (I wrote to you [Lorenzo Pietro de' Medici] at some length concerning my return from those new regions which we found and explored with the fleet, at the cost, and by the command of this Most Serene King of Portugal. And these we may rightly call a new world. Because our ancestors had no knowledge of them, and it will be a matter wholly new to all those who hear about them. For this transcends the view held by our ancients, inasmuch as most of them hold that there is no continent to the south beyond the equator, but only the sea which they named the Atlantic). Amerigo Vespucci, *Mundus Novus* (Augsburg: Johannes Otmar, 1504), n.p. For the English translation, see Amerigo Vespucci, *Mundus Novus: Letter to Lorenzo Pietro Di Medici*, trans. George Tyler Northup (Princeton, NJ:

Princeton University Press, 1916), 1. See also Romy Günthart, *Deutschsprachige Literatur im frühen Basler Buchdruck (ca. 1470–1510)* (Münster: Waxmann, 2007), 112.

5. The *Carta marina* of 1516, whose only extant copy is held in the Jay I. Kislak Collection at the Library of Congress, Washington, DC, has been extensively studied in Chet van Duzer, "A Northern Refuge of the Monstrous Races: Asia on Waldseemüller's 1516 Carta Marina," *Imago Mundi* 62, no. 2 (2010): 221–31.

6. Van Duzer, "A Northern Refuge," 221.

7. Following Ricardo Padrón, I use "America" to refer to the New World as a whole. Ricardo Padrón, *The Spacious Word: Cartography, Literature, and Empire in Early Modern Spain* (Chicago: University of Chicago Press, 2004), 239–40 [note 6].

8. I use the term "discovery" purposefully to point to the Eurocentric valence with which it was used in the sixteenth century by humanists and navigators alike. For the "cartographic discovery" of America, see Folker E. Reichert, "Die Erfindung Amerikas durch die Kartographie," *Archiv für Kulturgeschichte* 78, no. 1 (1996): 115–43.

9. See Padrón, *The Spacious Word*, 20.

10. Girolamo Fracastoro, *Syphilis sive Morbus Gallicus* (Verona: Stefano dei Nicolini da Sabbio, 1530). For the English translation, see Girolamo Fracastoro, *Syphilis sive Morbus Gallicus*, trans. Geoffrey Eatough (Liverpool: Francis Cairns, 1984). I use Eatough's bilingual edition of the poem if not otherwise indicated. More recently, *Syphilis* has been published in two independent French translations by Christine Dussin (Paris; Classiques Garniers, 2010) and Jacqueline Vons (Paris; Les Belles Lettres, 2011). The latest edition and translation into English is by James Gardner: Girolamo Fracastoro, *Latin Poetry*, trans. James Gardner (Cambridge, MA; Harvard University Press, 2013). The eminent sixteenth-century scholar and philologist Julius-Cesar Scaliger dedicated several pages to a discussion of Fracastoro's poem in his *Poetices libri VII*. In 1686, Nahum Tate translated *Syphilis* for the first time into English and, in 1693, the poem was reprinted along with Tate's address to the "great Son of Art," Thomas Hobbes, in the third part of John Dryden's *Miscellany Poems*. Nahum Tate, *Examen Poeticum: Being the Third Part of Miscellany Poems, Containing Variety of New Translations of the Ancient Poets, Together with many Original Copies, by the Most Eminent Hands* (London: Jacob Tonson, 1693), n.p.

11. Fracastoro's social network comprised noblemen of Verona and Venice and included many eminent intellectuals such as the Della Torre family, Andrea Navagero, Giovanni Battista Ramusio, Pietro Bembo, and Paolo Giovio. In 1545, at the height of his career, Fracastoro "was appointed by Pope Paul III as *medicus conductus et stipendiatus* at the Council of Trent." Miguel A. Granada and Dario Tessicini, "Copernicus and Fracastoro: The Dedicatory Letters to Pope Paul III, the History of Astronomy, and the Quest for Patronage," *Studies in History and Philosophy of Science* 36 (2005): 431–76: 439.

12. Jon Arrizabalaga, John Henderson, and Roger French, *The Great Pox: The French Disease in Renaissance Europe* (New Haven, CT: Yale University Press, 1997), 6.

13. I borrow this term from Tom Koch, *Cartographies of Disease: Maps, Mapping, and Medicine* (Redlands, CA: Esri Press, 2005).

14. Mariano Cuesta Domingo, "La cartografía en 'prosa' durante la época de los grandes descubrimientos americanos," in *Actas del Congreso de Historia del Descubrimiento, 1492–1556* (Madrid: Real Academia de la Historia, 1992), 299. Quoted in Padrón, *The Spacious Word*, 89.

15. Peter Martyr d'Anghiera, *De Orbe Novo decades* (n.p. [Alcalá de Henares]: n.d. [1516]); Gonzalo Fernández de Oviedo y Valdés, *Sumario dela natural hystoria delas Indias* (Toledo: Remo[n] de Petras, 1526); Francisco López de Gómara, *La historia general delas Indias, con todos los descubrimientos, y cosas notables que han acaescido enellas, donde que se ganaron hasta agora* (Antwerp: Joannes Steelsius, 1554); Bartolomé de Las Casas, *Brevíssima relacíon de la destruycíon de las Indias* (Seville: Sebastian Trugillo, 1552) and *História de las Indias* (Madrid: Miguel Ginesta, 1875–1876); José de Acosta, *De natura Novi Orbis libri duo* (Salamanca: Guillermo Foquel, 1588).

16. Francisco López de Gómara, *Historia general de las Indias y vida de Hernán Cortés*, ed. J. G. Lacroix (Caracas: Biblioteca Ayacucho, 1979), 20. Quoted by Padrón, *The Spacious Word*, 155.

17. Denis Cosgrove, *Apollo's Eye: A Cartographic Genealogy of the Earth in the Western Imagination* (Baltimore: Johns Hopkins University Press, 2001), 145.

18. In his detailed philological study of the manuscript (MS 250, Royal Library, Copenhagen), Karl Sudhoff established the hypothesis that syphilis existed already in the first half of the fifteenth century. See Karl Sudhoff, *Mal Franzoso in Italien in der ersten Hälfte des 15. Jahrhunderts; ein Blatt aus der Geschichte der Syphilis* (Gießen: Alfred Töpelmann, 1912), 8. More recent studies, such as *The Great Pox* by Arrizabalaga and colleagues dismiss Sudhoff's thesis on the grounds that he all too easily established an overlap between the new disease (syphilis) and a previously circulating terminology (*morbus gallicus*). Arrizabalaga, Henderson, and French, *The Great Pox*, 8.

19. See Joseph Grünpeck, the *Tractatus de pestilentiali Scorra sive mala de Franzos: Originem Remediaq[ue] eiusdem continens* (Nuremberg: Caspar Hochfeder, 1496 or 1497) and Grünpeck's own German translation, *Ein hubscher Tractat von dem ursprung des bosen franzos, das man nennet die wilden wartzen* (*A Fine Treatise on the Origin of the Evil French [Disease] Called the Wild Warts*) (Nuremberg: Caspar Hochfeder, ca. 1497). It includes Grünpeck's prose paraphrase of Sebastian Brant's poem *De pestilentiali scorra sive mala de Franzos, eulogium*. See also Niccolò Leoniceno, *De epidemia quam Itali morbum gallicum, galli vero Neapolitanum vocant* (1497).

20. Arrizabalaga, Henderson, and French, *The Great Pox*, 14.

21. Anna Foa, "The New and the Old: The Spread of Syphilis (1494–1530)," in *Sex and Gender in Historical Perspective*, ed. Edward Muir and Guido Ruggiero (Baltimore: Johns Hopkins University Press, 1990), 26–45: 36. I was not able to track down the sources where the Turkish terminology is used.

22. See Arrizabalaga, Henderson, and French, *The Great Pox*, 6 [note 11].

23. Etienne Balibar, *Politics and the Other Scene* (London: Verso, 2002), 76. Quoted in Sandro Mezzadra and Brett Neilson, *Border as Method, or, the Multiplication of Labor* (Durham, NC: Duke University Press, 2013), 4.

24. Thomas Nail, *Theory of the Border* (Oxford: Oxford University Press, 2016), 3.

25. Nail, *Theory of the Border*, 4.

26. Nail, *Theory of the Border*, 10.

27. Navagero, who edited Lucretius's *De rerum natura* for the Aldine press in 1515, died in 1529 on a diplomatic mission to France. His work as a historiographer of Venice was taken over by Pietro Bembo. See Pietro Bembo, *History of Venice*, ed. and trans. Robert W. Ulery Jr. (Cambridge, MA: Harvard University Press, 2007).

28. Like Peter Martyr, Navagero moved away from the belief that syphilis had a French origin and claimed that its origins were American instead. Eatough, *Syphilis*, Introduction, 22.

29. George B. Parks, "Columbus and Balboa in the Italian Revision of Peter Martyr," *Huntington Library Quarterly* 18, no. 1 (1954): 209–25: 209. According to Eatough, Fracastoro started working on *Syphilis* as early as 1510–12. Eatough, *Syphilis*, Introduction, 7.

30. Roberto Esposito, *Immunitas: The Protection and Negation of Life*, trans. Zakiya Hanafi (Cambridge: Polity Press, 2011), 122.

31. Esposito, *Immunitas*, 123.

32. Eatough, *Syphilis*, Introduction, 1.

33. Leona Baumgartner and John F. Fulton, *A Bibliography of the Poem* Syphilis sive Morbus Gallicus *by Girolamo Fracastoro of Verona* (New Haven, CT: Yale University Press/Oxford University Press, 1935), 11.

34. Pietro Bembo to Girolamo Fracastoro, November 26, 1525, *Delle Lettere di Messer Pietro Bembo Cardinale a' Prencipi e Signori, e suoi famigliari Amici scritte*, vol. 3 (Verona: Pietro Antonio Berno, 1743), 125.

35. Ulrich von Hutten, *De guaiaci medicina et morbo gallico liber unus* (Mainz: Joannis Scheffer, 1519). The treatise was published in several editions in Latin, in a German translation by Thomas Murner in 1519, and in a French translation by Jehan Cheradame in 1520. Sigmund Grimm, municipal physician and apothecary in Augsburg, published the earliest known printed treatise on the guaiac treatment: *Ain recept von ainem holtz zu brauchen für die kranckheit der Franzosen* (Augsburg: Sigmund Grimm and Marx Wirsung, 1518). Grimm ran his own press, partly financed by the wealthy merchant Marx Wirsung, and within a few days they published yet another treatise on the subject: *Lucubratiuncula de morbo gallico et cura ejus Novita reperta cum ligno Indico*, written by the Salzburg physician Leonard Schmaus. See John L. Flood and David J. Shaw, *Johannes Sinapius (1505–1560): Hellenist and Physician in Germany and Italy* (Geneva: Droz, 1997), 28.

36. On the dispute between von Hutten and Paracelsus regarding the effectiveness of the two cures see Christine R. Johnson, *The German Discovery of the World. Renaissance Encounters with the Strange and Marvelous* (Charlottesville: University of Virginia Press, 2008), 160–63.

37. Girolamo Fracastoro, *De Contagione, & Contagiosis Morbis, & eorum curatione, Libri Tres* (Lyon: Guillaume Gazeau, 1550), 512.

38. Denis Cosgrove, "Mapping New Worlds: Culture and Cartography in Sixteenth-Century Venice," *Imago Mundi* 44 (1992): 65–89: 80.

39. Fracastoro, *Syphilis*, 1.15, 38–39; 3.419, 106–7.

40. Bembo, *Lettere*, 125.

41. Bembo, *Lettere*, 125.

42. Fracastoro, *Syphilis*, 1.339–54, 54–55.

43. Fracastoro, *Syphilis*, 1.302, 53.

44. Pietro Bembo, *Lyric Poetry/Etna*, trans. Mary P. Chatfield (*Lyric Poetry*) and Betty Radice (*Etna*) (Cambridge, MA: Harvard University Press, 2005).

45. Bembo, *Lettere*, 125.

46. Fracastoro completed book three several years prior to book two, as we know from his correspondence with Pietro Bembo, who vehemently opposed the inclusion of the book on the mercury cure. Scholars have heretofore assumed that Fracastoro wrote *Syphilis* chronologically and that Bembo opposed the inclusion of the guaiac cure when mentioning "book three," not the mercury cure. However, for Bembo "book three" is the book Fracastoro completed last, that is book two in *Syphilis*. Bembo, *Lettere*, 125.

47. Fracastoro, *Syphilis*, 1.1–14, 39.
48. Fracastoro, *Syphilis*, 2.283–85, 76–77.
49. Fracastoro, *Syphilis*, 2.322–23, 78–79.
50. Fracastoro, *Syphilis*, 3.1–7, 87.
51. Bembo, *Lettere*, 125.
52. Fracastoro, *Syphilis*, 3.235, 96–97.
53. Fracastoro explicitly references the Antilles in the Portuguese spelling of Antilha (from "anti" and "ilha," "the island opposite"): curiously, Antilha was a fictitious island depicted on maps (the Pizzigano map) in the middle of the Atlantic as early as 1424, long before Columbus ventured across the ocean. On the Pizzigano nautical chart from 1424, the first to use the toponym Antilha, see Armando Cortesão, "The North Atlantic Nautical Chart of 1424," *Imago Mundi* 10 (1953): 1–13.
54. Fracastoro's and Ramusio's correspondence is included in the first volume of Ramusio's *Navigazioni e viaggi* (1550).
55. Giovanni Battista Ramusio, Dedication "All'eccellentiss. M. Ieronimo Fracastoro" in *Navigazioni e viaggi*, vol. 1, ed. Marica Milanesi (Turin: Einaudi, 1978), 3.
56. Fracastoro, *Syphilis*, 1.1–14, 38–39.
57. Eatough, *Syphilis*, Introduction, n.p.
58. Lucretius, *De rerum natura*, trans. W. H. D. Rouse (Cambridge, MA: Harvard University Press, 2014), 4.158, 288–89.
59. Pablo Maurette, "De rerum textura: Lucretius, Fracastoro, and the Sense of Touch," *Sixteenth Century Journal* 45, no. 2 (2014): 309–30: 316.
60. Lucretius, *De rerum natura*, 2.217–24, 112–13.
61. Fracastoro, *Syphilis*, 3.105, 90; 3.250, 98.
62. Fracastoro, *Syphilis*, 3.218, 96; "rex" (Spanish king) 3.208, 96; (native king) 3.259, 98 and 3.253, 98.
63. Fracastoro, *Syphilis*, 3.105, 90.
64. Fracastoro, *Syphilis*, 3.97; 90–91.
65. Lucretius, *De rerum natura*, 2.227–28, 112–13.
66. Concetta Pennuto, "La *natura* dei contagi in Fracastoro," in *Girolamo Fracastoro fra medicina, filosofia e scienze della natura*, ed. Alessandro Pastore and Enrico Peruzzi (Florence: Olschki, 2006), 57–71: 65.
67. J. Hillis Miller, *Topographies* (Stanford, CA: Stanford University Press, 1995), 7.
68. Fracastoro, *De sympathia et antipathia rerum liber I*, trans. Concetta Pennuto (Rome: Storia e Letteratura, 2008), 2.1–5. For the English translation, see Maurette, "De rerum textura," 325.
69. "Che tiene accostate le parti del cosmo secondo un ordine armonico di perfezione con l'esclusione del vuoto, fattore di imperfezione e di mancata connessione e comunica-

zione fra i corpi. Il cosmo è infatti concepito come un animale vivente, le cui membra sono in relazione grazie al vincolo della *sympathia*." Pennuto, "La *natura*," 59. The emphasis is Pennuto's.

70. Sympathy is correlated to antipathy, the power with which contrary elements are kept apart. As Pennuto points out, "Fracastoro coltiva i principi della simpatia neoplatonica come *amicitia similium et inimicitia contrariorum, consensus et dissensus rerum*, seguendo le leggi dell'*analogia*." ("Fracastoro cultivates the principles of neoplatonic sympathy as *amicitia similium et inimicitia contrariorum, consensus et dissensus rerum*, following the laws of *analogy*"). Pennuto, "La *natura*," 61.

71. Cheryll Glotfelty, "Introduction," in *The Ecocriticism Reader: Landmarks in Literary Ecology*, ed. Cheryll Glotfelty and Harold Fromm (Athens, GA: University of Georgia Press, 1996), xv-xxxvii: xix. For an in-depth study of ecocriticism, humanism, and early modern literature, see Phillip Usher, *Exterranean: Extraction in the Humanist Anthropocene* (New York: Fordham University Press, 2019).

72. Eatough, "Word Index," in Fracastoro, *Syphilis*, 287–88.

73. Charles and Dorothea Singer, "The Scientific Position of Girolamo Fracastoro [1478?-1553]. With Especial Reference to the Source, Character and Influence of his Theory of Infection," *Annals of Medical History* 1, no. 1 (1917): 1–34: 7. Ch. and D. Singer mistakenly spell Qazwini's name Kazrini.

74. Girolamo Fracastoro, *Homocentrica sive de stellis* (Venice: n.p., 1538), 1.12. Quoted in Singer and Singer, "Girolamo Fracastoro," 7.

75. "[Che]tutti i monti fossero stati fatti dal mare, ammassando, & accozzando insieme molta arena con l'onde sue, e che dove ora sono i monti, fosse già tempo stato il mare, i quali, partendosi quello a poco a poco erano restati in secca." Torello Saraina, *Dell'origine et ampiezza della città di Verona* (Verona: Gieronimo Discepoli, 1586), 9. Quoted by Luca Ciancio, "Un interlocutore fiammingo di Fracastoro: Il medico Iohannes Goropius Becanus (1518–1572) e la teoria dell'origine organica dei fossili," in *Girolamo Fracastoro fra medicina, filosofia e scienze della natura*, 141–55: 145.

76. Gilles Deleuze, "Desert Islands," in *Desert Islands and Other Texts, 1953–1974*, trans. Michael Taormina (Los Angeles: Semiotext(e), 2004), 9–14: 13.

77. Fracastoro, *Syphilis*, 2.402–11, 82–83.

78. Fracastoro, *Syphilis*, 2.25–26, 62–63.

79. Gilles Deleuze and Félix Guattari, *What Is Philosophy?*, trans. Hugh Tomlinson and Graham Burchell (New York: Columbia University Press, 1994), 194.

80. Deleuze and Guattari, *What Is Philosophy?*, 194.

81. Fracastoro, *Syphilis*, 2.376, 80.

82. See Yasmin Haskell and V. Haskell, "Round and Round We Go: The Alchemical *Opus Circulatorium* of Giovanni Aurelio Augurello," *Bibliothèque d'Humanisme et Renaissance* 59, no. 3 (1997): 583–606.

83. Saverio Campanini, "The Quest for the Holiest Alphabet in the Renaissance," in Nadia Vidro, Irene E. Zwiep, and Judith Olszowy-Schlanger, ed., *A Universal Art: Hebrew Grammar across Disciplines and Faiths* (Leiden: Brill, 2014), 196–217.

84. Joannes Augustinus (Antonius) Pantheus, *Ars et theoria transmutationis metallicae cum Voarchadumia, proportionibus, numeris, & iconibus rei accommodis illustrata* (Paris: Vivant Gaultherot, 1550).

85. The latitude is indicated in "miliaria germanica," the longitude in "miliaria italica." A text aligned with the circumference of the map, reminiscent of the *okeanos* hugging the *oikoumene* on T-O maps, explains the conversion of the two measuring units: "Circunferentia terrae continent miliaria germanica, 5400, italica vero 21600" (The circumference of the earth contains 5,400 German miles, that is, 21,600 Italian miles). Pantheus, *Ars et theoria*, frontispiece.

86. Giovanni Aurelio Augurelli, *Chrysopoeiae libri III et Geronticon liber primus* (Venice: Simone da Lovere, 1515). For Augurelli, see Roberto Weiss, "Augurelli," *Dizionario Biografico degli Italiani* (*DBI*), vol. 4 (Rome: Istituto dell'Enciclopedia Italiana, 1962), 578–81.

87. Ptolemy, *Liber Geographiae cum tabulis et universali figura et cum additione locorum que a recentioribus reperta sunt*, ed. Bernardo Sylvano (Venice: Jacopo Pencio, 1511), n.p. This is one of the first editions of Ptolemy's *Geography* to include the famous cordiform world map.

88. Fracastoro, *Syphilis*, 2.260–63; 270–72; 281–82; 74–77.

89. Fracastoro, *Syphilis*, 2.313, 76–77.

90. Fracastoro, *Syphilis*, 2.323, 78–79.

91. Fracastoro, *Syphilis*, 2.375–81, 80–81, with slight modifications to Eatough's translation.

92. Plato, *Phaedo*, ed. and trans. Chris Emlyn-Jones and William Preddy (Cambridge, MA: Harvard University Press, 2017), 111D-111E, 499; 501.

93. Fracastoro's reference to Diana as Trivia is programmatic in itself: used rarely in classical literature, the name is evocative of the Lucretian image of the sacrifice of Iphigenia on "the altar of our Lady of the Crossways" (*Triviai virginis aram*). See Lucretius, *De rerum natura*, 1.84, 10.

94. Theodore Cachey has explored the importance of Fracastoro's poem for Tasso's *Gerusalemme liberata*, especially the "ottave rifiutate" (rejected octaves) or "ottave estravaganti" (extravagant octaves) of the fifteenth canto in which Charles and Ubaldo venture to the New World. These octaves were eliminated from the first edition. See Theodore J. Cachey Jr., "Tasso's *Navigazione del Mondo Nuovo* and the Origins of the Columbus Encomium (GL, XV, 31–32)," *Italica* 69, no. 3 (1992): 326–44: 329.

95. Torquato Tasso, *Gerusalemme liberata* (Turin: Einaudi, 2000), 14.37, 430. For the English translation, see Torquato Tasso, *Jerusalem Delivered*, trans. Anthony Esolen (Baltimore: Johns Hopkins University Press, 2000), 14. 37, 277.

96. Tasso, *Gerusalemme liberata*, 14.38, 430/Tasso, *Jerusalem Delivered*, 14.38, 277.

97. Strabo, *Geography*, trans. Horace Leonard Jones (Cambridge, MA: Harvard University Press, 2004), 2.3.6, 391.

98. Francis Bacon, "New Atlantis. A worke unfinished," in *Sylva sylvarum, or, a Naturall History in Ten Centuries* (London: J.H. [John Haviland] for William Lee, 1626).

99. See John Monfasani, "For the History of Marsilio Ficino's Translation of Plato," *Rinascimento* 27 (1987): 293–99: 294. Simon Grynaeus, *Novus orbis regionum ac insularum veteribus incognitarum* (Basel: Johannes Hervagius, 1532).

100. Bartolomé De Las Casas, *Historia de las Indias*, ed. Juan Perez de Tudela Bueso (Madrid: Atlas, 1957), 36–39. Paraphrased in Vidal-Naquet, "Atlantis and the Nations," *Critical Inquiry* 18 (1992): 300–326: 312.

101. See Marco Ciardi, *Atlantide: Una controversia scientifica da Colombo a Darwin* (Rome: Carocci, 2002), 23.

102. The lengthy title of the map is "Situs Insulae Atlantidis, a pari olim absorpte ex mente Aegyptiorum et Platonis descriptio." Athanasius Kircher, *Mundus Subterraneus in XII Libros digestus* (Amsterdam: Joannes Janssonius & Elizeus Weyerstraten, 1665), n.p.

103. Kircher, *Mundus Subterraneus*, n.p.

104. In 1679, Kircher's contemporary the Swedish physician, botanist, and geographer Olof Rudbeck published a lengthy treatise describing Sweden as the formerly sunken island of Atlantis, the true repository of Japheth's faith. For the bilingual edition, see Olof Rudbeck, *Atland eller Manheim dedan Japhetz afkomne, de förnämste Keyserlige och Kungelige slechter ut till hela werlden / Atlantica sive Manheim vera Japheti posterorum sedes ac patria, ex qua non tantum Monarchae et Reges ad totum fere orbem reliquum regendum ac domandum* (Uppsala: Curio, 1679). I thank Charlotta Forss for this information.

105. Fracastoro, *Syphilis*, 3.265–81, 98–101.

106. Denis Cosgrove, *Geography & Vision: Seeing, Imagining and Representing the World* (London: I. B. Tauris, 2008), 53.

107. Virgil opens book two of his *Georgics* by referring to the "sacred groves" (*nemorumque sacrorum*). Virgil, *Georgics*, trans. H. Rushton Fairclough (Cambridge, MA: Harvard University Press, 1999–2000) 2.13, 136–37. Intimately entangled with transformation and movement, "glaucus" is a color-in-transition announcing liminal spaces, including the threshold between the living and the dead. Virgil uses "glaucus" to describe Aeneas' *translatio*, his movement "trans fluvium," across the river Styx in Charon's barge: "At once [Charon] takes aboard giant Aeneas. The seamy craft groaned under the weight, and through its chinks took in a marshy flood. At last, across the water, he lands seer and soldier unharmed on the ugly mire and grey sedge" ("simul accipit alveo / ingentem Aeneam. Gemuit sub pondere cumba / sutilis et multam accepit rimosa paludem. / tandem trans fluvium incolumis vatemque virumque / informi limo glaucaque exponit in ulva"). In the Tartarian swamps, "glaucus" marks the tension between life and death, between weight and weightlessness, and between the elements of earth and water that together create a "marshy flood." Virgil's *Aeneid*, trans. H. Rushton Fairclough (Cambridge, MA: Harvard University Press, 1999–2000), 6:412–16, 560–61.

108. Fracastoro, *Syphilis*, 2.213–22, 72–73.

109. Virgil, *Georgics*, 2.126–35, 144. See Luís Ramón-Laca, "The Introduction of Cultivated Citrus to Europe via Northern Africa and the Iberian Peninsula," *Economic Botany* 57, no. 4 (2003): 502–14: 506. The citron, referred to by ancient writers such as Theophrastus also as the "Median fruit" (*melon Medicon*) or "Persian fruit" (*melon Persicon*), was commonly associated with the "golden apples" of the Hesperides in Renaissance Neo-Latin poetry. Theophrastus, *Enquiry into Plants*, trans. Arthur Hort (Cambridge, MA: Harvard University Press, 1916), 4.4.3, 312–13.

110. See Giovanni Pontano, *De hortis Hesperidum sive de cultu citriorum* (Venice: Aldo Manuzio, 1505).

111. In Pontano's poem, Venus transforms Adonis into a citron, planting the citron tree in the so-called "Garden of the Hesperides." When he sees the tree, Hercules takes it from Africa and transplants it to Italy. Out of revenge, Juno destroys all the trees so that the citron tree remains forgotten in Italy—until Venus revives its memory by successfully

transplanting it again, this time from the remaining trees in Persia. Pontano, *De hortis Hesperidum*, 1.168–77, n.p.

112. Pontano, *De hortis Hesperidum*, 1.168–77, n.p.

113. Alchemists associated the transmutation of elements with the colors white (*albedo*) and black (*negredo*). Yet as Pantheus claimed, the colors oscillate around one specific chromatic shadow: "citrinus" (citric). In his description of the extraction of whiteness out of blackness, "citrinus" plays a significant and creative role: "Ego sum Albus nigri, & citrinus albi: & Rubeus citrini, veridicus sum certe, & non mentior. Notetis ergo albus nigri. Nam ipsum album, ex nigro extrahimus. . . . Viso ergo quod primus motus naturae est putrefacere. . . . Et ut perfecte sciamus facere, oportet nos habere cognitionem principiorum tam naturalium quam artificialium" (I am the white of the black, & the citric of the white, & the red of the citric, I am certainly truthful, & I do not lie. We extract white itself out of the black. . . . I see then that the first movement of nature is to putrefy. . . . And so that we know how to do it perfectly, we need the knowledge of the principles of both natural and artificial things). Pantheus, *Ars et theoria*, 31–32.

114. Fracastoro, *Syphilis*, 3.346; 3.347–49, 102–3.

115. "When I place before me and ponder, my most illustrious Queen, the antiquity of all the things that have been recorded for our memory and recollection, I find one thing and can conclude with certainty: that language was always the companion of empire, and followed it such that together they began, grew, and flourished—and, later, together they fell." Elio Antonio de Nebrija, "The Prologue to *Grammar of the Castilian Language* (1492)," trans. Magalí Armillas-Tiseyra, *PMLA* 131, no. 1 (2016): 202–8: 202.

116. The physician L. Núñez allegedly described the dictionary as "lleno de barbarismos y de arabismos." Micaela Carrera de la Red, "Lexicografía contrastiva castellano-catalana en el *Dictionarium medicum* de E. A. de Nebrija: campo léxico de los árboles," *Revista de filología románica* 14, no. 1 (1997) : 113–34: 115.

117. Elio Antonio de Nebrija, *Dictionarium medicum*, ed. Avelina Carrera de la Red (Salamanca: Ediciones Universidad de Salamanca, 2001), entry "Citrus," 52.

118. A hand-colored navigational chart published in Amsterdam in 1690 equates the archipelagoes already in its title: "Insulae de Cabo Verde Olim Hesperides, Sive Gorgades Belgice de Zoute Elanden" (Amsterdam: Gerard Valk and Peter Schenk, 1690).

119. Pliny the Elder, *Natural History*, trans. Harris Rackham (Cambridge, MA: Harvard University Press, 1962), 6.36, 486–87. I slightly changed Rackham's translation.

120. Álvaro Baraibar, "El mito de las Islas Hespérides en el discurso político de la monarquía hispánica: Gonzalo Fernández de Oviedo," *Romance Notes* 55 (2015): 15–23: 17.

121. Vidal-Naquet, "Atlantis and the Nations," 313.

122. Fracastoro, *Syphilis*, 3.1–7, 86–87.

123. Fracastoro, *Syphilis*, 3.405–13, 106–7.

124. Fracastoro, *Syphilis*, 3.337, 88–89. My emphasis.

125. Pseudo-Seneca, "Laus Caesaris," in Alfred Breitenbach, *Kommentar zu den Pseudo-Seneca-Epigrammen der Anthologia Vossiana* (Hildesheim: Weidmann, 2009), 271.

126. Marcus Terrentius Varro, *Rerum rusticarum libri tres*, 2.1.6. Quoted by A. T. Fear, "The Golden Sheep of Roman Andalusia," *Agricultural History Review* 40, no. 2 (1992): 151–55: 154.

127. Fracastoro, *Syphilis*, 3.120–35, 92–93.

128. In his letter to Pope Alexander VI from February 1502, Columbus writes, "Cotesta isola è Tarso, è Cetia, è Ofir, e Ofaz e Cipango; noi l'abbiamo chiamata Spagnuola" (This island is Tharsis, Cethya, Ophir, and Ophaz, and Cipango [Japan]; we have called it Hispaniola). Christopher Columbus, "Lettera di Cristoforo Colombo al Papa Alessandro VI," in *Raccolta completa degli scritti di Cristoforo Colombo* (Lyon: Lépagnez, 1864), 295.

129. "Auri terra ferax: sed longe ditior una / Arbore, voce vocant patrii sermonis Hyacum." Fracastoro, *Syphilis*, 3.34–35, 88–89.

130. This episode echoes, as David Quint has pointed out, the "violence that Aeneas and his men commit against the harpies in *Aeneid* 3 (234ff.)." David Quint, "Voices of Resistance: The Epic Curse and Camões's Adamastor," *Representations* 27 (1989): 111–41: 135. Reprinted in Stephen Greenblatt, ed., *New World Encounters* (Berkeley: University of California Press, 1993), 241–71, 265.

131. Timothy J. Reiss, "Wars, Birds, Cultural Origins, and Oceanic Exchanges," *Know: A Journal on the Formation of Knowledge* 1, no. 2 (2017): 353–72: 366.

132. Fracastoro, *Syphilis*, 3.218–19, 96–97.

133. "Tristitia" was equated with melancholy from early Christianity onward. For Gregory the Great, who lived in times of a great plague taken for God's punishment, "tristitia" was caused by pride (*superbia*) and negligence (*acedia*). See John Hamilton, *Security: Politics, Humanity, and the Philology of Care* (Princeton, NJ: Princeton University Press, 2013), 63; and Ann G. Carmichael, "Universal and Particular: The Language of Plague, 1348–1500," in *Pestilential Complexities: Understanding Medieval Plague* (*Medical History*, supplement 27), ed. Vivian Nutton (London: Wellcome Trust Centre for the History of Medicine at UCL, 2008), 17–52: 25.

134. Fracastoro, *Syphilis*, 3.267–68, 98–99.

135. Fracastoro, *Syphilis*, 3.347–49, 102–3.

136. I describe elsewhere syphilis as an amalgam of classical languages, a *pharmakon* created out of the Virgilian word "melisphyllon" (bruised balm) which appears in book four of the *Georgics*. See Katharina N. Piechocki, "Syphilologies: Fracastoro's Cure and the Creation of Immunopoetics," *Comparative Literature* 68, no. 1 (2016): 1–17.

137. Leoniceno, *De epidemia*, n.p.

138. Leoniceno, *De epidemia* n.p.

139. Fracastoro, *Syphilis*, 3.384–86, 104–5, punctuation slightly changed.

140. Walter Mignolo has investigated the use of Latin as a language of colonization in *The Darker Side of the Renaissance: Literacy, Territoriality, & Colonization* (Ann Arbor: University of Michigan Press, 1995).

141. Girolamo Fracastoro, *Naugerius sive de poetica dialogus*, in *Opera omnia* (Venice: Giunta, 1555). For an English translation, see Girolamo Fracastoro, *Naugerius sive de poetica dialogus / Navagero or A Dialogue on the Art of Poetry*, trans. Ruth Kelso (Urbana: University of Illinois Press, 1924).

142. Fracastoro, *Navagero: Della Poetica*, bilingual edition (Latin/Italian), ed. and trans. Enrico Peruzzi (Florence: Alinea, 2005), 78.

143. "Quae ex alio [tropo] in aliud velut viam praestat.... Est enim haec in metalempsi natura, ut inter id quod transfertur <et id quo transfertur> sit medius quidam gradus, nihil ipse significans sed praebens transitum." Quintilian, *The Orator's Education*, trans. H. E. Butler (Cambridge, MA: Harvard University Press, 2001), 8.6.37–38, 446–47. In the Middle Ages,

the terms *transumptivus* and *transumptus* were used by authors and church fathers such as Dante, Alexander of Hales, St. Thomas Aquinas, and Peter of Spain. William Purcell, "Transsumptio: A Rhetorical Doctrine of the Thirteenth Century," *Rhetorica* 5, no. 4 (1987): 369–410: 369.

144. Taking Dante's *Egloca altera* as an example, Mauda Bregoli-Russo identifies Dante's use of the unusual Greek verb "poymus" as a marker of the poet's *stilus transumptivus*: Within the Latin text, Dante's use of a rare Greek verb—"poymus," with the meaning "to show in poetic form"—hints at the process of poetic creation generating a tension between the time and space of the eclogue and that of its creation. See Mauda Bregoli-Russo, "Le Egloghe di Dante: Un'analisi," *Italica* 62, no. 1 (1985): 34–40: 38.

145. Leonard Barkan, *Transuming Passion: Ganymede and the Erotics of Humanism* (Stanford, CA: Stanford University Press, 1991), 45–46.

146. Barkan, *Transuming Passion*, 43–44.

CHAPTER FIVE

1. João de Barros, *Asia de Joam de Barros*, 3 vols. (Lisbon: Germão Galharde, 1552). Henceforth *Of Asia*. For Barros's impact on Camões's writing on Africa, see Josiah Blackmore, *Moorings: Portuguese Expansion and the Writing of Africa* (Minneapolis: University of Minnesota Press, 2009), 63–66.

2. The navigational chart (*carta de marear*) was carefully studied during those very years by the Portuguese mathematician and cosmographer Pedro Nunes, a chief touchstone for Gerhard Mercator. Pedro Nunes, *Tratado sobre certas dúvidas da navegação*, published with the *Tratado em defensam da carta de marear com o regimento da altura* (Lisbon: Germão Galharde, 1537). See also Henrique Leitão and Joaquim Alves Gaspar, "Globes, Rhumb Tables, and the Pre-History of the Mercator Projection," *Imago Mundi* 66, no. 2 (2014):180–95: 182.

3. Barros, *Asia*, 1:6, fol. 58 [68]. Curiously, in the 1628 edition the term "outro novo [mundo]" (another new world) was misspelled and rendered as "ouro novo" (new gold). João de Barros, *Decada Primeira da Asia* (Lisbon: Jorge Rodriguez, 1628), 1:6, fol. 109r.

4. Jörg Dünne, *Die kartographische Imagination: Erinnern, Erzählen und Fingieren in der Frühen Neuzeit* (Munich: Fink, 2011), 12.

5. Franco Farinelli, *Blinding Polyphemus: Geography and the Models of the World*, trans. Christina Chalmers (London: Seagull Books, 2018), 20.

6. Luís de Camões, *Os Lusíadas* (Lisbon: António Gonçalves, 1572). For a modern edition, see Luís de Camões, *Os Lusíadas* (Lisbon: Imprensa Nacional, 2005). For the English translation, see Luís de Camões, *The Lusíads*, trans. Landeg White (Oxford: Oxford University Press, 1997). I use these modern editions, if not otherwise stated.

7. Manuel de Faria e Sousa, "Lusíadas de Luís de Camões, com notas de Manoel de Faria e Sousa," 1621, Ms. Port 32, Houghton Library.

8. François Marie Arouet de Voltaire, "Essai sur la poésie épique," in *Oeuvres complètes*, vol. 8 (Paris: Garnier, 1877), 332. Quoted by Terry Cochran, *Twilight of the Literary: Figures of Thought in the Age of Print* (Cambridge, MA: Harvard University Press, 2001), 105–6.

9. See José Agostinho de Macedo, *Reflexões críticas sobre o episódio de Adamastor* (*Critical Reflections on the Adamastor Episode*) (Lisbon: Impressão Regia, 1811), 7. Quoted by Cochran, *Twilight of the Literary*, 140–41.

10. Camões, *Os Lusíadas*, 5.39, 172/Camões, *The Lusíads*, 5.39, 105.

11. Cochran, *Twilight*, 116.

12. Cochran, *Twilight*, 134.

13. Alexander von Humboldt, *Cosmos: A Sketch of a Physical Description of the Universe*, trans. E. C. Otté, vol. 2 (New York: Harper & Brothers, 1866), 69.

14. Humboldt, *Cosmos*, 68.

15. Bernhard Klein, "Camões and the Sea: Maritime Modernity in *The Lusíads*," *Modern Philology* 111, no. 2 (2013): 158–80: 164–65.

16. Bernhard Klein, "Mapping the Waters: Sea Charts, Navigation, and Camões's *Os Lusíadas*," *Renaissance Studies* 25, no. 2 (2010): 228–47: 232.

17. Humboldt, *Cosmos*, 70–71.

18. Ayesha Ramachandran, *The Worldmakers: Global Imagining in Early Modern Europe* (Chicago: University of Chicago Press, 2015), 122.

19. Michael Wintle, "Renaissance Maps and the Construction of the Idea of Europe," *Journal of Historical Geography* 25, no. 2 (1999): 137–65: 139.

20. Vincent Barletta, *Death in Babylon: Alexander the Great & Iberian Empire in the Muslim Orient* (Chicago: University of Chicago Press, 2010), 119.

21. Giancarlo Casale, *The Ottoman Age of Exploration* (Oxford: Oxford University Press, 2010), 15.

22. Jeremy Black, *The Power of Knowledge: How Information and Technology Made the Modern World* (New Haven, CT: Yale University Press, 2014), 75.

23. Josiah Blackmore, "Africa and the Epic Imagination of Camões," *Post-Imperial Camões*, special issue of *Portuguese Literary & Cultural Studies* 9 (2002): 107–15: 108; 112.

24. Camões, *Os Lusíadas*, 2.80, 63/Camões, *The Lusíads*, 2.80, 41.

25. Camões, *Os Lusíadas*, 3.20, 81/Camões, *The Lusíads*, 3.20, 52. White's English translation slightly changed.

26. See Walter Mignolo, *The Darker Side of the Renaissance: Literacy, Territoriality, and Colonization* (Ann Arbor: University of Michigan Press, 1995).

27. Klein, "Mapping the Waters," 245–46.

28. Camões, *Os Lusíadas*, 5.26, 167/Camões, *The Lusíads*, 5.26, 103.

29. Barros, *Décadas* (*Of Asia*), I, fol. 109r.

30. Camões, *Os Lusíadas*, 1.44, 15/Camões, *The Lusíads*, 1.44, 11.

31. Farinelli, *Blinding Polyphemus*, 175.

32. In my distinction between "route-enhancing" (for a navigational chart), "center-enhancing" (for a medieval T-O map), and "equipollent" (for a Ptolemaic map) I follow David Woodward, "Roger Bacon's Terrestrial Coordinate System," *Annals of the Association of American Geographers* 80, no. 1 (1990): 109–22: 119–20.

33. Jacques Le Goff, "The Medieval West and the Indian Ocean: An Oneiric Horizon," in *Facing Each Other: The World's Perception of Europe and Europe's Perception of the World*, ed. Anthony Pagden (Burlington, VT: Ashgate, 2000), 2. Quoted by Casale, *The Ottoman Age of Exploration*, 15.

34. Ramachandran, *The Worldmakers*, 48.

35. John A. Marino, "The Invention of Europe," in *The Renaissance World*, ed. John Jeffries Martin (New York: Routledge, 2007), 140–65: 150.

36. Farinelli, *Blinding Polyphemus*, 175.

37. "According to the charts and documents of experienced mariners who navigate the Indian Sea," Marco Polo contends, "it is a fact that in this sea of India there are 12,700 islands, inhabited or uninhabited." Marco Polo, *Il Milione* (Venezia: Giunti, 1559). Quoted by Piero Falchetta, *Fra Mauro's World Map with a Commentary and Translations of the Inscriptions* (Turnhout: Brepols, 2006), 31, 186.

38. The description of the map's legends is published in Falchetta, *Fra Mauro's World Map*, 31, 187.

39. Casale, *The Ottoman Age of Exploration*, 15.

40. Fernão Lopes de Castanheda, *História do descobrimento & conquista da India pelos portugueses*, ed. Pedro de Azevedo (Coimbra: Imprensa da Universidade, 1924), 14. Quoted in English by Cochran, *Twilight*, 151.

41. Cochran, *Twilight*, 149.

42. Giovan Battista Giraldi Cinzio, *Dell'Hercole* (Modena: Gadaldini, 1557), 25.61; 25.67, 328. For Giraldi Cinzio's importance within the Renaissance epic system, see Hélio J. S. Alves, *Camões, Corte-Real e o sistema da epopeia quinhentista* (Coimbra: Por Ordem da Universidade, 2001), 66.

43. Camões, *Os Lusíadas*, 8.73, 281/Camões, *The Lusíads*, 8.73, 171. I modified White's translation.

44. *Oxford Latin Dictionary*, ed. P. G. W. Glare, vol. 2 (Oxford: Oxford University Press, 2012), entry "stagnum."

45. Humboldt, *Cosmos*, 69.

46. Carlo Saccone, "Percorsi indo-mediterranei: trascorsi e ricorsi: Una riflessione sulla genesi, lo spirito e la finalità dei Quaderni di Studi Indo-Mediterranei," *Rivista di Studi Indo-Mediterranei* 4 (2014), http://kharabat.altervista.org/index.html (accessed on December 4, 2017).

47. "'Indo-mediterraneo' è una espressione che richiama immediatamente i due grandi mari del mondo antico, i mari che la cultura mesopotamica del secondo millennio a.C. . . . chiama già rispettivamente il 'mare superiore' e il 'mare inferiore'." Saccone, "Percorsi indo-mediterranei."

48. *An Eleventh-Century Egyptian Guide to the Universe: The Book of Curiosities*, ed. Yossef Rapoport and Emilie Savage-Smith (Leiden: Brill, 2014), 479.

49. "Il Mare Nostrum . . . è sempre stato in effetti percepito sin dalla più remota antichità come un 'mare superiore' che fronteggiava in un ambiguo rapporto di confronto-scontro l'altro grande mare, posto più a sud, quello che conosciamo come Oceano Indiano." Saccone, "Percorsi indo-mediterranei."

50. "Gran parte della storia del mondo antico . . . si potrebbe leggere come una successione di eventi che hanno sullo sfondo questo confronto tra i due mari e le varie civiltà che vi sono cresciute intorno." Saccone, "Percorsi indo-mediterranei."

51. Barros, *Decada primeira da Asia* (1628), fol. 175r.

52. *The Qur'ān*, trans. Tarif Khalidi (New York: Penguin, 2008), 18.60–61, 237. Saccone, "Percorsi indo-mediterranei."

53. In the Islamic cartographic tradition, the place holding the two seas together was associated with the Arabian Peninsula (Al-Jazeera), in particular a black square symbolizing the Ka'aba in Mecca and indicating the radiating power of Islam across the two seas. Karen Pinto, *Medieval Islamic Maps: An Exploration* (Chicago: University of Chicago Press, 2016), 36.

54. Pinto, *Medieval Islamic Maps*, 276.

55. Pinto, *Medieval Islamic Maps*, 276.

56. Casale, *The Age of Ottoman Exploration*, 25.

57. Alves, *Camões, Corte-Real e o sistema da epopeia*, 324.

58. Jerónimo Corte-Real, *Felicíssima victoria concedida del cielo al señor don Juan d'Austria, en el golfo de Lepanto de la poderosa armada Othomana. En el año de nuestra salvacion de 1571* (Lisbon: Manuel Antunez, 1578). Alves, *Camões, Corte-Real e o sistema da epopeia*, 190. Tradition has it that the *Felicíssima victoria* was influenced by Camões's poem, already completed in 1553, before Camões left for India. According to Hélio Alves, however, Camões completed the final cantos of *Os Lusíadas* after his return from India and was probably to a certain extent influenced by Corte-Real, who was completing portions of his *Felicíssima victoria* already around 1569.

59. James S. Romm, *The Edges of the Earth in Ancient Thought* (Princeton, NJ: Princeton University Press, 1992), 131.

60. Romm, *The Edges of the Earth in Ancient Thought*, 131.

61. Barros, *Decadas (Of Asia)*, I, fol. 109r.

62. Camões, *Os Lusíadas*, 8.44, 271/Camões, *The Lusíads*, 8.44, 165.

63. Simone Pinet, *The Task of the Cleric: Cartography, Translation, and Economics in Thirteenth-Century Iberia* (Toronto: University of Toronto Press, 2016), 21.

64. Camões, *Os Lusíadas*, 3.6, 77/Camões, *The Lusíads*, 3.6, 49.

65. Camões, *Os Lusíadas*, 3.20, 81/Camões, *The Lusíads*, 3.20, 52.

66. Camões, *Os Lusíadas*, 7.22, 234/Camões, *The Lusíads*, 7.22, 143.

67. Barros, *Décadas (Of Asia)*, vol. 2 (Lisbon: Regia Officina Typografica, 1777), 175.

68. Camões, *Os Lusíadas*, 3.20, 81/Camões, *The Lusíads*, 3.20, 52. White's translation has been modified.

69. Camões, *Os Lusíadas*, 8.78, 283/Camões, *The Lusíads*, 8.78, 172. Translation slightly adapted.

70. Camões, *Os Lusíadas*, 5.14, 163/Camões, *The Lusíads*, 5.14, 100. White's translation has been slightly modified.

71. Camões, *Lusíadas de Luís de Camões comentadas por Manuel de Faria e Sousa* (1639), vol. 1 (Lisbon: Imprensa Nacional-Casa da Moeda, 1972), 543. Quoted by Blackmore, "Africa and the Epic Imagination of Camões," 114 [note 2].

72. Blackmore, "Africa and the Epic Imagination of Camões," 114 [note 2]. As James S. Romm observes, in ancient times "the southerly continent could also be referred to as Antichthon or 'Counterworld' and its supposed inhabitants as Antichthones—a term which attests to their remoteness, since it derives from the Pythagorean term for an alien *planet* located behind the 'dark side' of the earth." Romm, *The Edges of the Earth*, 131. The emphasis is Romm's.

73. Camões, *Os Lusíadas*, 6.8; 6.10, 195–96. The ekphrasis extends from 6.7 to 6.14, 195–97/Camões, *The Lusíads*, 6.8; 6.10, 120. The ekphrasis extends from 6.7 to 6.14, 120–21.

74. Corte-Real, *Felicíssima victoria*, 8, 109. I thank Hélio Alves for pointing this parallel out to me.
75. Camões, *Os Lusíadas*, 9.42, 305/Camões, *The Lusíads*, 9.42, 185.
76. Camões, *Os Lusíadas*, 7.47, 242/Camões, *The Lusíads*, 7.47, 148.
77. Camões, *Os Lusíadas*, 7.48, 243/Camões, *The Lusíads*, 7.48, 148.
78. Michel Foucault, "Of Other Spaces," *Diacritics* 16, no. 1 (1986): 22–27: 22.
79. David Quint, "Voices of Resistance: The Epic Curse and Camões's Adamastor," *Representations* 27 (1989): 111–41: 131–32.
80. Camões, *Os Lusíadas*, 5.50, 175/Camões, *The Lusíads*, 5.50, 108. I slightly modified White's translation.
81. Cochran, *Twilight*, 157.
82. David Woodward, "Medieval *mappaemundi*," in *The History of Cartography*, vol. 1: *Cartography in Prehistoric, Ancient, and Medieval Europe and the Mediterranean*, ed. J. B. Harley and David Woodward (Chicago: University of Chicago Press, 1987), 286–370: 287.
83. The image of a continuous body of land is repeated in canto ten, when Thetis discloses to Gama the territorial discoveries to come by framing them within a Ptolemaic cosmology and cartography: "Vedes a grande terra que contina / Vai de Calisto ao seu contrário Pólo" (Behold a vast continent which stretches / From the Great Bear to the opposite pole). Camões, *Os Lusíadas*, 10.139, 369/Camões, *The Lusíads*, 10.139, 224.
84. Camões, *Os Lusíadas*, 2.27–28, 46/Camões, *The Lusíads*, 2.27–28, 30.
85. Nonnus, *Dionysiaca*, trans. W. H. D. Rouse (Cambridge, MA: Harvard University Press, 1940).
86. Plato, *Phaedo*, ed. and trans. Chris Emlyn-Jones and William Preddy (Cambridge, MA: Harvard University Press, 2017), 109B, 491; 493. Also quoted by J. B. Harley, David Woodward, and Germaine Aujac, "The Foundations of Theoretical Cartography in Archaic and Classical Greece," in *The History of Cartography*, vol. 1, 130–47: 136.
87. Ptolemy's *Geography* informed Islamic cartography throughout the Middle Ages: among others, the twelfth-century Islamic cartographer Al-Idrisi, who was "unaware of the proposed connection [by other cartographers] of the Indian Ocean with the Atlantic through channels to the south of the sources of the Nile." S. Maqbul Ahmad, "Cartography of al-Sharīf al-Idrīsī," in *The History of Cartography*, ed. J. B. Harley and David Woodward, vol. 2: *Cartography in the Traditional Islamic and South Asian Societies*, part 1 (Chicago: University of Chicago Press, 2007), 156–74: 161.
88. Camões, *Os Lusíadas*, 1.51, 18/Camões, *The Lusíads*, 1.51, 13. White's translation has been modified.
89. Camões, *Os Lusíadas*, 5.41; 5.43, 172/Camões, *The Lusíads*, 5.41; 5.43, 106. White's translation modified.
90. Dante, *Inferno*, ed. Anna Maria Chiavacci Leonardi (Milano: Mondadori, 1991), 26.108–9, 785/Dante, *Inferno*, trans. Robert Hollander and Jean Hollander (New York: Random House, 2012), 26.108–9, 482–83.
91. Camões, *Os Lusíadas*, 5.39, 172/Camões, *The Lusíads*, 5.39, 105.
92. Blackmore, "Africa and the Epic Imagination of Camões," 109.
93. Lawrence Lipking, "The Genius of the Shore: Lycidas, Adamastor, and the Poetics of Nationalism," *PMLA* 111, no. 2 (1996): 205–21: 215.

94. Camões, *Os Lusíadas*, 5.39, 172/Camões, *The Lusíads*, 5.39, 105.
95. Barletta, *Death in Babylon*, 102–3.
96. Gomes Eanes da Zurara, *Crónica da Tomada de Ceuta* (= *Chronicle of the Conquest of Ceuta*), ed. Reis Brasil (Mira-Sintra: Publicações Europa-América, 1992), 226.
97. Camões, *Os Lusíadas*, 5.40, 172/Camões, *The Lusíads*, 5.40, 106.
98. Camões, *Os Lusíadas*, 6.1, 193/Camões, *The Lusíads*, 6.1, 119. I modified White's translation.
99. Quint, "Voices of Resistance," 135.
100. Jacques Derrida, *Specters of Marx: The State of the Debt, the Work of Mourning and the New International*, trans. Peggy Kamuf (New York: Routledge, 1994), 20.
101. Camões, *Os Lusíadas*, 5.50, 175/Camões, *The Lusíads*, 5.50, 108.
102. Derrida, *Specters of Marx*, 20.
103. Camões, *Os Lusíadas*, 5.51, 176/Camões, *The Lusíads*, 5.51, 108.
104. Camões, *Os Lusíadas*, 5.56, 177/Camões, *The Lusíads*, 5.56, 109.
105. For a discussion of King Manuel's dream as a "space of history itself," see Timothy Hampton, "Virgil in India: Epic, History, and Military Tactics in the *Lusiads*," *MLN* 130, no. 2 (2015): 169–82: 181.
106. Camões, *Os Lusíadas*, 4.74, 147/Camões, *The Lusíads*, 4.74, 91.
107. Camões, *Os Lusíadas*, 4.74, 147/Camões, *The Lusíads*, 4.74, 91.
108. Klein, "Mapping the Waters," 236.
109. Klein, "Mapping the Waters," 239.
110. Klein, "Mapping the Waters," 239.
111. Torquato Tasso, *Discorsi del poema eroico* (Naples: Stigliola, 1594). Quoted in Ramachandran, *The Worldmakers*, 109.
112. Leonardo Dati, *In questo libro se contien la forza di pianeti che governano el mondo qual si chiama la Spera, cosa bellissima etc.* (Venice: Battista Sella, 1503).
113. Woodward, "Medieval *Mappaemundi*," 301 [note 83].
114. See George Tolias, "*Isolarii*, Fifteenth to Seventeenth Century," in *The History of Cartography*, ed. David Woodward, vol. 3: *Cartography in the European Renaissance*, part 1 (Chicago: University of Chicago Press, 2007), 263–84. More recently, Tom Conley has explored the relationship between poetic form (the sonnet) and cartographic isolation in sixteenth-century French and Italian literature. Tom Conley, "A Cartography of Exile: Du Bellay's *France, mere des arts*," in *Cartographies of Exile. A New Spatial Literacy*, ed. Karen Elizabeth Bishop (New York: Routledge, 2016), 44–66.
115. "Historic deeds such as theirs [the Portuguese navigators] / Transcend fables, and would eclipse / Boiardo's Orlando, and Ariosto's too, / Even if all they wrote of him were true" (As verdadeiras vossas são tamanhas / Que excedem as sonhadas, fabulosas, / Que excedem Rodamonte e o vão Rugeiro / E Orlando, inda que fora verdadeiro). Camões, *Os Lusíadas*, 1.11, 4/Camões, *The Lusíads*, 1.11, 5.
116. Georges Bataille, *The Accursed Share. An Essay on General Economy*, vol. 1, trans. Robert Hurley (New York: Zone Books, 1988), 29. Quoted, in Portuguese translation, by Alves, *Camões, Corte-Real e o sistema da epopeia*, 503 [note 61].
117. Alves, *Camões, Corte-Real e o sistema da epopeia*, 502–3.
118. Camões, *Os Lusíadas*, 8.65; 8.67, 278–79/Camões, *The Lusíads*, 8.65; 8.67, 170. See Jorge de Sena, "Camões: novas observações acerca da sua epopeia e do seu pensamento,"

Ocidente Número Especial (1972): 3–24: 18. Quoted by Alves, *Camões, Corte-Real e o sistema da epopeia*, 503.

119. Alves, *Camões, Corte-Real e o sistema da epopeia*, 503.
120. Bataille, *The Accursed Share*, 29.
121. Camões, *Os Lusíadas*, 7.78, 253/Camões, *The Lusíads*, 7.78, 154.
122. Camões, *Os Lusíadas*, 8.94–95, 288/Camões, *The Lusíads*, 8.94–95, 175–76.
123. Alves, *Camões, Corte-Real e o sistema da epopeia*, 496.
124. Alves, *Camões, Corte-Real e o sistema da epopeia*, 502–3.
125. For the two etymologies, see K. V. Krishna Ayyar, "The Importance of the Zamorins of Calicut," *Proceedings of the Indian History Congress* 37 (1976): 252–59: 252, and Glenn J. Ames, ed., *Em nome de Deus: The Journal of the First Voyage of Vasco da Gama to India, 1497–1499* (Leiden: Brill, 2009), 74 [note 19].
126. Camões, *Os Lusíadas*, 9.60, 311/Camões, *The Lusíads*, 9.60, 189.
127. Klein, "Mapping the Waters," 243–44.
128. While the first English translator of *The Lusiads* (1655), Richard Fanshawe, still preserves Camões's image of a tapestry, it is lost in Landeg White's most recent English translation of the poem. Here, the more metaphorical verb "to carpet" is substituted for Fanshawe's noun, "the carpet": "The fine and noble Carpets then (which there / Lye to be trod on by the meanest Plant) / Make those of Persia, course; and pleasanter / These of the gloomy Valley All will grant. / Narcissus, there, over the water cleere / Hangs his sick head, who what he had, did want. / There flaunts the Grand-child-Son of Cynaras, / For whom Thou, Paphian Queen, cry'st yet, alas!" Luís de Camões, *The Lusiads*, trans. Richard Fanshawe (London: Humphrey Moseley, 1655), 9.60, 186.
129. Alves, *Camões, Corte-Real e o sistema da epopeia*, 496.
130. While there are ekphrastic instances in the first half of the poem—descriptions of battle shields in cantos one and three, evocative of Virgil's description of the Battle of Actium on Vulcan's shield in book eight of the *Aeneid*—those moments are rare when compared to the almost uninterrupted production of ekphrases in the second half of *Os Lusíadas*.
131. Camões, *Os Lusíadas*, 8.43, 271/Camões, *The Lusíads*, 8.43, 165. I slightly modified White's translation.
132. Barbara Karl, *Embroidered Histories: Indian Textiles for the Portuguese Market during the Sixteenth and Seventeenth Centuries* (Wien: Böhlau, 2016), 16.
133. Karl, *Embroidered Histories*, 16.
134. Gaspar Correia, *Lendas da Índia* (*Legends from India*) (Lisbon: Academia Real das Sciencias, 1858–66), 287. Quoted in Karl, *Embroidered Histories*, 29.
135. "Se colhe tanto algodão, e ha tantos Officiães, que tecem finíssimos pannos, que póde dar de vestir com elles a toda Europa." Barros, *Décadas* (*Of Asia*), Fourth Decade, Part 2, 9.1 (Lisbon: Regia Officina Typographica, 1777), 456–57. Quoted in Karl, *Embroidered Histories*, 30.
136. Camões, *Os Lusíadas*, 9.44, 305/Camões, *The Lusíads*, 9.44, 185.
137. Jörg Dünne, "Luís de Camões und der globale Blick. Die bewegte Welt der Lusíadas," in *Mobile Eyes: Peripatetisches Sehen in den Bildkulturen der Vormoderne*, ed. David Ganz und Stefan Neuner (Munich: Fink, 2013), 295–318: 299.

138. Gilles Polizzi, "Politiques de Cythère: *topothesia* et *topographia* chez René d'Anjou, Colonna et Rabelais," in *Paysage Politique. Le regard de l'artiste*, ed. Isabelle Trivisani-Moreau (Rennes: Presses Universitaires de Rennes, 2011), 229–42.

139. Frank Lestringant, *Le Livre des îles: Atlas et récits insulaires de la Genèse à Jules Verne* (Geneva: Droz, 2002), 14.

140. Camões, *Os Lusíadas*, 9.53, 308/Camões, *The Lusíads*, 9.53, 187.

141. Camões, *Os Lusíadas*, 9.52–53, 308/Camões, *The Lusíads*, 9.52–53, 187.

142. Camões, *Os Lusíadas*, 9.64, 312/Camões, *The Lusíads*, 9.64, 189.

143. Camões, *Os Lusíadas*, 9.74, 315/Camões, *The Lusíads*, 9.74, 191.

144. Camões, *Os Lusíadas*, 9.78, 317/Camões, *The Lusíads*, 9.78, 192. "Between the grain and my hand what wall is set?" Petrarch, *Lyric Poems*, trans. Robert M. Durling (Cambridge, MA: Harvard University Press, 1974), 134.

145. "O soneto [de Petrarca] assinala o momento em que a experiência de vida é co-extensiva da experiência de leitura: Petrarca percebe que a sua vida só pode ser descrita por meio de um conjunto de modelos que aprendeu." João Ricardo Figueiredo, "*Os Lusíadas* e a vaidade da poesia," *Revista Colóquio/Letras*, 155/156 (2000): 9–38: 32.

146. "Se alguma coisa há de pornográfico na Ilha dos Amores, não são as 'alvas carnes' das Ninfas . . . , mas sim a absoluta e explícita nudez com que é exibido o verso '*Tra la spica e la man, qual muro he messo*'." Figueiredo, "*Os Lusíadas* e a vaidade da poesia," 32.

147. Alves, *Camões, Corte-Real e o sistema da epopeia*, 601–2 [note 96]. Here, Alves refers to Guido Almansi, *L'Estetica dell'Osceno* (Turin: Einaudi, 1974), 192, and Stephen Marcus, who coined the term "pornotopia" in *The Other Victorians, A Study of Sexuality and Pornography in Mid-Nineteenth Century England* (New Brunswick, NJ: Transaction Publishers, 1966).

148. Camões, *Os Lusíadas*, 9.55, 309/Camões, *The Lusíads*, 9.55, 188.

149. Franco Farinelli, *La crisi della ragione cartografica* (Turin: Einaudi, 2009), 77.

150. Dünne, "Luís de Camões und der globale Blick," 300.

151. A further detail regarding the "mesa" illustrates that its function is to mark the "inverted" world of the Southern Hemisphere. In a gesture that reverses the biblical image of Paradise and its four rivers flowing out of its contained space, on the Isle of Love, rivers flow *into* the "mesa."

152. Joaquim Alves Gaspar, "Revisiting the Mercator World Map of 1569: An Assessment of Navigational Accuracy," *Journal of Navigation* 69 (2016): 1183–96: 1193.

153. Jerry Brotton, "A 'Devious Course': Projecting Toleration on Mercator's 'Map of the World', 1569," *Cartographic Journal* 49, no. 2 (2012): 101–6: 101.

154. Walter Ghim, in Gerhard Mercator, *Atlas sive Cosmographicae Meditationes de Fabrica Mundi et Fabricati Figura* (Duisburg: Albertus Busius, 1595). Quoted by Joaquim Alves Gaspar and Henrique Leitão, "Squaring the Circle: How Mercator Constructed His Projection in 1569," *Imago Mundi* 66, no. 1 (2014): 1–24: 20 [note 1].

155. Dipesh Chakrabarty, *Provincializing Europe: Postcolonial Thought and Historical Difference* (Princeton, NJ: Princeton University Press, 2000), 4.

156. I allude here to Ngũgĩ wa Thiong'o, *Decolonising the Mind: The Politics of Language in African Literature* (London: Heinemann, 1986), along with James R. Akerman,

ed., *Decolonizing the Map: Cartography from Colony to Nation* (Chicago: University of Chicago Press, 2017).

157. Frantz Fanon, *The Wretched of the Earth* (New York: Grove Press, 1968), 102. Quoted by Raymond B. Craib, "Cartography and Decolonization," in Akerman, *Decolonizing the Map*, 11–71: 11.

CONCLUSION

1. Nicholas de Genova, "Introduction: The Borders of 'Europe' and the European Question," in Nicolas de Genova, ed., *The Borders of "Europe": Autonomy of Migration, Tactics of Bordering* (Durham, NC: Duke University Press, 2017), 1–35: 1.

2. Kenneth White and Joan Brandt are among the early scholars (and poets, in the case of White) who forged and mobilized the term "geopoetics" from the 1980s on. Kenneth White, *Le plateau de l'albatros: Introduction à la géopoétique* (Paris: Grasset, 1994); and Joan Brandt, *Geopoetics: The Politics of Mimesis in Poststructuralist French Poetry and Theory* (Stanford, CA: Stanford University Press, 1997). For the performative utterance as a normative and normalizing act, see the classic study of J. L. Austin, *How to do Things with Words* (Oxford: Clarendon Press, 1975).

3. Kenneth White, *Le plateau de l'albatros: Introduction à la géopoétique* (Marseille: Le Mot et Le Reste, 2018), Preface, 6. The emphasis is White's.

4. Oscar J. Martínez, *Border People: Life and Society in the U.S.–Mexico Borderlands* (Tucson: University of Arizona Press, 1994), 20.

5. Willibald Pirckheimer, *Germaniae tam superioris quam inferioris descriptio* (Strasbourg: Lazarus Zetzner, 1610), f. 713. Quoted by Carlos Alberto Campos, *Technology, Scientific Speculation and the Great Discoveries* (Coimbra: Imprensa de Coimbra, 1985), 518.

6. Etienne Balibar, "The Borders of Europe," in Etienne Balibar, *Politics and the Other Scene* (London: Verso, 2002), 87–103: 89; 92. Quoted by Mezzadra and Neilson, *Border as Method*, 29.

7. Nigel Thrift, "Space," in *Theory, Culture & Society* 23, nos. 2–3 (2006): 139–55: 140.

8. Amir Eshel, *Futurity: Contemporary Literature and the Quest for the Past* (Chicago: University of Chicago Press, 2013), 4.

9. J. B. Harley, "Deconstructing the Map," *Cartographica* 26, no. 2 (1989): 1–20: 2.

10. Harley, "Deconstructing the Map," 3.

INDEX

Abbasid court, 137
abstraction (in mathematics), 16–20, 34–35, 72, 78–79, 102, 113, 126, 129, 186
Abyla. *See* Hercules: Abyla and Calpe
Achilles, 67
Acosta, José de, 150
Actium, Battle of, 294n130
Adamastor, 187, 208–11, 213–18. *See also* monsters
Adam of Bremen, 269n138
Adelard of Bath, 133
adolescence. *See* youth
Adonis, 175, 223, 285n111
Aegean Sea, 6–9, 40, 76, 271n27
Aeneid (Virgil), 140, 169, 276n115, 285n107, 287n130, 294n130. *See also* Virgil
Aethicus Ister, 90
Afranius, 28, 250n11
Africa, 132, 176, 190, 196; *Chamesia*, 109; as colonized by Europeans, 30, 190, 193, 214, 224, 248n78, 251n23; in continental thinking, 6, 10, 14, 73, 109, 153, 172, 176, 196–97, 215, 218, 220, 243n19; in geographical and historical writing, 8, 42, 185, 197, 202, 208–11, 215, 218, 244n29, 247n65, 276n115, 285n111, 288n1; in geography, 99–100, 140, 159, 174, 180, 276n115; Libya, 8, 158–59, 180, 208, 276n114; in maps, 1–2, 18, 21, 52, 107, 166, 212
Agenor, 2, 107, 121
Ailly, Pierre d', 115, 272n31
Alans, 82–83, 94–95
Alberti, Leandro, 29

Alberti, Leon Battista, 114, 129, 253n54
Albertus Magnus, 24, 28–29
Album Castrum, 74, 76
alchemy, 25, 156, 165–66, 168, 171, 175, 181, 183, 286n113. *See also* mining
Aldus Manutius, 272n36, 285n110
Alexander of Hales, 288n143
Alexander the Great, 57, 96, 106; Alexander romance, 57; Altars of, 74, 82, 93, 263n54
Alexander VI (pope), 180, 266n97, 287n128
Alicante. *See* Castrum Album
Al-Idrisi, 199, 292n87
Alif, 136–37
Al-Istakhrī, 199
Al-Jazeera. *See* Arabian Peninsula
Al-Khwārazmī, 132–33
allegory, 2, 17, 22, 27–28, 107, 109, 200, 225, 270n9
alphabetic writing, 14, 71, 101, 113, 117–21, 125, 129–32, 136–40, 142; Arabic, 50, 71, 113, 130–33, 136–37, 139–40, 267n104, 275n105; Cyrillic, 71, 101; Greek, 10, 71, 101, 112, 119, 121, 132, 136; Hebrew, 71, 101, 109, 113, 132, 166, 275n105; Latin, 25, 71, 101, 111–12; pseudoscript, 275n105; Utopian alphabet, 117, 139, 140, 143
Alps, 28, 50, 60
Alt, Georg, 34, 249n1
Alves, Hélio, 220, 291n58, 295n147
America. *See* New World
anadiplosis, 64
analogy, 89–90, 210
Anaximander, 6–7
Anghiera, Peter Martyr d', 150, 152, 281n28

297

Annius of Viterbo, *Antiquities*, 114, 177, 271n29
Antarctica, 16, 209, 211, 243n19
antemurale Christianitatis. See *propugnaculum*
anthropomorphization, 2, 27, 67, 145, 190, 208, 217. See also body, human
antichthones. See antipodes
Antilha. See Antilles
Antilles, 157–58, 161, 177–79, 282n53
antimetabole, 205
antipodes, 201–2, 205, 291n72
apeiron. See under *epeiron*
Apollo, 39–41, 44–47, 67, 120, 142, 190, 226, 255nn79–80
apple: as globe (*apfel*), 30, 251n24; Golden Apples, 174, 179–80, 285n109; as sheep, 179–80
Aquinas, Thomas, 288n143
Arabian Peninsula, 1, 291n53
Arabic alphabet. See alphabetic writing
arae Alexandri. See Alexander the Great: Altars of
archipelago, 6–7, 286n118. See also *isolario* (Island Book)
Arctic region, 16, 104, 202
Argus, 124, 126, 225
Ariosto, Ludovico, 68–69, 90, 95, 220, 222, 259n3, 260n6, 260n15, 267n107, 293n115
Aristophanes, *The Frogs*, 211
Aristotle, 82–83, 114, 116–17, 233, 272n39; Aristotelian cosmos, 34; *Politics*, 116–17
arithmetic. See mathematics: arithmetic
Asia, 11, 107, 109–10, 117, 148, 176, 185, 247n65, 261n17, 272n39; colonization of, 72, 79–80, 93–94, 96–97; in continental thinking, 4, 6, 8–10, 14, 18, 25, 69–70, 72, 75, 77–80, 92, 107–9, 116, 119–20, 124–27, 137, 142, 146–47, 149, 153, 158–59, 166, 176, 185, 203, 218, 220, 233, 243n19, 244n29, 263n54; Eurasia, 6, 18, 80, 115, 149–50, 231; and Europa (mythological figure), 4, 107, 148, 278n2; in geography, 25, 42, 61, 68, 70, 72, 74–75, 77–80, 82–84, 86–90, 93–96, 99–100, 116, 118–19, 121, 146, 268n122; on maps, 2, 18, 166; Minor, 6, 9, 118–19, 124, 245n52; and nomadism, 78–79, 83, 87–96; *Semia*, 109. See also Barros, Jõao de: *Of Asia*; continents; Piccolomini, Enea Silvio (later Pius II): *De Asia*

astronomy, 22, 27, 29–30, 49–50, 71, 101, 252n34, 278n1. See also Copernicus, Nicolaus
Athens, 6–7, 118, 141–42, 155, 244n34
Athos, Mount, 7, 262n40
Atlantic Ocean, 13–14, 16, 74, 88, 148, 150, 152–53, 156, 161, 168, 172, 176, 179–80, 202, 209, 278n4, 282n53, 292n87
Atlantis, 154, 160, 165, 171–74, 176, 181, 184, 285n104; *New Atlantis* (Bacon), 172
Atlas: mountains, 176; mythological figure, 17, 172–74
atlas (genre), 2, 29, 79, 104, 172, 189, 247n65, 263n50; antiAtlas, 79
Augsburg, 18, 70, 115, 267n114, 281n35
Augurelli, Giovanni Aurelio, 166, 168–69
Augustine, 18, 201
Aulus Gellius, 112, 250n11
aurea vetula, 102–4, 106, 269n138, 269n141
Aurillac, Gerbert of (Pope Sylvester II), 133
Austria, 2, 60, 63, 252n34
Azov, Sea of, 6, 73, 76–77, 85, 265n89, 268n122

Bacchus, 45, 56, 66, 205–6, 211, 224. See also Dionysus
Bachelard, Gaston, 69
Bacon, Roger, 247n66
bahrayn (the two seas), 199
Balboa, Vasco Núñez de, 149
Balibar, Étienne, 151, 232
Baltic Sea, 2, 37, 50–51, 53, 60, 71, 85, 100
Barbara Codonea (Baltic Barbara), 50, 53, 55
barbarism, 8, 42, 53, 55–56, 95, 176, 286n116. See also civilization
Barkan, Leonard, 184
Barletta, Vincent, 189
Barros, Jõao de, 185–87, 189, 191–92, 199, 201, 203, 216, 224–25; *Of Asia*, 185–87, 199, 203, 224–25, 288n1, 288n3, 294n135
bashlyk (Tatar headdress), 90
Bataille, Georges, 220
Becanus, Johannes Goropius, 163
Behaim, Martin, 29–30, 251n23, 251n24
Beirut, 115–16
Belorussian language, 71, 267n104
Bembo, Pietro, 152–57, 168, 279n11, 282n46
Beneventano, Marco, 92, 94, 261n20
Berlin Congo Conference, 17, 232, 248n78
Berlinghieri, Francesco, 145, 251n19, 277n136
Bianchi, Emanuela, 146

Bible, 1, 11, 18–19, 35–36, 64, 94, 109, 114, 149, 153, 174, 179, 248n86, 264n73, 295n151; biblical prophecy, 11, 180, 272n31; Genesis, 35, 248n83, 264n73. *See also* Christianity
Bilhorod-Dnistrovskyi. *See* Album Castrum
Biondo, Flavio, *Italia illustrata*, 26, 29
birth, 36, 39–41, 86, 171, 183, 275n109; place of, 40–41, 207, 218, 226, 253n52; of territory, 16, 144, 146. *See also* chôra; matrix
Blackmore, Josiah, 190
Black Sea, 6, 60, 71, 73–74, 76–77, 83, 85, 99, 268n122
Boccaccio, Giovanni, 12, 219, 245n53, 249n98
body, human, 1–2, 43–45, 56, 111–12, 121–22, 126, 137, 145, 175, 186, 214, 230; as measurement, 16–17, 121–22, 129; in medicine, 155–56, 165, 168; mystical, 66. *See also* anthropomorphization; Eucharist
Boethius, 28, 114
Bohemia, 2, 34, 60–63, 252n30, 252n34, 258n126, 267n114
Boiardo, Matteo Maria, 27, 219–20, 293n115
Book of Curiosities, 73
border, 4–5, 13, 15, 24, 27, 43, 49–50, 53, 56–57, 60–61, 67, 77, 82, 100, 102–4, 106, 223, 230–33, 263n54; bordering, 78, 152, 230, 233, 263n46; borderland, 18, 71–72, 78–80, 87, 92, 96, 99–102, 232, 256n91; borderline, 72, 78–80, 87, 92, 233, 256n100; cartographic, 17, 87; *grammen*, 112, 118–20, 122, 137; *linea*, 16, 77, 112; *raya*, 16, 153; studies, 79, 130, 151–52; studies manifesto, 25, 68–69; zone, 4–5, 79, 92, 151, 232–33. *See also* limits; *propugnaculum*
Bordone, Benedetto, 8, 76–77. *See also isolario* (Island Book)
Borysthenes, 83
Bosphorus, 73, 77
boustrophedon, 142
Bovelles, Charles de, 112, 134–36. *See also* nothing; zero
Brandis, Lucas, *Rudimentum Novitiorum*, 50–51, 256n94
Braude, Benjamin, 18, 248n83, 248n86
Braudel, Fernand, 244n34
Bregoli-Russo, Mauda, 288n144
Breydenbach, Bernhard von, *Peregrinatio in Terram Sanctam*, 50

Buckinck, Arnold, 18
Budé, Guillaume, 112, 143
bulwark. *See propugnaculum*
Buondelmonti, Cristoforo, 7, 76, 271n27. *See also isolario* (Island Book)

Cachey, Theodore J., Jr., 29, 41, 243n20, 284n94
Cádiz, 6, 74–75, 78, 116, 121; as *Gades*, 6, 74–75, 78, 121, 197
Caesar: emperor, 96, 119; *Laus Caesaris* (Pseudo-Seneca), 179
Caffa, 77, 101
Calicut, 188, 202–3, 221, 294n125
calligraphy, 132, 137
Callirhoe, 156
Calpe. *See* Hercules: Abyla and Calpe
Camões, Luís de, 2, 25, 187–92, 196–98, 200–208, 210–11, 215, 218–21, 223–25, 227–29, 241n8, 288n1, 291n58, 294n128. See also *Lusíadas, Os*
Canzoniere. *See* Petrarca, Francesco
Cão, Diogo, 30
Capella, Martianus, 114, 129–30
Cape of Good Hope, 13, 187–88, 191–92, 196, 208–9, 218
Cape of Torments. *See* Cape of Good Hope
Cape Verde Islands. *See* Hesperides
cardinal directions. *See* directions, cardinal
Caria, 8, 9, 118
Caribbean, 152–53, 156–57, 161, 174–75, 177–79, 182
Carpathian mountains, 61
Carroll, Lewis: *Alice in Wonderland*, 57; *The Hunting of the Snark*, 219
cartographeme, 111, 117, 119, 122, 124–25, 130, 142, 144–45, 147
cartography: cartographic anxiety, 92; cartographic writing, 22, 29, 73, 243n20; and early modernity, 4, 16–17, 20–22, 39–40, 43–45, 73–76, 87–88, 92–94, 148–50, 189–93, 208, 216, 228, 243n20, 249n98, 262n125, 274n79; and epistemology, 14, 28, 88, 113, 128–29, 134, 138–42, 145, 181, 184–85, 189, 191, 203, 209–14, 216–18, 225, 228–29, 292n86; and geography, 12, 17–19, 21, 28–29, 34–36, 42, 62, 102–4, 106, 120, 152–54, 161, 164, 171, 184, 231, 256n98, 261n20, 263n54, 268n125, 292n87; and literature, 5, 27, 49, 55, 57–58, 60, 63–69, 75, 109–15, 117–20, 139,

cartography (cont.)
145, 150–51, 157–59, 165–66, 184, 187, 198–99, 201, 205, 208, 218–23, 227, 246n56, 252n34, 278n1; as a new discipline, 4, 6–7, 11, 13–15, 22, 24–27, 35, 42, 45, 49, 57, 65, 71, 86, 113, 115, 117, 134, 164, 172, 196, 232–34, 242n10, 243n20, 249n98, 250n7, 252n34, 279n8, 292n87; prose, 70, 104, 109. *See also* Islam: Islamic maps; latitudes and longitudes; *mappa mundi*; nautical charts; Ptolemy; T-O map; zonal map (Macrobian map)

Casas, Bartolomé de las, 150, 172

Caspian Sea, 84, 86, 88–89, 99

Castanheda, Fernão Lopes de, *História do descobrimento & conquista da India*, 187, 196–97

Castilian language, 114, 286n115

Castrum Album, 73

Cathay, 68–69, 100

Caucasia, 73, 84, 262n31

Cave, Terence, 43

Celtis, Conrad, 22–24, 26–29, 34, 36–43, 45–51, 53–58, 60–67, 232, 249n1, 249n3, 249n6, 250n11, 250n14, 252n34, 253n50, 253n52, 254n63, 255nn79–80, 255n85, 256n91, 257n118, 258n122, 258n128; *Germania generalis* (and *Germania illustrata*), 26, 58, 249n6, 255n85, 257n118, 258n122. See also *Quatuor Libri Amorum secundum Quatuor Latera Germanie*

Ceylon, 199

Chakrabarty, Dipesh, 4, 229

Chaldea, 28, 101, 113

Cham. *See* Ham (son of Noah)

Chamesia. See Africa

Champ fleury, 111–14, 117, 119–22, 124, 126–27, 129, 131, 134, 136–40, 142–43, 145, 147, 271n22, 271n26

Chancellor, Richard, 104, 269n139

Charles V (emperor), 96, 177, 274n79

Charles VIII (of France), 107, 149

Châtillon, Walter of, *Alexandreis*, 57

Chekin, Leonid, 74

Chenoweth, Katie, 121

chiasmus, 204–5, 220

China. *See* Cathay

chôra, 142, 144–47, 277nn139–40. *See also* birth; matrix

chorography, 9, 26, 43–44, 58, 107. *See also* scale (in cartography): regional

Christ, 1, 7, 103, 259n147; *mappa mundi* and, 1, 62, 67, 259n147; mystical body of, 66–67. *See also* Christianity; Eucharist

Christianity, 1, 15, 34, 42, 64–66, 68, 87, 100–101, 103, 133, 151, 170, 200, 207, 215, 231, 245n52, 248n82, 248n86, 287n133; and Europe, 11–12, 18, 62, 65–68, 87, 95–96, 245n52, 259n147, 287n133. *See also* Bible; Christ; Church fathers; Eucharist; Paradise

Chronica Polonorum. *See* Miechowita, Maciej

chronology, 28, 39, 86, 216, 282

Chrysoloras, Manuel, 5

Church fathers, 18, 42, 287n143. *See also* Christianity

Cicero, 24, 28, 42, 89, 114, 140, 202

cipher, 132–34, 136, 139, 275n96. *See also* nothing; zero

citron, 174–77, 188, 285n109, 285n111. *See also* Hesperides

civilization, 7, 36, 55, 113–14, 137, 160, 189, 199, 257n101. *See also* barbarism

Claudian, Nikolaus, 252n30

colonialism, 5, 7–8, 11, 14, 16, 24–25, 30, 72, 80, 177–81, 187, 189–90, 232; colonization, 48, 78, 92, 96, 102, 109, 174, 190, 192, 198, 214, 226, 228–29, 287n140; colony, 115, 179. *See also* postcolonialism

Colonna, Francesco, 114, 225

Columbus, Christopher, 11, 13, 70, 126, 149, 152, 172, 180, 186, 272n31, 282n53, 287n128

comparative literature, perspective of, 24, 55, 90, 117, 243n20, 246n56

comparison (as disciplinary method), 89–90, 96, 98–99, 116–17, 141–42, 160, 172, 198–99, 201, 203–5, 210. *See also* analogy

compass, 30, 120, 142–43, 219

Conley, Tom, 22, 39–40, 56, 109, 243n20, 252n28, 257n110, 293n114

Constantinople, 7, 11, 76, 101, 231

continents: as *continens*, 9–11, 13, 94, 148; continental belonging, 88, 92; continental divide, 6, 8, 16, 69, 82, 87, 99–100, 119–20, 137–38, 140, 153, 190, 196–97, 213; continental thinking, 6–13, 15, 75, 78–80, 102, 151, 172, 196–97, 246n59.

See also Africa; Asia; Europe; mainland; New World
Copernicus, Nicolaus, 13, 29, 49, 71, 261n18
Correia, Gaspar, *Lendas da Índia (Legends from India)*, 224
Corte-Real, Jerónimo, *Felicíssima victoria*, 200, 206, 291n58
Cortés, Hernán, 30, 172, 251n25
Cortesão, Armando, 282n53
Cosgrove, Denis, 34, 154, 174
Cosmas of Prague, 61
Cosmographia. See *Geography* (Ptolemy)
cosmography, 9, 12, 26, 43–44, 58, 73, 82–83, 107, 115, 145, 148, 163–64, 220, 246n58, 251n24, 252n34, 254n69, 288n2. *See also* scale (in cartography): global
Council: of Constance, 84; of Trent, 279n11. *See also* Nuremberg: Nuremberg City Council
Cracow. *See* Kraków
Crimea, 74–77, 79, 97, 231, 262n41, 266n103
Croca. *See* Kraków
crusade, 11, 94
Cuba, 149
Cusanus, Nicolaus, 30, 31, 114
Cyprus (island), 200–201; as Venus's birthplace, 207
Cyrillic alphabet. *See* alphabetic writing

dactyl, 63, 65–67, 258n140; catalectic dactylic tetrameter, 258n140; dactylic hexameter, 63, 150
Dalché, Patrick Gautier, 251n22, 276n114
Dalmatia, 60
Dante Alighieri, 12, 145, 205, 213, 287n143, 288n144
Dantiscus, Johannes, 92
Danube (river), 37, 48, 50, 60, 63, 83, 171, 265n88
Dardanelles, 7, 73
Dati, Leonardo, 220
Davies, Surekha, 55
death: of Adonis, 175; deathscape, 230; of Dom Sebastião, 187; geographic north as, 36–37, 41; and poetic creation, 181, 183, 285n107
Deleuze, Gilles, 56–57, 120, 163, 165, 193
Delos, 40, 226. *See also* Aegean Sea; Apollo
democracy, 16
Derrida, Jacques, 138, 216

Descriptio Sarmatiarum. *See* Miechowita, Maciej
Diana (mythological figure), 153, 156, 169–70, 226, 284n93
Dias, Bartolomeu, 196
Diderot, Denis, 72
Dionysus, 44, 67, 120, 211, 263n54. *See also* Bacchus
directions, cardinal, 15, 24, 34, 36–37, 39–41, 49–50, 53, 57, 59, 62, 64–65, 138, 142, 219, 225, 258n122; directionality, 41, 57, 64, 78, 138, 225
Długosz, Jan, 84–86, 261n24, 264n72
Don (river). *See* Tanais (river)
Donatus, 42
dragons, 44–45, 90
Drusus, 59
Dryden, John, 279n10
Dünne, Jörg, 186, 228, 265n79
Dürer, Albrecht, 30, 39, 137, 145, 253n43; *Philosophia* (woodcut), 22–23, 27–29
Dutch. *See* Netherlands

Eatough, Geoffrey, 159, 279n10, 281n29
Ebsdorf Map, 67, 259n147
Eck, Johannes von, 70
Eden. *See* Paradise
Egypt, 8, 28, 73, 101, 113–14, 171, 193, 207. *See also* Nile
Eitzinger, Michael, 2–4; *De Europae Virginis, . . . liber*, 2–3, 241n5
ekphrasis, 57, 189, 203, 205, 207, 222–28, 291n73, 294n130
Elbe, 59, 63, 86
elegance (in cartographic writing), 97, 115
elegy, 24, 26–28, 37, 39, 49–51, 53, 62, 67, 256n91, 257n109, 263n57
Elsula Alpina, 50, 62
encyclopedism, 10, 115, 199. *See also* Isidore of Seville
England, 2, 70, 72, 269n139; English Channel, 179; English travelers, 70, 104, 269n139
Enlightenment, 36, 72, 234
epeiron, 6–7, 9–10, 171–72; *apeiron*, 7, 244n29
equator, 19–20, 98, 137, 201–2, 278n4; torrid zone, 98, 201. *See also* zonal map (Macrobian map)
equipollent: logic, 134; map, 16–17, 117, 129, 192, 221, 229, 289n32

Erasmus, Desiderius, 112, 121–22, 140, 266n98, 276n116
Eratosthenes, 8, 201
Erebos. *See* Würzburg, as *Erebos*
Esposito, Roberto, 5, 152–53, 242n11, 243n21, 272n39
Estado da Índia. *See* Portuguese Empire
Etzlaub, Erhard, 30, 37, 252n30
Eucharist, 66–67, 259n147
Euclid, 134
Eurasia. *See* Asia: Eurasia
Europa (mythological figure), 2–4, 17, 107–10, 121, 145–46, 241n5, 244n29, 270n9, 277n141. *See also* Asia: and Europa (mythological figure)
Europe, 7, 12, 14, 22, 30, 50, 74, 112–14, 116–21, 130, 132, 137, 142, 148–49, 154, 174, 178–79, 182, 186, 189, 191–93, 200, 228–30, 243n18, 246n61, 249n97, 270n2, 270n6, 272n39; in continental thinking, 1–6, 8–16, 24, 29, 36, 68–69, 72, 77, 82, 92, 146–47, 165–66, 176, 196–97, 220, 223, 243n19, 244n29, 245n52, 246n64, 263n54; European (as europeus, Europico, and europianus), 11–12; in geography, 21, 25, 34, 39, 41–43, 48, 56, 58–60, 67, 70, 73, 85–90, 93–102, 159, 177, 245n51, 252n34, 256n98, 261n17, 261n20, 268n122; idea of Europe, 11, 20, 190, 198, 231–32, 242n11; *Japetia*, 109; limits of, 24, 26–27, 53, 57, 61, 67, 73, 75, 78–80; in maps and representations, 2, 4, 17–18, 31, 51–52, 68, 81, 84, 107, 109, 111, 145, 202–5, 215–16, 225, 241n8, 246n56, 247n65, 268n125, 270n9; and the New World, 152–53, 172, 181; *Respublica christiana* and, 11, 96. *See also* Africa; Asia: Eurasia; continents; Eitzinger, Michael: *De Europae Virginis, . . . liber*; Piccolomini, Enea Silvio (later Pius II): *De Europa*
Euxinus, Pontus. *See* Black Sea

Familiares. *See* Petrarca, Francesco: *Familiares*
Farinelli, Franco, 16–17, 19, 44–45, 75, 113, 126–27, 129, 136, 139, 141, 146, 192, 228, 242n10, 247n72, 259n147
Fenni (Finns), 56
Fibonacci, Leonardo, *Liber Abaci*, 133
Fichtel Mountain, 58–59, 62–63
Ficino, Marsilio, 113–14, 144–45, 172

Filelfo, Francesco, 95–96
flood, biblical, 18, 45, 47, 109, 285n107. *See also Gal*
Fortunate Islands, 24
Foucault, Michel, 208
Fracastoro, Girolamo, 25, 149–66, 168–76, 179–84, 279nn10–11, 281n29, 282n46, 282n53, 282n54, 283n70, 284nn93–94. *See also Syphilis sive Morbus Gallicus*
Fra Mauro, 37, 193, 246n58
France, 2, 5, 22, 41, 50, 72, 109, 112–13, 115, 117, 119, 121, 159, 250n17, 280n27, 281n28, 293n114. *See also* French language
Franconia, 60–62, 253n52
Frege, Gottlob, 127, 274n74
French language, 25, 111, 113–14, 117–18, 121, 133, 143–44, 151, 256n94, 279n10, 281n35. *See also* alphabetic writing; Tory, Geoffroy
Fridericus, Brother, 50
Frisians, 60
frontier-letter ("I"), 119–20, 124, 130, 137–38
frontispiece, 2, 37–39, 90–91, 122–23, 166. *See also* humanism; printing technology
Fugger (family), 267n114
Fusoris, Jean, 251n22

Gades. *See* Cádiz
Gal, 109, 270n5. *See also* flood, biblical; Postel, Guillaume
Galilei, Galileo, 146
Gallia. *See* Gaul (France)
Gallipoli, 7
Gama, Vasco da, 13, 187–93, 196–98, 202, 205, 207–8, 210–11, 213, 215–17, 219–22, 224, 292n83
Garden of the Hesperides. *See* Hesperides: as mythological figures
gardens, 36, 59, 120, 153, 159, 165, 174–75, 177, 179, 265n89, 276n116, 285n107, 285n111. *See also* Hesperides; Paradise
Gasché, Rodolphe, 86–87
Gaspar, Joaquim Alves, 288n2
Gastaldi, Giacomo, 30
Gaul (France), 50; as *Galli*, 39, 158–59, 181, 270n5; as *Gallia Belgica*, 48; the Gauls, 40, 109. *See also* France; French language; Hercules: Gallic; syphilis
gemination (in rhetoric), 178–79

geography (discipline), 5, 7, 12–13, 28, 42, 62, 74, 87, 93, 113, 115, 118, 121, 143, 150, 157–59, 168, 202, 243n21, 252n34
Geography (Ptolemy), 5, 9–10, 15–16, 18–19, 26, 43, 72, 80, 82, 84–85, 94, 107, 113, 115, 192–93, 233, 245n51, 247nn65–66, 264n68, 292n87; editions and translations of, 30, 48–50, 52, 84, 92, 94, 108–9, 145, 168, 193–94, 261n20
geological thinking, 156, 158, 162–64, 168
geometry, 14, 16–18, 22, 27, 43, 73, 113, 127, 129–30, 132, 137, 199; and religion, 35, 45, 253n39
George, Saint, 44–45
Georgia. *See* toponymic symmetry (Iberia and Hiberia)
Georgics (Virgil), 157, 175, 285n107, 287n136
Germania. *See* Tacitus
German lands. *See* Germany
German Sea. *See* Baltic Sea
Germany, 2, 5, 17, 24, 26–30, 34, 36–37, 39–43, 47–51, 53, 55–65, 67, 80, 83–85, 99, 121, 148, 151, 153, 172, 188, 232, 243n18, 253n54, 254n63, 255n85, 256n91, 257n118, 258n121, 258nn128–29; German language, 70–71, 127, 133, 182, 249n1, 252n33, 260n13, 265n79, 265n86, 280n19, 281n35. *See also* Celtis, Conrad; *Quatuor Libri Amorum secundum Quatuor Latera Germanie*; Tacitus
Gervase of Tilbury, *Otia Imperialia*, 85
Gibraltar, Straits of, 20, 73, 120, 213. *See also* Hercules: Abyla and Calpe; Hercules: Pillars of
Giles of Viterbo, 277n141
Giovio, Paolo, 279n11
GIS, 21
Glaber, Andrzej, 90, 265n86
global scale. *See* scale (in cartography): global
global thought, 5, 12, 17, 24–25, 58, 79, 102, 151–52, 159, 174, 188, 191–92, 198, 203, 232, 253n54, 267n114
globes, 13, 29–30, 34, 37, 154, 251n22, 251n24; use of term in poetry, 165, 169. *See also* apple
Gniezno, 93
Golden Apples. *See* apple; Hesperides
Golden Old Woman, 102–4, 106, 269n138, 269n141. *See also* *aurea vetula*; *złota baba*
Gómara, Francisco López de, 150, 172

GoogleMaps, 21
Goths, 60, 83, 86, 94–95, 268n125
Grafton, Anthony, 114
Greece, 2, 6, 8–9, 28, 41, 48, 71, 116, 118–19, 141, 172, 180, 201, 233, 253n52, 272n39; Greek language, 6–10, 14, 19, 28, 36–37, 40, 42, 57, 66, 73, 75, 90, 101, 112, 116, 119, 121, 136, 180, 182, 205, 211, 216, 231, 246n61, 254n63, 262n40, 265n86, 288n144
grid (cartographic), 16–17, 34, 56–57, 62, 72, 78, 80, 87–88, 94, 102–3, 113, 130, 136, 140, 257n110. *See also* latitudes and longitudes; Ptolemy
Grimm, Sigmund, 281n35
groves. *See* gardens
Grünpeck, Joseph, 151, 280n19
Grynaeus, Simon, 70, 172, 260n13; *Novus orbis regionum ac insularum* (with Johann Hutten), 70, 172
Gutenberg, Johannes, 50

Hakluyt, Richard, 70, 104
Ham (son of Noah), 18–19, 109. *See also* Africa: *Chamesia*; Noah
hapax legomenon, 64
Harley, J. B., 234, 242n10
Hasilina Sarmatica, 49
Hay, Denys, 11, 242n11, 245n52, 246n60, 248n82
Hebrew alphabet. *See* alphabetic writing: Hebrew
Heidegger, Martin, 17, 88, 247n76
heliocentrism, 71
hemispheres, 36, 97–98, 137, 148, 150, 161, 189–92, 197–98, 201–5, 207–8, 211, 215, 221, 224–25, 295n151
Herberstein, Sigismund von, 70, 104–5, 260n9, 263n9, 263n57, 269n138
Hercules, 45, 118–22, 124, 137, 140, 180, 185–86, 196–98, 215–17, 255n80, 263n54, 273n56, 285n111; Abyla and Calpe, 120, 140, 196–97, 213, 216–17; Gallic, 118–22; Melqart, 121; Pillars of, 1, 6, 74–75, 96, 174, 177, 213
Hercynian Forest, 2, 34, 58–63, 67, 258n124, 258n128
Hereford map, 1, 2, 18, 230, 241n2, 241n4
Herodotus, 7–8, 83, 201, 265n89, 266n97
Hesiod, 40, 120

Hesperides: Hesperus (Evening Star), 176–77; as islands, 16, 119, 174–80, 285n109, 285n111, 286n118; as mythological figures, 119–20, 174, 180. *See also* Caribbean; Pontano, Giovanni, *De hortis Hesperidum* (*The Gardens of the Hesperides*)
hexameral literature, 26, 34, 50
hexameter, 27, 53, 63, 150
Hiberia. *See* toponymic symmetry (Iberia and Hiberia)
hieroglyphs, 113, 168; *Hieroglyphica* (Horapollo), 113, 271n26; *Monas Hieroglyphica* (Dee), 166. *See also* alphabetic writing: pseudoscript; Ficino, Marsilio
Hinduism, 207
Hirschvogel, Augustin, 104–5, 269n141
Hispaniola, 149, 153–54, 173–74, 179–81, 287n128
historiography. *See* history
history: as discipline, 5, 14, 24, 84–86, 107, 116, 118, 150, 191, 196, 199, 223–24, 228, 234, 243nn18–19, 244n29, 256n91, 256n98, 264n65, 280n27, 288n2, 293n105; as genre, 42, 71–73, 116, 140, 152, 185, 187, 224, 258n126, 261n20
Hobbes, Thomas, 279n10
Hodgson, Marshall, 1, 21, 242n11
Hölderlin, Friedrich, 36
Holy League, 200
Holy Roman Empire, 26, 29–30, 62, 71, 96; Holy Roman Emperor (Ferdinand I), 107. *See also* Germany
Homer, 6, 24, 67, 188, 255n80
Honter, Johannes, 73
Horace, 66, 114
Horapollo, 113–14, 271n26. *See also* Ficino, Marsilio
Houghton Library, 23, 31–32, 38, 46, 52, 54, 77, 105, 110, 131, 135, 138, 167, 194–95, 288n7
human body. *See* body, human
humanism: and continental thinking, 7, 11, 20, 24, 34, 41, 58, 84–85, 87–88, 110–11, 152, 161, 172, 187, 232–33, 283n71; critique of, 4, 11–13, 15, 17–18, 26–27, 29–30, 36, 41–42, 55, 57, 80, 95–96, 113–17, 142, 246n61, 249n98, 252n34, 267n114; and organization of knowledge, 4, 22, 24, 28, 42, 112–14, 122, 170, 243n20, 249n97; and printing, 14, 29, 41–42, 192, 252n54, 285n111; Renaissance, 11–12, 15–16, 22, 28, 116–17, 172, 176, 232–33, 243n20. *See also* philology; printing technology; translation
Humboldt, Alexander von, 83, 188, 198, 223, 225
humor, 56–57
Hungary, 2, 71, 94–96, 252n34. *See also* Ladislas III (of Poland and Hungary)
Huns, 95
Husserl, Edmund, 117
Hutten, Johann. *See* Grynaeus, Simon
Hutten, Ulrich von, 153, 281n35, 281n36
Hyperborean mountains, 68, 80, 82–83, 93

Iazyges, 48, 61
Iberia (Hiberia). *See* toponymic symmetry (Iberia and Hiberia)
Iberian Peninsula, 2, 6, 50, 73, 116, 262n31, 285n109
Ibn al-Wardī, 199
Ibn Muqla al-Shirazi, 137
Iliad (Homer), 6, 67
immunology, 150, 153
India, 21, 97–101, 141–42, 187, 196, 211, 220–24; Indian wood (guaiac), 153, 291n58
Indian Ocean, 10, 187–89, 192–93, 196–203, 205–11, 213, 215–19, 221, 223–25, 228, 290n37, 290n49, 292n87
Indicopleustes, Cosmas, 248n82
Indonesia, 21
Io, 124–25, 146
Ionia, 6–7, 9, 118, 124–27, 146
Isidore of Seville, 10, 18, 24, 50, 90, 114–15, 265n88
Islam, 94, 137, 185–86, 188–89, 193, 200, 210–11, 214, 247n66; Islamic calligraphy, 137; Islamic maps, 37, 73, 133, 189, 199–200, 291n53, 292n87; and Tatars, 94
Island Book. *See isolario* (Island Book)
islands, 1, 6–8, 40–41, 48–50, 74–76, 78, 119, 129, 139–43, 148–50, 152–54, 157, 161, 163, 166, 171–74, 176–77, 180–82, 184, 188, 193, 199–201, 203, 220, 223, 225–28, 230, 244n24, 244n29, 244n34, 262n40, 271n27, 274n79, 276n114, 282n53, 285n104, 287n128, 290n37, 295n146. *See also* Grynaeus, Simon; *isolario* (Island Book)
Isle of Love, 188, 201, 206, 223–28, 295n151
isolario (Island Book), 7–8, 29, 70, 76, 220, 225, 271n27

INDEX 305

Istanbul. *See* Constantinople
Ister. *See* Danube (river)
Italy, 2, 5, 14, 18, 29, 31, 41–42, 48, 50, 95, 141, 158, 175, 192, 199, 233, 245n51, 249n98, 252n34, 277n141, 285n111

Jagiellonian dynasty. *See* Władysław II Jagiełło (founder of Jagiellonian dynasty)
Jagiellonian University. *See* University of Kraków
Jan. *See* Javan
Japetia. *See* Europe
Japetus. *See* Japheth
Japheth, 18, 94, 109, 248n83, 285n104
Javan, 94
Jenkinson, Anthony, 104, 106, 269n143
Jerome, 18
Jerusalem, 11, 15, 18, 49, 66–67, 170
Josephus, 18, 118
Judaism, 101
Jupiter (mythological figure), 2, 107, 124, 207–8, 269n138
Jupiter (planet), 146

Karl, Barbara, 224
Kircher, Athanasius, *Mundus Subterraneus*, 172–73
Kivelson, Valerie, 72, 94, 246n56
Klein, Bernhard, 188, 218–19
Klibansky, Raymond, 28
Korenjak, Martin, 258n128
Kraków, 27, 37, 49, 69–71, 73, 103, 256n91, 261n18
Kraye, Jill, 116
Kulmbach, Hans Süss von, 253n43

Lactantius, 35–36, 41–42, 276n113
Ladislas III (of Poland and Hungary), 96, 267n111
Łaski, Jan, 93, 95, 266n98
Latin alphabet. *See* alphabetic writing
latitudes and longitudes, 5, 15–16, 18, 88–89, 129–30, 199, 247n66, 284n85. *See also* grid (cartographic); Treaty of Tordesillas
Lefèvre d'Etaples, Jacques, 112
Le Goff, Jacques, 192
Leitão, Henrique, 288n2
Lemaire de Belges, Jean, 112
Leoniceno, Niccolò, 151, 181–82
Lepanto, Battle of, 200
Lestringant, Frank, 243n20

Liber Chronicarum (*Nuremberg Chronicle*). *See* Schedel, Hartmann
Liber de nichilo. *See* Bovelles, Charles de
libraries and collections, early modern, 72, 107, 253n55, 262n25, 266n98
Libya. *See* Africa: Libya
limitlessness, geographic, 57. *See also epeiron*
limits: in calculus, 151; of Europe, 4, 48, 50, 53, 55–56, 61, 72–75, 78, 111, 120, 137, 204, 213, 233; geographic and cartographic, 10, 15, 18, 24, 47, 60, 82, 85, 104, 106, 142, 152, 174, 216, 220–21, 254n66, 263n54; linguistic, 137–38, 220–21, 275n107
literary criticism. *See* comparative literature, perspective of
Lithuania, 71, 84–85, 88–90, 92–93, 100, 259n3, 267n104. *See also* Poland: Poland-Lithuania
Livy, *Ab urbe condita* (*History of Rome*), 73
Llull, Ramon, 114
locus amoenus, 64–65
Low Countries. *See* Netherlands
loxodromes. *See* rhumb lines
Lübeck, 27, 37, 50, 100
Lublin. *See* Union of Lublin
Lucan, 118; *Pharsalia*, 74, 253n52, 265n89
Lucretius, 155, 158–61, 181, 280n27, 284n93
Lupset, Thomas, 143
Lusíadas, Os, 2, 187–93, 196–98, 200–211, 213–29. *See also* Camões, Luís de
Lusíads, The. *See Lusíadas, Os*

Machiavelli, Niccolò, *Discourses of the First Decade of Titus Livius*, 95
Macrobian map. *See* zonal map (Macrobian map)
Macrobius, *Commentary on the Dream of Scipio*, 202
Magellan, Ferdinand, 13
Maggi, Annibal, 70
Magnus, Johannes, 268n125, 269n138
Magnus, Olaus, 256n98, 268n125, 269n138
Main (river), 59, 62
mainland, 6–8, 10, 75–77, 139, 142–43, 172, 203, 240nn40–41. *See also* continents; *epeiron*
Mainz, 27, 37, 41, 50
Mamluk Sultanate, 193, 200
manual (genre), 25, 111, 127, 132, 137, 182, 193, 276n112. *See also Champ fleury*

Manuel I (of Portugal), 96–97, 99, 217
Manuzio, Aldo, 272n36, 285n110
map (synonyms for): *descriptio*, 21, 210, 228; *figura* (and *figuratio*), 12, 168, 187, 210, 214, 216, 218; *forma*, 210; *orbis pictus*, 210; painting (*pintura*), 185–86, 192; tapestry, 223–28, 294n128. See also cartography; *mappa mundi*
mappa mundi, 1–2, 11, 15, 34, 37, 62, 65–67, 74, 134, 166, 170, 193, 196, 224, 241n2, 241n4, 246n58, 248n80, 259n147
Marco Polo, 70, 99–100, 126, 193, 247n72, 251n24, 290n37
mare nostrum. See Mediterranean
Marino, John, 11–12
Marmara, Sea of, 73
Martellus, Henricus, 31
Martínez, Oscar, 71, 232
mathematics, 16, 19, 22, 29, 43, 112–13, 129, 132–34, 136, 145, 151, 249n97, 252n34, 288n2; algebra, 133; arithmetic, 16, 22, 27, 113. See also geometry; numbers
matrix, 16, 142–44, 146, 214. See also birth; chôra
Maurette, Pablo, 160
Maximilian I (emperor), 49–50, 263n57
measurement (in cartography), 16, 19–20, 30, 35, 44–45, 47, 51, 53, 56, 62, 64, 67, 121–22, 126, 129–30, 247n72, 255n77, 258n129; digits, 112, 133
Mediterranean: Indo-, 199, 290n47; *mare nostrum*, 196, 199, 230, 290n49; Sea, 1–2, 40, 73–74, 120–21, 153, 156, 170, 179, 196–203, 206–8, 216, 225, 230, 271n27; symmetry of, 73–74; vegetation, 188, 223–24, 227; world, 14, 55, 94, 115–16, 157, 174, 188, 211, 219, 225, 244n34, 248n80, 268n125
Mehmed II (Ottoman sultan), 200
Melinde, king of, 224
Mercator, Gerhard, 5, 20–21, 70, 104, 223, 228–29, 232, 263n50, 269n141, 288n2; *Nova et aucta orbis terrae descriptio*, 21, 228
merchants, 8, 26, 30, 99, 101–2, 193, 271n27, 281n35
mercury (metal), 150, 153–54, 156–57, 164–65, 168–69, 174, 282n46
Mercury (mythological figure), 45, 124–27
metabolē. See antimetabole

metalepsis. See *transumptio*
metaphor, 5, 117, 138, 141–42, 144, 164–65, 180, 190, 200, 211, 229, 233, 294n128
metonymy, 34, 75, 121, 126–27
Mexico City. See Tenochtitlan
Mezzadra, Sandro (and Brett Neilson), 92, 102
Miechowita, Maciej, 12, 24, 69–73, 75–80, 82–104, 106, 260n7, 260n9, 261n17, 261n19, 261n24, 262n25, 263n57, 264n60, 265n89, 266n103, 267n114, 267n116, 268n125, 268n136, 269n138, 269n142; *Chronica Polonorum*, 71–72, 261n24; *Descriptio Sarmatiarum*, 70, 90–91, 103, 268n125, 268n136. See also *Tractatus de duabus Sarmatiis Asiana et Europiana*
Miernowski, Jan, 134
Mignolo, Walter, 4–5, 14, 191, 248n82, 287n140
migration, 48, 88, 92–94, 104, 106, 109, 230, 248n78. See also nomadism
Miletus, 6, 244n24
Minerva, 45
mining, 141, 153–54, 156, 164–66, 168–70, 174, 267n114, 283n71, 283n84. See also alchemy
Moguntia. See Mainz
Moluccas, 100, 128–29, 274n79
Monmonier, Mark, 21
monsters, 1, 47, 55–56, 84–85, 90, 208, 257n101. See also Adamastor; dragons
morbus gallicus. See syphilis
More, Thomas, *Utopia*, 75, 139–43, 275n109, 276n112
Moschus, 2, 244n29
Müller, Gernot Michael, 63
Müller, Johann. See Regiomontanus, Johannes
mundus novus. See New World
Münster, Sebastian, 70, 104, 110, 190, 193–95, 209, 270n9
Münzer, Hieronymus, 30
Murad II (Ottoman sultan), 96
Muscovy, 70–71, 83, 89, 92–93, 97, 100, 103–4, 259n3, 263n57, 265n89, 269n138. See also Russia
Muslims. See Islam
mysticism: in Christianity, 62, 65–67; Sufi, 137
mythology, 2, 6, 17, 44, 83, 93, 103, 107, 109, 118–21, 124, 127, 137, 140, 145–46, 154, 160, 165, 172–75, 177, 192, 207, 217, 224, 277n141

Nagel, Alexander, 275n105
nautical charts, 15, 30, 34, 142, 148, 185, 189, 218–19, 222, 247n70, 282n53, 286n118, 288n2, 289n32
Navagero, Andrea, 152, 183, 279n11, 280n27, 281n28
Navarre, Marguerite de, 39
navel, 24, 30, 66–67, 170–71, 232, 259n144; umbilication, 30, 143, 232, 252n28, 259n144
navigation, 13, 21, 30, 98–100, 104, 148, 161, 176, 187–88, 192–93, 196, 203, 211, 216, 218–19, 279n8, 286n118, 288n2, 289n32, 290n37. *See also* Hakluyt, Richard; nautical charts; Ramusio, Giovanni Battista
Nebrija, Elio Antonio de, 114, 175–76, 286n115
neo-Platonic thought, 162, 283n70. *See also* Ficino, Marsilio; Plato
Neptune, 107, 205–7, 224
nesos. *See* islands
Netherlands, 2, 72; Dutch, 70
New World, 11, 13, 16, 25, 30, 70, 88, 104, 148, 150, 152–53, 157–58, 161, 165, 172–73, 178, 180, 182–84, 186, 191, 243n19, 249n6, 278n1, 278n4, 279n7, 281n28, 284n94, 288n3; *mundus novus*, 148–49, 170, 175, 184, 249n6, 278n4. *See also* Grynaeus, Simon
Niedźwiedź, Jakub, 267n104, 268n117
Nile, 8, 10, 157, 171, 196, 248n80, 292n87. *See also* Egypt
Noah, 1, 18–19, 109, 115, 163, 172, 174, 248n83, 248n86
nomadism, 78–80, 83, 87–90, 92–96, 102, 265n89
Nonnus, *Dionysiaca*, 211
Norway, 85, 100
nothing, 57, 133–34, 136; *nichil*, 134–36, 139, 287n143; *nulla*, 132. *See also* Bovelles, Charles de; zero
numbers, 15, 111, 114, 117, 120, 130, 132–34, 136, 142; Pythagorean number (perfect number ten), 114, 129–30, 136
Nunes, Pedro, 228, 288n2
Nuremberg, 26–27, 29–30, 34, 58, 232; Nuremberg City Council, 30, 251n24
Nuremberg Chronicle. *See* Schedel, Hartmann

ocean, 10, 41, 48–49, 74, 84, 97, 120–21, 134, 163, 165, 172–73, 179–80, 183, 188, 190, 197, 202, 206, 214, 220, 225, 228. *See also* Atlantic Ocean; Indian Ocean; *okeanos*; Pacific Ocean
oikoumene, 1–2, 8–11, 14–15, 18, 20, 37, 67, 84, 101–2, 109, 111, 142, 148, 150, 152–54, 158–59, 165–66, 169–75, 177–78, 182, 184, 188, 198, 201, 211, 220, 230, 246n58, 284n85
okeanos, 1, 34, 248n80, 284n85
old age, 37, 39, 50, 56, 257n109
Olsson, Gunnar, 87
Ophir. *See* Hispaniola
Orkney Islands, 50
Ortelius, Abraham, 70, 79, 104, 263n50, 264n60
Os Lusíadas. *See Lusíadas, Os*
Österreichische Nationalbibliothek, 3, 91, 270n7
ottava rima, 197, 284n94; and cartography, 219–20
Ottoman Empire, 11, 94, 96, 98, 193, 199–200, 245n52, 268n122
Ovid, 2, 27, 45, 47, 56, 83, 114, 124, 145, 257n109
Oviedo y Valdés, Gonzalo Fernández de, 150, 152, 154, 177–78

Pacific Ocean, 14, 148–50
padrão, 196, 198
Padrón, Ricardo, 114, 279n7
Padua, 261n18
Palestine, 50–51
Pannartz, Arnold, 41, 253n55
Panofsky, Erwin, 28
Pantheus, Joannes Augustinus (Antonius), 166–67, 284n85, 286n113
Paracelsus, 153, 166, 281n36
Paradise, 1, 59, 64–65, 218, 258n121, 295n151
Paris, 112, 119, 128, 141–42, 166, 256n94, 271n16; Parrhisians (Parrhasians), 119, 137
pars pro toto, 14, 44
patera, 66–67
Pechriggl, Alice, 145
Peirce, Charles, 88, 265n79
peninsulas, 1–2, 31, 49–50, 73–77, 79, 139, 149, 231, 262n40, 291n53. *See also* Arabian Peninsula; Crimea; Iberian Peninsula; Italy
peraiai, 6–7

Perekop, Isthmus of. *See* Perekopian Strait
Perekopian Strait, 75–77
Petrarca, Francesco, 22, 27, 41, 168, 226–27, 295n145; *Canzoniere*, 27, 226–27, 295n146; *Familiares*, 24, 41
Peutinger, Konrad, 252n34
pharmakon, 16, 232, 287n136
philology, 12–15, 22, 24–25, 80, 112, 114–19, 124, 136, 139, 146–47, 153–54, 166, 170, 179–80, 182–84, 232, 279n10, 280n18. *See also* humanism
philosophy, 6–8, 24, 28–29, 42, 72, 75, 114, 116–17, 133, 140, 155, 201, 242n11, 243n21, 272n39; visual representation of, 22–24, 27–29, 39. *See also* Dürer, Albrecht (*Philosophia* woodcut)
Phoebus. *See* Apollo
Piccolomini, Enea Silvio (later Pius II), 11–12, 61, 87, 114–15, 233, 245n52, 249n1, 256n91, 258n126; *De Asia*, 11, 115; *Cosmographia*, 114–15; *De Europa*, 11–12, 115, 233
Pico della Mirandola, Giovanni, 42, 114
Piemontese, Angelo Michele, 136–37
pilgrimage, 27, 49–50
Pinet, Simone, 141–42
Pinifer Mons. *See* Fichtel Mountain
Pinto, Karen, 200, 291n53
Pirckheimer, Willibald, 30, 48–49, 70, 232
Pius II. *See* Piccolomini, Enea Silvio (later Pius II)
place, definitions of, 19, 57, 87, 133, 139–41, 145–46, 257n115, 276n112, 276n113, 276n116
Plato, 28, 89, 114, 144–46, 160, 169–72, 211. *See also* Ficino, Marsilio; neo-Platonic thought
Pliny the Elder, 15, 20, 24, 73, 85, 140, 176, 209, 251n24, 262n31, 263n54
Plutarch, 73, 262n31
Poland, 2, 5, 69–71, 78, 83–85, 90, 92–96, 100, 103, 252n34, 261n20, 264n60, 264n65, 264n72, 267n107; Poland-Lithuania, 71, 92–94, 97; Polish Kingdom, 37, 71, 79, 83, 92–93, 96–98, 256n91, 261n18; Polish language, 70, 83, 95, 101, 103, 268n136
Polo, Marco. *See* Marco Polo
Pomponius Mela, 15, 22, 73, 85, 115, 118, 123, 196, 209, 258n124, 258n129
Pontano, Giovanni, *De hortis Hesperidum* (*The Gardens of the Hesperides*), 174–75, 285n111. *See also* citron; Hesperides

Porcacchi, Tommaso, 70, 76–77, 260n15
portolan charts. *See* nautical charts
Portugal, 5, 16, 25, 30, 69, 72, 186–90, 201–2, 222, 224; and colonialism, 96–97, 99–100, 129, 148, 153, 193, 209, 214–15, 247n70, 251n23, 274n79, 278n4, 293n115
Portuguese Empire, 187, 202–3
Poseidon. *See* Neptune
postcolonialism, 4–5, 102. *See also* colonialism
Postel, Guillaume, 107, 109–10, 270n5, 270n6
Pravda, Ján, 111
printing technology, 10, 13, 15, 17–19, 26, 29–30, 41–42, 50–51, 64, 71, 76, 94, 103, 107, 111, 115, 132, 148, 219, 248n86, 249n6, 250n18, 251n21, 252n30, 253n54, 253n55, 260n6, 267n114, 281n35. *See also* frontispiece; humanism
Probus, Valerius, 115
Propertius, 27, 66
propugnaculum, 95–96
Prudentius, 258n140; *Cathemerinon*, 64–67
Prussia, 68–69, 84–85, 93, 261n18
pseudo-Berosus. *See* Annius of Viterbo, *Antiquities*
pseudoscript. *See* alphabetic writing: pseudoscript
Pseudo-Seneca, 179
Ptolemy, 9–10, 13, 15–16, 18–19, 22, 26, 28, 43–45, 48–49, 61–62, 70, 72–73, 208–11, 245n51, 247n65; and cartography, 5, 20, 29, 34, 53, 56–58, 67, 78–80, 113, 140, 165, 196, 198, 213, 218–21, 232–33, 261n20, 289n32, 292n83; reception of, 7, 15, 25, 30, 70, 72, 82–88, 92, 94, 107, 109, 115, 117, 145, 168, 188–89, 192–93, 210, 216, 247n66, 251n21, 251n24, 260n7, 292n87. *See also Geography* (Ptolemy); grid (cartographic); latitudes and longitudes
Putsch, Johannes, 109
Pythagoras, 114, 130, 136, 291n72. *See also* numbers: Pythagorean number (perfect number ten)

quadrivium. *See* humanism: and organization of knowledge
Quatuor Libri Amorum secundum Quatuor Latera Germanie, 22–23, 26–27, 29, 36–40, 45–46, 49–51, 54, 58–59, 62–64, 67, 249n7, 253n52, 255n85, 257n109, 258n121

quicksilver. *See* mercury (metal)
Quint, David, 208, 215, 287n130
Quintilian, 55, 57, 183–84, 257n115, 287n143
Qur'ān, 199

Rabelais, François, 275n109
Ramachandran, Ayesha, 193, 243n20, 253n39
Ramusio, Giovanni Battista, 70, 98–100, 152, 157, 172, 183, 260n14, 261n17, 262n41, 268n122, 279n11; *Navigazioni e viaggi* (also *Navigationi et viaggi*), 70, 98, 157, 172, 260n14, 261n17, 262n41
Ratisbona. *See* Regensburg
Red Sea, 1, 98, 200
Regensburg, 27, 37, 50, 258n121
Regiomontanus, Johannes, 30
Reiss, Timothy J., 113, 180, 249n97, 275n96
Reuchlin, Johann, 114
rhetoric, 22, 24, 27–28, 42, 57, 64, 111, 113, 121, 140, 178, 183–84, 187, 189, 205, 210, 220, 224, 276n112; rhetorical expenditure, 220–21
Rhine (river), 37, 48, 50, 59
Rhodes, 7, 214–15
rhumb lines, 21, 30, 34, 65, 142, 219
Rime Sparse. *See* Petrarca, Francesco: *Canzoniere*
Ringmann, Matthias. *See* Waldseemüller, Martin
Riphean mountains, 61, 74, 80, 82–83, 263n57
Robert, Jörg, 64
Rome, 14, 28, 30, 34, 92, 116, 141–42, 177, 179, 250n17
Romm, James S., 291n72
Romweg map. *See* Etzlaub, Erhard
Roxolania. *See* Ruthenia
Rudbeck, Olof, 285n104. *See also* Atlantis
Russia, 72, 79, 82, 100–101, 103–4, 231, 259n3, 268n126, 268n136, 269n139, 269n142, 269n143. *See also* Muscovy; Ruthenia
Ruthenia, 68–69, 83, 85, 88–89, 92–93, 100–101, 259n3, 261n24, 268n126

Saint Emmeran, monastery of, 50
Samorin (king of Calicut), 202–3
Saragossa. *See* Treaty of Saragossa
Sarmatia, 39–41, 47–49, 61, 68–70, 72, 74, 76, 78–80, 83–91, 93–98, 102–3, 256n91, 259n3, 264n65, 264n72, 265n89; Sarmatian Gates, 74; *Sauromatae*, 83, 265n89
Sarmatian Sea. *See* Baltic Sea

Saruth, 86, 264n73
Saxl, Fritz, 28
Saxons, 60, 73
scale (in cartography), 9, 15, 19, 30, 44, 51, 53, 58–59, 112, 150, 256n96; continental, 9, 58, 150–51, 159, 192; global, 8, 12, 58–59, 192, 232; national, 58–59, 150–51; regional, 8–9, 12, 14, 27, 29, 31, 49, 58, 62, 104, 107, 247n65. *See also* chorography; cosmography; topography
scansion, 63
Scarperia, Jacopo Angelo da, 9, 84
Schedel, Hartmann, 26, 29, 34–37, 39, 41–42, 50, 253n55; *Liber Chronicarum (Nuremberg Chronicle)*, 26, 29–30, 34–37, 41–42, 50, 134, 248n86, 249n1, 252n33
Schnitzer von Armsheim, Johannes, 50–51
Schöner, Johann, 29
Schreyer, Sebald, 26
Scotland, 2
Scythia, 68–69, 79, 83, 87, 90, 93–96, 102, 263n54, 265n89, 266n97, 266n102. *See also* Tartaria
Semia. *See* Asia
seminarium, 143, 145, 161
Seneca, 255n80; Pseudo-, 179
senectus. *See* old age
Shell, Marc, 75, 244n29, 256–57n100, 258n135, 262n40
Shem, 18, 86, 109
Sidon, 141–42
Siegert, Bernhard, 87
sifr. *See* cipher
Sigismund I (of Poland), 71, 79, 92–93, 96, 99
situs. *See* topography
Slata Baba. *See* złota baba
Slavic: languages, 75, 268n122; peoples, 60, 71, 85–86, 93–94, 103–4, 259n3, 268n122. *See also* Muscovy; Poland; Russia; Ruthenia; Sarmatia; Scythia; Tartaria
Solomon (king), 174
Sousa, Manuel de Faria e, 205
space, definitions of, 15–17, 19–21, 45, 47, 51, 63, 65, 78, 126, 128, 141–42, 145, 229, 232–33, 243n21, 247n67, 247n72
Spain, 14, 72–74, 114, 119, 121, 140, 150–52, 173, 175, 180–82, 187, 200, 247n66, 262n31, 282n62; and empire, 16, 96, 100, 129, 148, 153, 156, 161, 172, 177, 179, 251n25, 274n79
Spain, Peter of, 288n143

speech, 42, 53, 55–56, 64, 111, 122, 173, 178, 180, 258n135. *See also* barbarism; tongue
spondees, 63, 66, 258n140
standardization, cartographic, 18–20, 25, 111, 229
Statius, 276n113
Stockhammer, Robert, 265n79
Strabo, 6–9, 15, 19–20, 73–74, 145, 171–72, 201, 209, 251n24, 258n129, 262n31, 270n7
Stuchs, Georg, 29, 251n21
Subiaco, 18, 41–42, 253n54
Sufism, 137
Sumatra, 199
surface (in cartography), 16, 19, 21, 44–45, 49, 57, 62, 65, 111, 113, 120, 125, 130, 136, 140, 156, 158, 161, 163–65, 171, 184, 186, 192, 198, 219, 228, 265n79
Sweden, 85, 100, 256n98, 268n125, 285n104
Sweynheym, Konrad, 17, 41, 253n55
Sylvester II (pope), 133. *See* Aurillac, Gerbert of (Pope Sylvester II)
symbol, 126–28, 133
symmetry, cartographic, 18, 72–75, 96–97, 137, 201, 203, 205
syphilis, 148–49, 151–52, 154, 157, 159, 162, 165, 168–69, 175, 180–84, 280n18, 281n28, 281n35, 287n136; syphilitic limology, 150, 152, 184
Syphilis sive Morbus Gallicus, 148–54, 156–59, 161–62, 165, 168, 171, 173–77, 179–80, 182, 184, 279n10, 281n28, 282n46

table (cartographic), 43–45, 49, 56–57, 62, 67, 80–81, 84–85, 113, 117, 126, 128–30, 132, 134, 136, 140, 143–44, 146, 219, 228; *mensa*, 227–29; *Tabula Cebetis* (pseudo-Lucian), 114; *tabula rasa*, 45, 65, 192, 229
tabula and *tabula rasa*. *See* table (cartographic)
Tacitus, 42, 47, 49, 55–56, 83, 114, 121, 255n85, 273n56; *Germania*, 47, 49
Tana, 99, 171, 268n122
Tanais (river), 10, 20, 61, 73–74, 77, 80, 82–85, 88–89, 93, 99, 171, 196, 248n80, 263n54, 265n89, 268n122
tapestry. *See* map (synonyms for): tapestry
Tartaria, 70, 94–95, 99, 131–32, 266n102, 269n142, 285n107. *See also* Scythia
Tasso, Torquato, 170–71, 220, 284n94
Tatars, 76, 78, 80, 87–90, 94–95, 97, 104, 266n103; Cosanenses (Kazan), 266n103; Lithuanian, 71, 267n104; Nohacenses (Noghay), 266n103; Perekopian (Crimean), 75–76, 97, 266n103; Zavolhenses (Trans-Volga), 266n103
Tate, Nahum, 297n10
Tenochtitlan, 8, 30, 251n25
territory, 10, 21, 43, 53, 56, 59, 62–64, 68–69, 71–72, 80, 83, 89, 93, 107, 115, 137, 142–43, 145–47, 149, 166, 168, 170, 218–19, 222, 233, 258n128, 258n129, 259n3; territorial boundaries, 13, 15–17, 19, 35, 47–48, 63, 69, 78, 103, 106, 121, 140, 143, 146, 150–51, 204–5, 247n70, 248n78, 263n54, 269n138; territorial expansion, 30, 72–73, 79, 82, 88, 92, 94, 96–97, 172, 187, 200, 292n83; terror and territory, 47–48, 255n82, 255n83
terror. *See* territory: terror and territory
tertium aliquid. *See* analogy; Cicero
Thaurica Chersonesus. *See* Crimea
Thessaly, 124
Thetis, 217, 224, 292n83
Thor (Nordic mythology), 269n138
Thrift, Nigel, 51, 78, 233
Thucydides, 7
Thule, 24, 41, 50, 85
Tibullus, 27
T-O map, 18, 50, 115, 166, 196, 220, 248n80, 284n85, 289n32
tongue, 53, 56, 61, 121–22. *See also* barbarism; speech
topography, 2, 12, 26–27, 42, 49, 57–58, 75, 83, 93, 139–41, 146, 168, 276nn112–13, 276nn115–16; *situs*, 42, 254n64, 285n102
toponymic symmetry (Iberia and Hiberia), 73–74, 262n31
toponyms, 1–2, 15, 18, 73, 83, 85–87, 89–90, 93, 139–40, 152, 162, 241n4, 262n31, 262n40, 275n109, 282n53; ambiguous, 148–49, 168, 179, 231, 262n40; choice of, 148–49, 278n2; and colonization, 179; and continental thinking, 78–79, 84, 86–87, 89, 115, 152, 176; emergence of, 125–27, 148–49; and etymology, 40, 87; and poetry, 69
topothesia, 139–41, 146, 276n112, 276n115. *See also* topography
Tordesillas. *See* Treaty of Tordesillas
torrid zone. *See* equator
Toruń, 261n18

Tory, Geoffroy, 25, 111–18, 120–22, 124–25, 127–32, 134, 136–43, 145–47, 271n16, 272n36. See also *Champ fleury*
Toscanelli, Paolo Dal Pozzo, 186
Tractatus de duabus Sarmatiis Asiana et Europiana, 68–73, 75, 78–80, 82, 85–90, 92, 94–96, 100, 102–3, 260n7, 267n115, 268n125; editions and translations of, 70, 261n17, 268n119, 268n125, 280n19
translatio cartographiae, 28–30
translation, 9, 13, 15, 22, 24, 28, 30, 34, 48–49, 57, 70, 80, 84, 89–90, 95, 102, 107, 111–12, 114, 116, 121, 132–33, 138, 142, 145, 151–52, 168, 172, 178–79, 182–84, 187, 208, 227, 241n8, 249n1, 256n94, 256n100, 257n110, 259n1, 260n13, 277n127, 278n4, 280n19, 281n35, 294n128
transumptio, 183–84, 287n143, 288n144
triton genos. See chôra
Treaty of Saragossa, 129
Treaty of Tordesillas, 13, 16–17, 72, 129, 153, 232. *See also* border: *linea*; latitudes and longitudes
Treaty of Villier-Cotterêt, 121
triangulation, 17, 120, 125–27, 129
trivium. See humanism: and organization of knowledge
Tsardom of Russia, 96
Turzo, Stanisław, 96, 267n114, 268n119
Tyre, 121

Ulans. See Tatars: Perekopian (Crimean)
Union of Lublin, 71
Universitas Cracoviensis. See University of Kraków
University of Kraków, 69, 101, 262n25, 265n86, 267n114
Ursula Galla, 50
Usher, Phillip, 283n71
Utopia. See More, Thomas, *Utopia*
Utopian alphabet. See *under* alphabetic writing

Vandals, 94–95
Varro, Marcus Terrentius, 180

Venice, 8, 30, 37, 99, 132, 152, 157, 166, 168, 193, 200, 268n122, 279n11, 280n27
Venus (mythological figure), 45, 56, 175, 200–201, 206–7, 225–26, 228, 285n111
Vespucci, Amerigo, 70, 88, 148–50, 152, 248n79, 249n6, 278n1, 278n4
Vidal-Naquet, Pierre, 177
Villier-Cotterêt, Treaty of. *See* Treaty of Villier-Cotterêt
Virgil, 24, 28, 140, 155, 157, 169, 175, 188, 285n107, 287n136, 294n130. *See also Aeneid* (Virgil); *Georgics* (Virgil)
Vistula (river), 37, 48–49, 84, 86, 89, 256n91
Vitruvius, 112, 129, 137, 140
Volga (river), 82–83, 99; Zavolhenses (Trans-Volga), 266n103. *See also* Tatars
Voltaire (François-Marie Arouet), 72, 187

Waldseemüller, Martin, 80–82, 148–49, 153, 165, 211–12, 278nn1–2
Walter of Châtillon. See Châtillon, Walter of, *Alexandreis*
Wapowski, Bernard, 70–71, 92, 261n20
Wied, Anton, 104
winds, 59, 143, 186, 193, 196, 206, 221, 256n96, 258n122
Wirsung, Marx. See Grimm, Sigmund
Władysław II Jagiełło (founder of Jagiellonian dynasty), 71
Wolff, Larry, 260n9
Woodward, David, 15, 117, 129, 247n66, 247n67, 248n80
Würzburg, as *Erebos*, 40–41, 253n52
Wuttke, Dieter, 250n14, 252n34

youth, 37, 39, 49–50, 257n109

Zárate, Augustín de, *Historia del descubrimiento y conquista de la provincia del Perú*, 172
zero, 132–34, 136, 139, 275n96. *See also* cipher
złota baba, 102–4, 106, 269n138. See also *aurea vetula*; Golden Old Woman
zonal map (Macrobian map), 15, 98, 189, 202, 233. *See also* equator

www.ingramcontent.com/pod-product-compliance
Lightning Source LLC
Chambersburg PA
CBHW021935290426
44108CB00012B/843